Heterocycles from Double-Functionalized Arenes
Transition Metal Catalyzed Coupling Reactions

RSC Catalysis Series

Editor-in-Chief:
Professor James J Spivey, *Louisiana State University, Baton Rouge, USA*

Series Editors:
Professor Chris Hardacre, *Queen's University Belfast, Northern Ireland*
Professor Zinfer Ismagilov, *Boreskov Institute of Catalysis, Novosibirsk, Russia*
Professor Umit Ozkan, *Ohio State University, USA*

Titles in the Series:

1: Carbons and Carbon Supported Catalysts in Hydroprocessing
2: Chiral Sulfur Ligands: Asymmetric Catalysis
3: Recent Developments in Asymmetric Organocatalysis
4: Catalysis in the Refining of Fischer–Tropsch Syncrude
5: Organocatalytic Enantioselective Conjugate Addition Reactions: A Powerful Tool for the Stereocontrolled Synthesis of Complex Molecules
6: *N*-Heterocyclic Carbenes: From Laboratory Curiosities to Efficient Synthetic Tools
7: *P*-Stereogenic Ligands in Enantioselective Catalysis
8: Chemistry of the Morita–Baylis–Hillman Reaction
9: Proton-Coupled Electron Transfer: A Carrefour of Chemical Reactivity Traditions
10: Asymmetric Domino Reactions
11: C-H and C-X Bond Functionalization: Transition Metal Mediation
12: Metal Organic Frameworks as Heterogeneous Catalysts
13: Environmental Catalysis Over Gold-Based Materials
14: Computational Catalysis
15: Catalysis in Ionic Liquids: From Catalyst Synthesis to Application
16: Economic Synthesis of Heterocycles: Zinc, Iron, Copper, Cobalt, Manganese and Nickel Catalysts
17: Metal Nanoparticles for Catalysis: Advances and Applications
18: Heterogeneous Gold Catalysts and Catalysis
19: Conjugated Linoleic Acids and Conjugated Vegetable Oils
20: Enantioselective Multicatalysed Tandem Reactions
21: New Trends in Cross-Coupling: Theory and Applications
22: Atomically-Precise Methods for Synthesis of Solid Catalysts

23: Nanostructured Carbon Materials for Catalysis
24: Heterocycles from Double-Functionalized Arenes: Transition Metal Catalyzed Coupling Reactions

How to obtain future titles on publication:
A standing order plan is available for this series. A standing order will bring delivery of each new volume immediately on publication.

For further information please contact:
Book Sales Department, Royal Society of Chemistry, Thomas Graham House, Science Park, Milton Road, Cambridge, CB4 0WF, UK
Telephone: +44 (0)1223 420066, Fax: +44 (0)1223 420247
Email: booksales@rsc.org
Visit our website at www.rsc.org/books

Heterocycles from Double-Functionalized Arenes
Transition Metal Catalyzed Coupling Reactions

Xiao-Feng Wu
Leibniz Institute for Catalysis, Rostock, Germany and Zhejiang Sci-Tech University, Hangzhou, China
Email: xiao-feng.wu@catalysis.de

Matthias Beller
Leibniz Institute for Catalysis, Rostock, Germany
Email: matthias.beller@catalysis.de

ROYAL SOCIETY OF CHEMISTRY

RSC Catalysis Series No. 24

Print ISBN: 978-1-78262-136-2
PDF eISBN: 978-1-78262-312-0
ISSN: 1757-6725

A catalogue record for this book is available from the British Library

Published by The Royal Society of Chemistry,
Thomas Graham House, Science Park, Milton Road,
Cambridge CB4 0WF, UK

Registered Charity Number 207890

For further information see our web site at www.rsc.org

Printed in the United Kingdom by CPI Group (UK) Ltd, Croydon, CR0 4YY, UK

Dedicated to my wife and my kids:
Qing-Yuan Wei, Nuo-Yu Wu, and Nuo-Lin Wu
Thanks for their understanding and support!

Preface

The preparation of heterocyclic compounds is one of the main branches of organic chemistry due to the prevalence of heterocycles in various areas. More than four out of five of the best-selling drugs in the USA in 2013 have contained at least one heterocycle in their structure. Hence, efforts to develop efficient procedures to synthesize heterocycles are in demand. Nowadays, it is clear that current research is focused on mono-functionalized arenes, which are employed in C–H activation–functionalization reactions. But most of the developed C–H activation methodologies have met with problems such as harsh reaction conditions, high catalyst loading *etc.* As the topic of heterocycle synthesis has become more related to fine chemical synthesis and pharmaceuticals, there is more of a focus on the reaction efficiency, functional group tolerance and related parameters. All these points lead to one conclusion: a book on heterocycles produced from double functionalized arenes is necessary.

The main contributions to the synthesis of heterocycles based on double-functionalized arenes as substrates have been summarized and discussed. The contents are catalogued by the size of the ring formed and sub-divided by the substrates applied. The reactions included here usually have high efficiency and selectivity. However, the main disadvantage for most of the procedures mentioned is the requirement of pre-activation of the starting materials, which sometimes means that the substrate scope is narrow and limited. In the future, the development of highly efficient methodologies without pre-activation of substrates, such as C–H activation, will be in demand.

Here, it is important to claim that this book is not intended to be comprehensive but representative. On the other hand, we did our best to collect all related literature and to correct our writing; however, the possibility of

RSC Catalysis Series No. 24
Heterocycles from Double-Functionalized Arenes: Transition Metal Catalyzed Coupling Reactions
By Xiao-Feng Wu and Matthias Beller

Published by the Royal Society of Chemistry, www.rsc.org

missing literature and mistakes still exists. We apologize in advance for any possible omissions and mistakes.

We do wish this book to be useful for chemists who are working hard to look for a bright future for human beings! Thanks!

Xiao-Feng Wu

Contents

RSC Catalysis Series No. 24
Heterocycles from Double-Functionalized Arenes: Transition Metal Catalyzed Coupling Reactions
By Xiao-Feng Wu and Matthias Beller

Published by the Royal Society of Chemistry, www.rsc.org

CHAPTER 1

Introduction

1.1 The Importance of Heterocycles

In Asia, the life expectancy of a human was 40 years in 1960, which increased to 68 years in 2013 and will increase to 78 years in 2050.[1] In all the possible positive effects to increase lifetime, the development of pharmaceuticals plays a crucial role. Importantly, most of the best selling drugs contain a heterocyclic moiety as their core structure.[2] Additionally, more than 90% of naturally occurring compounds have a heterocyclic structure. Based on this importance of heterocycles, their preparation has become one of the main branches in modern organic synthesis.[3] Although the current trend in methodology development is C–H activation, the requirements of reaction efficiency and functional group tolerance mean that the pharmaceutical industry still have to look at double-functionalized aromatic compounds.

In this book, we are going to discuss the application of transition metal catalysts in the synthesis of heterocycles by using double-functionalized arenes as substrates. The chapters are organized by the size of the rings formed and sub-divided by the substrates applied. In order to make this book more applicable and readable, the preparation of the double-functionalized arenes applied will be mentioned first.

1.2 The Preparation of Double-functionalized Arenes

For the double-functionalized arenes applied, in general, they can be divided into three main analogues (Table 1.1). One is 1,2-dihaloarenes, also called 1,2-di-electrophilic arenes (DEA); the next is *ortho*-halogen activated arenes, also called 1,2-electrophilic-nulceophilic coexisted arenes (ENA); the third is called 1,2-di-nulceophilic arenes (DNA).

RSC Catalysis Series No. 24
Heterocycles from Double-Functionalized Arenes: Transition Metal Catalyzed Coupling Reactions
By Xiao-Feng Wu and Matthias Beller

Published by the Royal Society of Chemistry, www.rsc.org

Table 1.1 List of frequently used double-functionalized arenes.[a]

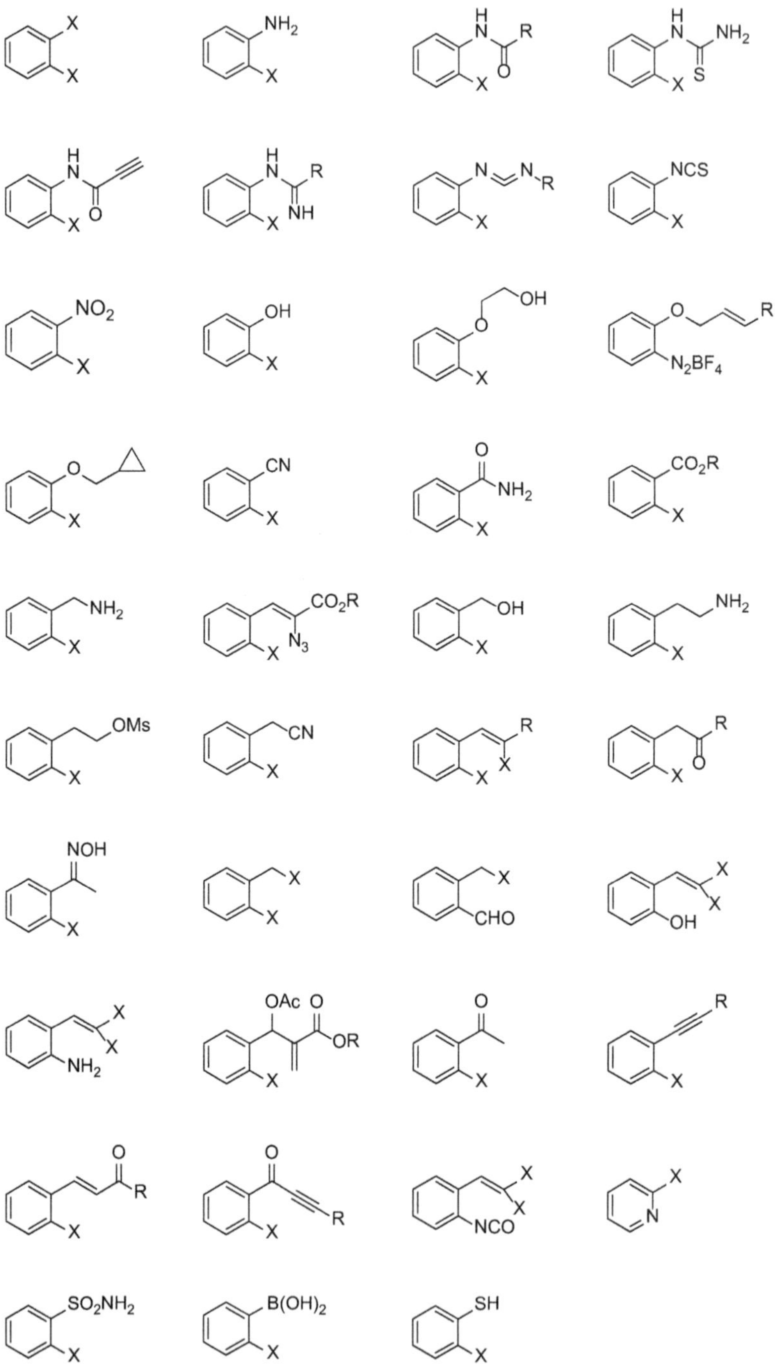

[a]X = I, Br, Cl, F; R = alkyl, aryl.

1.2.1 1,2-Dihaloarenes

1,2-Dihalogenized arenes are an important class of compounds that have been reported with broad applications in coupling reactions. As the halogen mentioned here normally refers to fluoride, chloride, bromide and iodide, the combination of these elements offers several possibilities for the 1,2-dihaloarenes formed. The symmetrical 1,2-dihaloarenes (such as 1,2-diiodobenzene, 1,2-dibromobenzene, 1,2-dichlorobenzene and 1,2-difluorobenzene) can be easily prepared by the reaction of the corresponding arenes with a halogen atom (I_2, Br_2, Cl_2, F_2) in acidic media. Regarding the non-symmetrical arenes, they can be synthesized by the halogenation of the pre-monohalogenated arenes.

1.2.2 2-Halophenol Derivatives

2-Halophenols have broad applications in the synthesis of oxygen-containing heterocycles. Typically, 2-halophenol can be prepared by the halogenation of phenols. Then, 2-halophenol can be used for the preparation of the other derivatives. Notably, 2-halophenol also acts as the precursor for arynes by activation of the C–O bond.

1.2.3 2-Haloaniline Derivatives

2-Haloaniline derivatives are widely available from chemical suppliers and have been applied extensively in nitrogen-containing heterocycle synthesis. Additionally, 2-haloanilines can act as 1,2-dihaloarene precursors as well in the Sandmeyer reaction. As bulky chemicals, anilines are usually applied as substrates for the synthesis of 2-haloanilines after halogenation.

1.2.4 1-Carbon-2-haloarenes

For 1-carbon-2-haloarenes, such as 2-bromoacetophenone, 2-bromotoluene, 2-bromobenzyl amine, 2-bromobenzaldehyde and so on, in general, they can all be prepared by the halogenation of their parent molecules.

Here, we can conclude that halogenation can effectively activate the parent molecules of double-functionalized arenes.

References

1. H. Kaplan, K. Hill, J. Lancaster and A. M. Hurtado, *Evol. Anthropol.*, 2000, **9**, 156–185.
2. http://en.wikipedia.org/wiki/List_of_largest_selling_pharmaceutical_products.
3. (a) T. Eicher, S. Hauptmann, A. Speicher, *The Chemistry of Heterocycles: Structures, Reactions, Synthesis, and Applications*, Wiley-VCH, 3rd edn, 2012; (b) J. A. Joule, K. Mills, *Heterocyclic Chemistry*, Wiley-VCH, 2010; (c) X.-F. Wu, M. Beller, *Economic Synthesis of Heterocycles: Zinc, Iron, Copper, Cobalt, Manganese and Nickel Catalysts*, RSC, 2014.

CHAPTER 2

Five-membered Heterocycle Synthesis

In this chapter, the applications of double-functionalized arenes in the synthesis of five-membered heterocycles will be discussed. The contents are divided according to the different types of substrates applied. The sub-chapters are organized based on the types of nucleophiles *ortho*-substituted to aryl halides. For the heterocycle synthesis based on C–X bond activation, in general, transition metal catalyst promoted activation of the C–X bond initiated the reaction sequence and was followed by intramolecular or intermolecular cyclization.

2.1 1,2-Dihaloarenes

The application of 1,2-dihaloarenes in organic synthesis has experienced long-term development. The most common application is the *in situ* generation of benzyne derivatives. However, using transition metal catalysts with 1,2-dihaloarenes as substrates offers more diversity for the outcome.

In 1991, Perry and Turner reported the preparation of *N*-substituted phthalimides by palladium-catalyzed carbonylation of 1,2-dihaloarenes with primary amines.[1] Various desired products were produced in moderate to good yields under CO pressure (7 bar) in the presence of a palladium catalyst (Scheme 2.1a). In this procedure, both aromatic and aliphatic primary amines were successfully applied. In the case of 1,2-diiodobenzenes, the nitro group cannot be tolerated and no desired product could be detected. 1,2-Dibromocyclopentene was tested as a substrate as well; 20% of the desired imide was formed and several by-products were formed. This transformation was studied further by Alper's group,[2] Kollár's group[3] and our group.[4] In the report from Alper and co-worker, they found PSIL 102

RSC Catalysis Series No. 24
Heterocycles from Double-Functionalized Arenes: Transition Metal Catalyzed Coupling Reactions
By Xiao-Feng Wu and Matthias Beller

Published by the Royal Society of Chemistry, www.rsc.org

(trihexyl(tetradecyl)phosphonium bromide) is a particularly effective general reaction media for the palladium-catalyzed double carbonylation reactions of dihaloarenes and amines. The desired products were afforded in excellent yields. Remarkably, the catalyst system can be reused. After the reaction, the ionic liquid was partitioned with hexane, and the substituted phthalimide product was extracted into hexane. The IL phase containing active palladium catalyst was dried under vacuum and recharged with starting material and base. After 24 h, the target products were isolated in 65 and 73% yield. The nature of the base played an important role in the reaction. DBU was found to be the best base, while NEt_3 gave no desired product. In our report, we studied the use of 2-aminopyridines as coupling partners with 1,2-dibromobenzenes. Some very sterically hindered amines were checked as well, such as 2,6-diisopropylaniline and 1-adamantylamine; in addition to the corresponding phthalimides, some isomers were isolated too (Scheme 2.1b).[5]

In 2011, Bhanage and co-workers reported a carbon monoxide-free one-step synthesis of phthalimides by using formamides as an amine and CO source.[6] With $POCl_3$ as the activator, various phthalimides were produced in moderate to excellent yields (Scheme 2.2). In addition to 1,2-dihalobenzenes, 2-iodobenzoic acid and methyl 2-iodobenzoate can be applied as substrates for phthalimide preparation with formamides as an amine and CO source as well. This methodology was extended and applied in the synthesis of isobenzofuran-1(3*H*)-one (70% yield) by using *o*-iodobenzyl alcohol as the substrate.

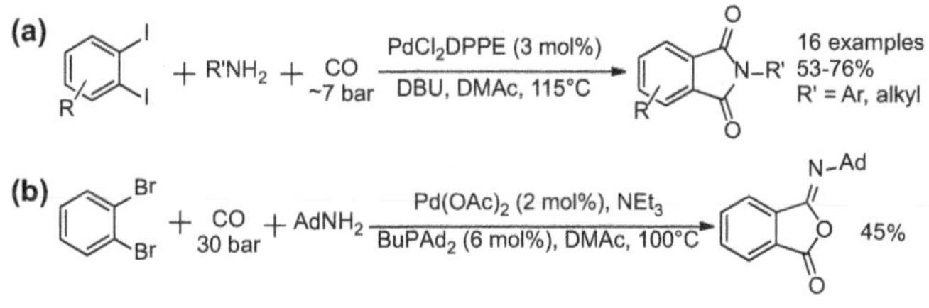

Scheme 2.1 Palladium-catalyzed phthalimide synthesis with CO.

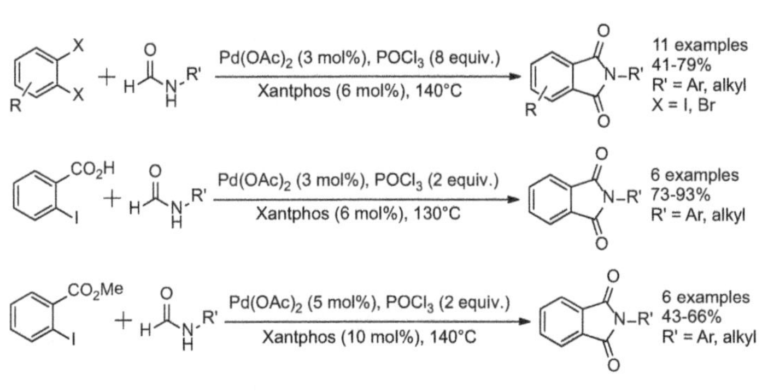

Scheme 2.2 Palladium-catalyzed phthalimide synthesis with formamides.

Reaction Procedure (Scheme 2.2): A mixture of *o*-diiodoarene (0.5 mmol), $Pd(OAc)_2$ (3 mol%) and Xantphos (6 mol%) in formamide (5 mL) was placed in a 25 mL two-necked round-bottomed flask equipped with a condenser under nitrogen at room temperature. $POCl_3$ (4 mmol) was added, and the reaction mixture was stirred at 140 °C for 20 h. After completion, the reaction mixture was cooled to room temperature and poured into a saturated solution of $NaHCO_3$ (50 mL). The product was extracted into diethyl ether (4×20 mL). The combined organic layers were dried with Na_2SO_4 and evaporated to afford the crude product, which was purified by column chromatography on silica gel (petroleum ether–ethyl acetate) to afford the pure product.

Meanwhile the same group found palladium on carbon can be used as an efficient and reusable heterogeneous catalyst for the carbonylative synthesis of *N*-substituted phthalimides.[7] 1,2-Diiodoarenes were dicarbonylated with amines and provided the corresponding *N*-substituted phthalimides in excellent yields. *o*-Halobenzoates and *o*-halobenzoic acid can be applied as substrates as well. During their optimization process, they observed that the nature of the solvent affected the yield of the reaction. Solvents such as DMF, DMSO and THF provided lower yields of the expected product whereas acetonitrile did not furnish the desired product. The best result was observed in toluene. Remarkably, no phosphine ligands were needed here and the catalyst can be reused for up to eight consecutive cycles with maintenance of high activity and selectivity (Scheme 2.3). The authors performed ICP-AES (inductively coupled plasma atomic emission spectroscopy) analysis of the 1st and 8th recycle runs, where below 0.01 ppm of palladium in solution was estimated thus indicating no significant leaching of the palladium catalyst. In 2014, they immobilized a palladium catalyst on an ionic liquid and applied it in the double carbonylation of 1,2-diiodobenzene as well.[8] Good yields of the desired products can be isolated and the catalyst can be reused.

At the end of 2013, our group described a palladium-catalyzed carbonylative synthesis of phthalimides from 1,2-dibromoarenes.[9] Molybdenum hexacarbonyl was applied as a CO source here. In this easy and convenient way, various *N*-substituted phthalimides were produced in a one-pot manner. A wide range of different primary amines as well as a variety of miscellaneous 1,2-dibromobenzenes can be applied as substrates, and the corresponding phthalimides were obtained in moderate to excellent yields (Scheme 2.4).

Miura and co-workers reported a palladium-catalyzed cross-coupling of benzyl ketones and 1,2-dibromobenzenes to benzofurans.[10] By using already available substrates and with the assistance of a palladium catalyst, various desired benzofurans were produced in moderate to good yields (Scheme 2.5). Regarding the reaction mechanism, the initial step was proposed to be the C–C coupling of the C–Br bond with the benzylic position of benzyl

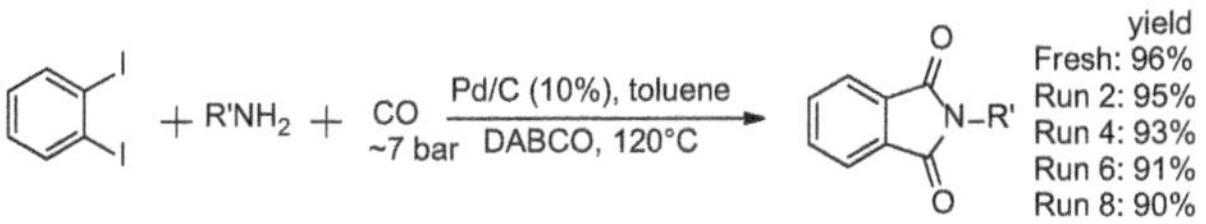

Scheme 2.3 Pd/C-catalyzed phthalimide synthesis with CO.

Reaction Procedure (Scheme 2.3): To a 100 mL stainless steel autoclave, *o*-iodo derivative (0.5 mmol), amine (0.7 mmol), 10% Pd/C (2.5 mol%), toluene (7 mL) and DABCO (1.5 mmol) were added. The autoclave was closed, purged three times with carbon monoxide, pressurized with 90 psi of CO and then heated at 120 °C for 2 h. The reactor was cooled to room temperature and the remaining CO gas was carefully vented and the reactor was opened. The reactor vessel was thoroughly washed with ethyl acetate (2×10 mL) to remove any traces of product and catalyst if present. The catalyst was filtered and the reaction mixture was evaporated under vacuum. The residue obtained was purified by column chromatography (petroleum ether–ethyl acetate, 95 : 5) to afford the desired product.

Procedure for recycling the catalyst: The catalyst obtained after filtration was thoroughly washed with distilled water (3×5 mL) and then with methanol (3×5 mL) to remove any traces of organic material if present. The resulting catalyst was dried in an oven at 80 °C for 4 h and was then used for the next cycle.

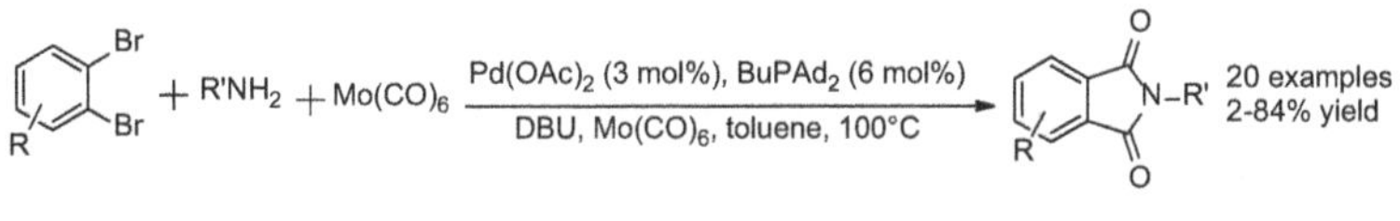

Scheme 2.4 Pd-catalyzed phthalimide synthesis with $Mo(CO)_6$.

Reaction Procedure (Scheme 2.4): 1,2-Dibromobenzene (1 mmol), 1.2 mmol of butylamine (110 mL), 6.7 mg of $Pd(OAc)_2$ (3 mol%), $BuPAd_2$ (6 mol%), 1 mmol of $Mo(CO)_6$ (264 mg), 2 mmol of DBU (300 mL) and 5 mL toluene were charged in an argon-flushed pressure tube which was subsequently sealed. The mixture was stirred for 16 h at 100 °C. Afterwards, the mixture was diluted with CH_2Cl_2 and washed with water. The aqueous phase was extracted twice with CH_2Cl_2, the organic layers were combined and dried over Na_2SO_4. The solvent was removed under reduced pressure. Column chromatography (hexane–ethyl acetate, 1 : 0–8 : 2) gave *N*-butylphthalimide as the final product.

ketones followed by intramolecular cyclization to give the desired product. Interestingly, 1-naphthol and 2-*tert*-butylphenol can be applied as coupling partners with 1,2-dibromobenzene as well. The corresponding

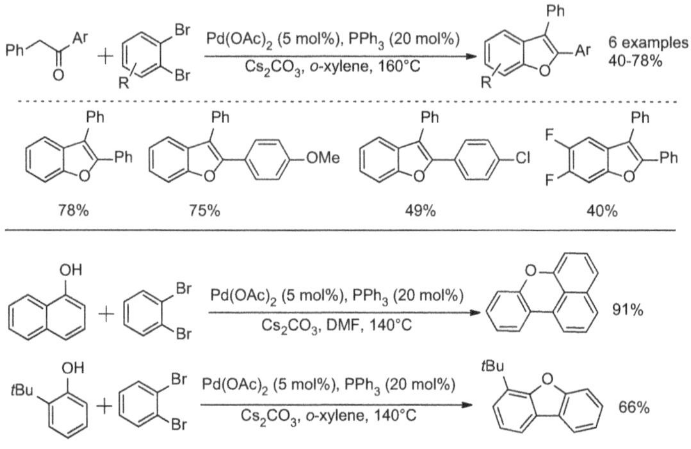

Scheme 2.5 Pd-catalyzed benzofuran synthesis from benzyl ketones.

oxygen-containing heterocycles were isolated in moderate yields under identical conditions. Later on, Domínguez and co-workers studied this transformation further and polymer-anchored palladium (FibreCat™) was applied in the benzofuran synthesis in addition to a homogeneous catalyst.[11]

A palladium-catalyzed intramolecular cyclization of enolate *O*-arylation and thio-enolate *S*-arylation was reported by Willis and co-workers in 2006.[12] A catalyst generated from $Pd_2(dba)_3$ and the ligand DPEphos effects the key bond formation to deliver a variety of substituted products from both cyclic and acyclic precursors. The analogous thio-ketones undergo C–S bond formation using identical reaction conditions and were converted to benzothiophene products. A cascade sequence that produces the required α-aryl ketones *in situ* has also been developed (Scheme 2.6).

In addition to palladium catalysts, copper as a cheaper transitional metal catalyst was explored in the synthesis of benzofurans as well. In 2007, a CuI-catalyzed coupling of 1-bromo-2-iodobenzenes with β-keto esters in THF at 100 °C leading to 2,3-disubstituted benzofurans was developed by Ma and co-workers.[13] This domino transformation involves an intermolecular C–C bond formation and a subsequent intramolecular C–O bond formation process. Benzofurans with different substituents at the 5- and 6-position were accessible in good yields by employing the corresponding 1-bromo-2-iodobenzenes as substrates (Scheme 2.7a). This transformation was further studied by Beifuss and co-workers in 2012.[14] They found the reaction conditions from Ma's group failed in the case of 1,3-diketones. In their reported procedure, the reaction of 1-bromo-2-iodobenzenes and other 1,2-dihalobenzenes with 1,3-cyclohexanediones proceeded selectively in DMF at 130 °C using Cs_2CO_3 as a base and pivalic acid as an additive. The corresponding 3,4-dihydrodibenzo[*b*,*d*]furan-1(2*H*)-ones were isolated with yields ranging from 47 to 83% (Scheme 2.7b). The highly regioselective domino process is

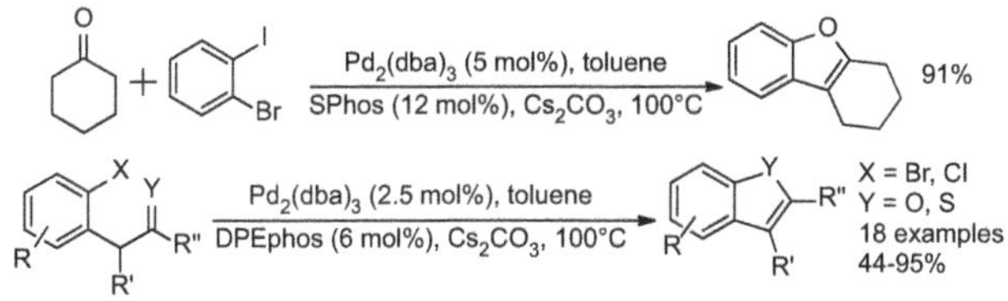

Scheme 2.6 Pd-catalyzed benzofuran synthesis from ketone.

Reaction Procedure (Scheme 2.6): Caesium carbonate (691 mg, 2.212 mmol) was added to a flask charged with $Pd_2(dba)_3$ (32 mg, 0.035 mmol) and SPhos (17 mg, 0.042 mmol) under nitrogen. The reagents were suspended in anhydrous toluene (1.00 mL) and 1-bromo-2-iodobenzene (200 mg, 0.707 mmol, 0.09 mL) and cyclohexanone (80 mg, 0.848 mmol, 0.09 mL) were added under nitrogen and the reaction was heated to 100 °C for 48 h. After cooling, the reaction mixture was diluted with diethyl ether (*ca.* 10 mL), filtered through Celite and reduced *in vacuo*. The residue was purified *via* flash column chromatography (petrol) to yield the benzofuran (0.111 g, 91%) as a colourless oil.

based on an intermolecular Ullmann-type *C*-arylation followed by an intramolecular Ullmann-type *O*-arylation. Substituted products are accessible by employing substituted 1-bromo-2-iodobenzenes and substituted 1,3-cyclohexanediones as substrates. Reaction with an acyclic 1,3-diketone leads to the corresponding benzo[*b*]furan.

Benzoxazoles are an important class of compounds containing one oxygen atom and one nitrogen atom, and have been reported with many applications in various areas. Traditionally, they can be prepared by a reaction between 2-aminophenols and aldehydes or carboxylic acids under relatively harsh conditions. In 2004, Glorius's group reported a copper-catalyzed C–N and C–O coupling procedure for the synthesis of benzoxazoles.[15] Readily available 1,2-dihaloarenes and primary amides were applied as substrates, the desired products were isolated in good yields in a single step (Scheme 2.8). This procedure was later applied by our group in the successive synthesis of benzoxazoles from aryl bromides and 1,2-dibromobenzenes.[16] By the combination of palladium-catalyzed aminocarbonylation and copper-catalyzed coupling reactions, a variety of substituted benzoxazoles were produced in moderate to good yields.

In 2013, Xiang, Wang and co-workers developed a copper-catalysed annulation reaction of 1,2-dihaloarenes with nitriles.[17] In this procedure, the nitriles were hydrated by a base into the corresponding primary amides *in situ* and then coupled with 1,2-dihaloarenes. In this system, combined bases (Cs_2CO_3 and KOH) and a relative high temperature were needed. Later on, they found that the reaction conditions can be much milder by adding acetaldoxime as an additive. In this new procedure, the role of acetaldoxime

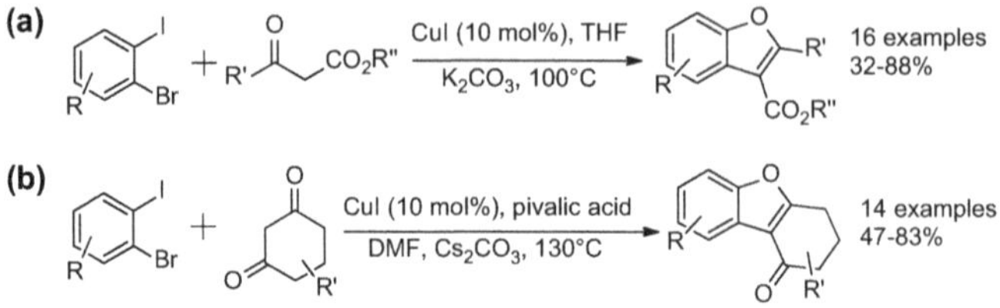

Scheme 2.7 Cu-catalyzed benzofuran synthesis.

Reaction Procedure (Scheme 2.7a): An oven-dried Schlenk tube was charged with CuI (10 mg, 0.05 mmol), potassium carbonate (207 mg, 1.5 mmol), and 1-bromo-2-iodobenzenes (0.5 mmol). The tube was evacuated and backfilled with argon (three times), and then ethyl acetoacetate (130 mg, 1 mmol) and 1.5 mL of THF were added. The reaction mixture in this sealed tube was stirred at 100 °C until the starting aryl iodide was consumed. The cooled solution was partitioned between ethyl acetate (40 mL) and saturated brine (10 mL). The organic layer was isolated, and the aqueous layer was extracted with ethyl acetate. The combined organic layers were dried over Na_2SO_4 and concentrated in vacuum. The residue thus obtained was purified by silica gel chromatography to give the product.

Reaction Procedure (Scheme 2.7b): An oven-dried sealed tube was flushed with argon and charged with pivalic acid (122 mg, 1.2 mmol), Cs_2CO_3 (651 mg, 1.5 mmol), a 1,2-dihaloarene (1 mmol), a 1,3-diketone (1.5 mmol), and CuI (19 mg, 0.1 mmol). The tube was evacuated and backfilled with argon twice, and dry DMF (2 mL) was added. The reaction mixture was stirred at 130 °C until the 1,2-dihaloarene was consumed. After being cooled to room temperature, the reaction mixture was partitioned between EtOAc (30 mL) and saturated ammonium chloride (20 mL). The organic layer was isolated, and the aqueous layer was extracted with EtOAc (3×30 mL). The combined organic layers were dried over anhydrous $MgSO_4$ and concentrated in vacuum. The residue was purified by flash chromatography on silica gel to afford the desired product.

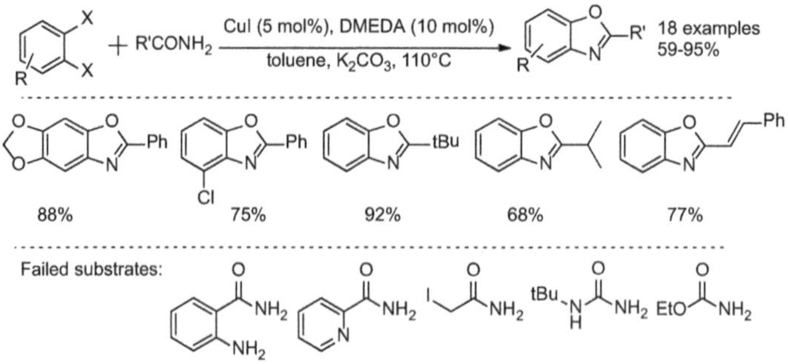

Scheme 2.8 Cu-catalyzed synthesis of benzoxazoles from amides.

Reaction Procedure (Scheme 2.8): 1,2-Dibromobenzene (120 mL, 1.0 mmol), benzamide (133 mg, 1.1 mmol), K_2CO_3 (414 mg, 3.0 mmol) and CuI (10 mg, 0.05 mmol) were weighed into a vial under air. The vial was evacuated and filled with argon, followed by the addition of *N,N′*-dimethylethylenediamine (11 mL, 0.1 mmol) and toluene (3 mL). The vial was closed and the reaction mixture stirred at 110 °C for 24 h. After cooling to room temperature, the reaction mixture was poured into 25% aqueous NH_4OH, extracted with EtOAc, dried over Na_2SO_4, filtered and concentrated. Chromatography over silica gel (EtOAc–hexane, 1 : 10) yielded the pure product.

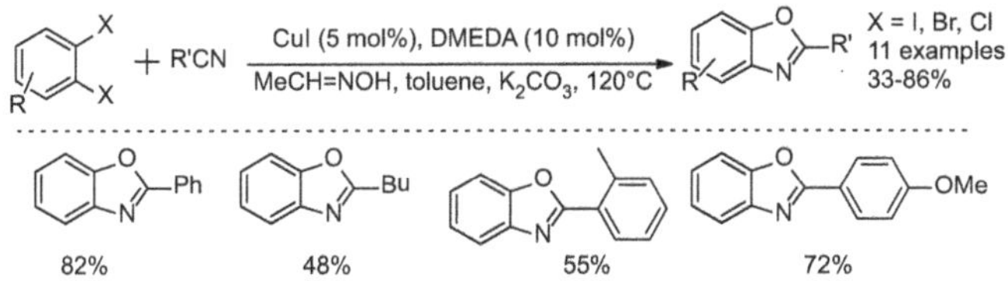

Scheme 2.9 Cu-catalyzed synthesis of benzoxazoles from nitriles.

Reaction Procedure (Scheme 2.9): To a suspension of CuI (9.5 mg, 0.05 mmol) and K_2CO3 (207.3 mg, 1.5 mmol) in toluene (2.0 mL) were added aryl *o*-dihalides (0.5 mmol), nitrile (1.0 mmol), DMEDA (8.0 μL, 0.075 mmol) and acetaldoxime (92.0 μL, 1.5 mmol) under argon. The reaction mixture was stirred for 15 h at 120 °C under argon. The solution was cooled to room temperature and the solvent was removed under vacuum. The crude product was purified by column chromatography on silica gel (eluent: petroleum ether–ethyl acetate, 30 : 1) to afford the benzoxazoles.

is to hydrolyse nitriles to the corresponding amides. The desired products were formed in moderate to good yields (Scheme 2.9).

Nitrogen-containing heterocycles are important heterocyclic compounds with broad applications in various areas. Among the known compounds, carbazoles are a representative example. In 2007, a procedure based on palladium-catalyzed N–H/C–H activation was developed by Ackermann and co-workers.[18] By using 1,2-dichloroarenes and anilines as substrates and palladium as the catalyst, the desired products were isolated in moderate to good yields (Scheme 2.10). Additionally, naturally occurring murrayafoline A was synthesized in a good yield by this protocol as well. In their further study, they found that IPr · HCl and PPh_3 can be effective ligands as well if 1,2-dibromobenzene is applied. The reaction temperature can be decreased to 105 °C by using NaO*t*Bu as a base.

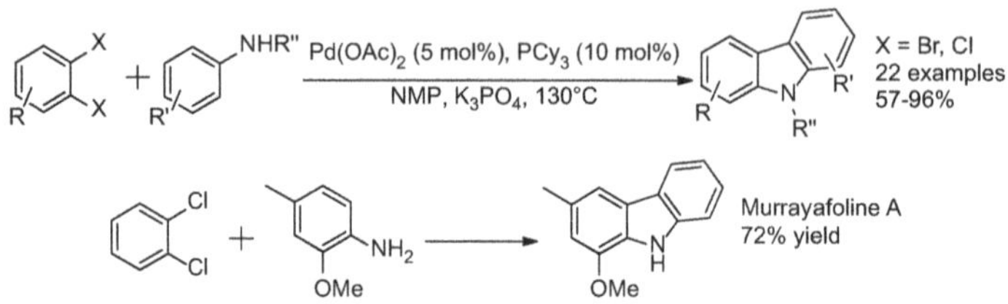

Scheme 2.10 Pd-catalyzed synthesis of carbazoles from anilines.

Reaction Procedure (Scheme 2.10): Synthesis of murrayafoline A: A solution of $Pd(OAc)_2$ (11.2 mg, 0.05 mmol, 5.0 mol%), PCy_3 (28.9 mg, 0.10 mmol, 10 mol%), finely powdered K_3PO_4 (467 mg, 2.20 mmol), 2-methoxy-4-methylaniline (165 mg, 1.20 mmol), and 1,2-dichlorobenzene (424 mg, 1.20 mmol) in dry NMP (10.0 mL) was stirred for 18 h at 130 °C under N_2. Et_2O (25 mL) and H_2O (25 mL) were added to the reaction mixture at ambient temperature. The separated aqueous phase was extracted with Et_2O (2×75 mL). The combined organic layers were washed with brine (50 mL), dried over $MgSO_4$, and concentrated *in vacuo*. The remaining residue was purified by column chromatography on silica gel (*n*-pentane–Et_2O, 50 : 1–30 : 1) to yield murrayafoline A as an off-white solid (152 mg, 72%).

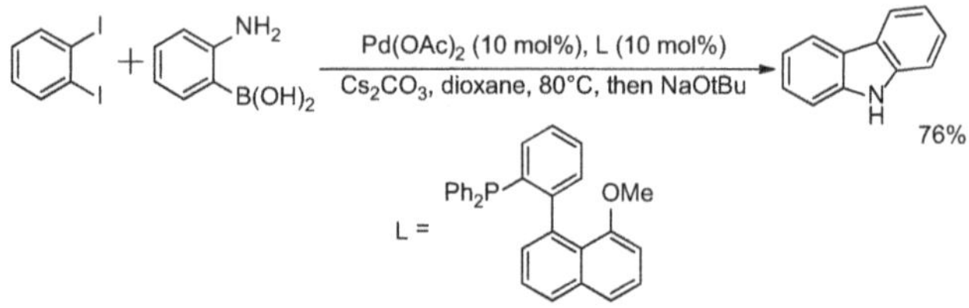

Scheme 2.11 Pd-catalyzed synthesis of carbazoles from 2-aminophenylboronic acid.

Another procedure was developed by Kan and co-workers in 2008 by applying Suzuki–Miyaura and amination reactions as the two key elemental steps.[19] 2-Aminophenylboronic acid and 1,2-diiodobenzene were applied as the starting materials; the corresponding carbazole was formed in a one-pot two-step manner in good yield (Scheme 2.11). In their optimizations, they found that the choice of suitable palladium precursor and solvent had a very important effect; $Pd_2(dba)_3$ or toluene resulted in much lower yields.

The reaction between vinylogous amides and aryl halides was studied by Edmondson and co-workers in 2000.[20] In this report, the first tandem Hartwig–Buchwald–Heck cyclization was described as well and applied to the synthesis of 2,3-disubstituted indole derivatives. The reaction between enaminone and 2-bromobenzaldehyde could lead to a quinoline derivative under their reaction conditions (Scheme 2.12).

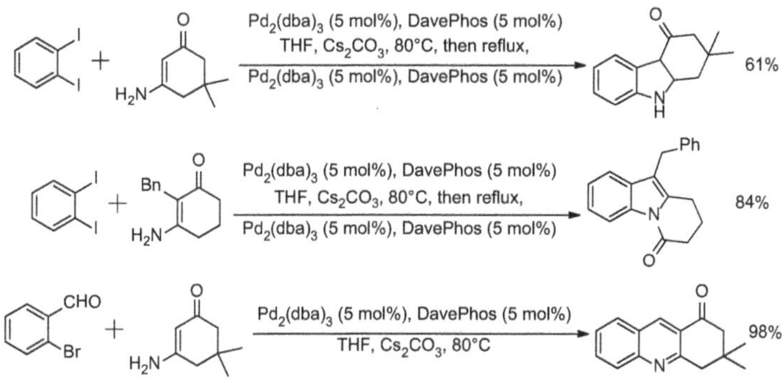

Scheme 2.12 Pd-catalyzed synthesis of indoles from enaminone.

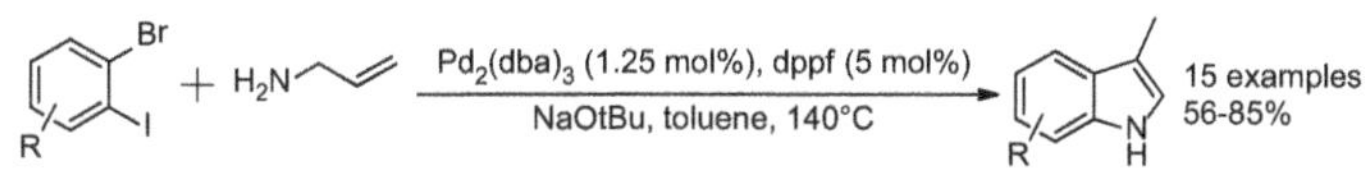

Scheme 2.13 Pd-catalyzed synthesis of indoles from allylamine.

Reaction Procedure (Scheme 2.13): [Pd_2dba_3] (1.25 mol%), DPPF (5 mol%), NaO*t*Bu (3.75 mmol), the aryl halide (1.5 mmol), toluene (4 mL), and allylamine (1.5 mmol) were placed in a 7 mL screw-cap vial. The resulting mixture was heated from room temperature to 140 °C over approximately 0.5 h and then stirred at 140 °C for 5 h.

Due to the importance of indoles, Joergensen and co-worker reported a palladium-catalyzed amination, followed by a Heck reaction of 1,2-dihaloarenes in 2008.[21] Various 3-substituted indoles were isolated in moderate to good yields by using allylamine as a coupling partner (Scheme 2.13). In their study, they showed that the first step for this transformation was C–N coupling and the intermediate was isolated as well.

Kurth and co-worker developed a versatile one-pot three component assembly route to highly substituted indoles in 2012.[22] Their procedure was based on palladium-catalyzed coupling reactions and 1,2-dihaloarenes, primary amines and ketones or aldehydes were applied as the substrates. Indole formation can be accomplished using suitable reaction conditions that favour *in situ* generation of an aniline, condensation to an arylenamine, and a subsequent arene–alkene coupling reaction to close the ring. A single palladium catalyst/ligand system mediated both coupling reactions. A detailed mechanistic study and DFT calculation for the palladium-catalyzed ring closing step was presented as well. Moderate to good yields of the desired indoles were produced (Scheme 2.14).

In 2014, Xi and co-workers developed an interesting indole synthesis based on a palladium-catalyzed four-component coupling reaction.[23]

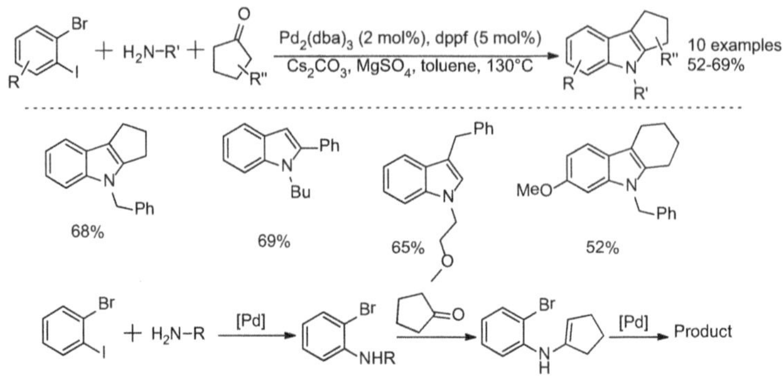

Scheme 2.14 Pd-catalyzed synthesis of indoles from amines and ketones.

Reaction Procedure (Scheme 2.14): A 10 mL vial was charged with Cs_2CO_3 (0.720 g, 2.2 mmol), DPPF (0.028 g, 0.05 mmol), $Pd_2(dba)_3$ (0.018 g, 0.02 mmol), $MgSO_4$ (0.600 g, 5 mmol), dry toluene (2 mL) and a small stir bar. The vial was sealed with a Teflon cap and the solution was degassed by bubbling Ar through for 5 min. The appropriate *o*-bromoiodoarene (1 mmol) and primary amine (1 mmol) were added to the reaction vessel *via* a syringe and the reaction was stirred at 130 °C for 6 hours. The appropriate aldehyde or ketone (3 mmol) was then added *via* a syringe and the reaction was again stirred and heated at 130 °C for 48 hours. The reaction was allowed to cool to room temperature, diluted with EtOAc (50 mL) and filtered through a bed of Celite. The crude mixture was isolated under reduced pressure and purified *via* column chromatography. All reactions were run by strictly avoiding oxygen.

By applying 1,2-dihaloarenes, alkynes, and amines as substrates, the corresponding indole derivatives were formed in moderate to excellent yields with the cleavage of $C(sp^3)$–N bonds (Scheme 2.15).

Barluenga and co-workers presented a new efficient method for the synthesis of indoles from readily available starting materials, through palladium-catalyzed domino and three-component/cascade processes in 2007.[24] This methodology represented a competitive alternative for the preparation of structurally diverse indoles by incorporation of an azaallylic anion—obtained by deprotonation of an imine—into the target heterocycles. By using imines or imines formed *in situ* and 1,2-dihaloarenes as substrates, the desired indoles were isolated in moderate to good yields (Scheme 2.16). A more detailed study of the scope of this new palladium-catalyzed synthesis of indoles was investigated and reported later on. They succeeded in extending their substrates from 1,2-dihaloarenes to *o*-halobenzene sulfonates. Mechanistically, this cascade reaction comprises an imine α-arylation followed by an intramolecular C–N bond-forming reaction promoted by the

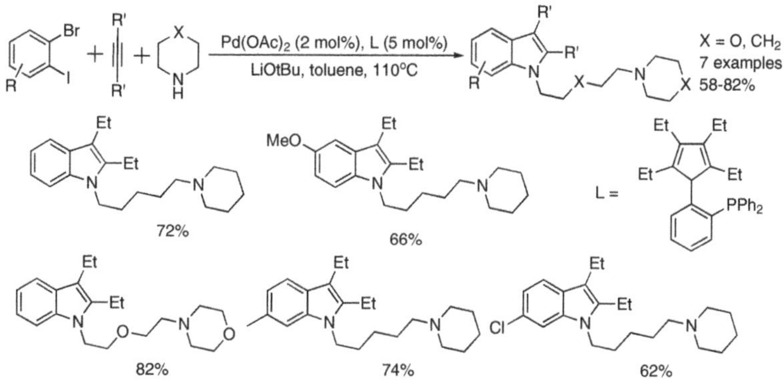

Scheme 2.15 Pd-catalyzed four-component synthesis of indoles.

Reaction Procedure (Scheme 2.15): Under an atmosphere of nitrogen, $Pd(OAc)_2$ (2 mol%), L (5 mol%), and Li*Ot*Bu (1.5 mmol) were added to toluene (2 mL). After this reaction mixture was stirred at room temperature for 5 min, 1,2-dihaloarene (0.5 mmol), alkyne (0.6 mmol), and amine (2.5 mmol) were added, and this reaction mixture was stirred at 110 °C for 12 h. The reaction mixture was quenched with water and extracted with Et_2O. The extraction was washed with brine and dried over $MgSO_4$. The solvent was then evaporated under vacuum, and the residue was purified by using a silica gel column with petroleum ether and ethyl acetate as eluent to afford the final products.

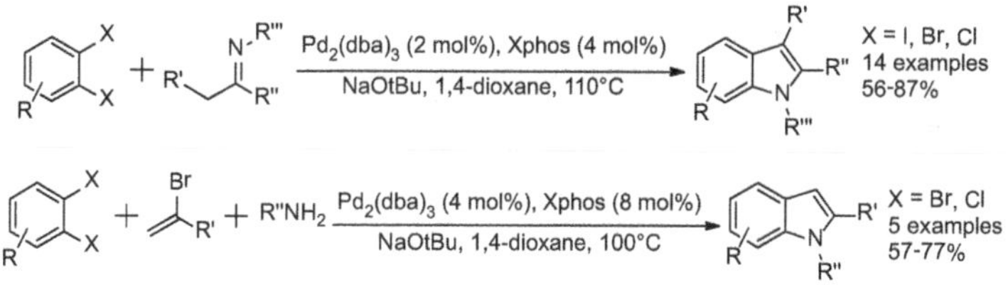

Scheme 2.16 Pd-catalyzed synthesis of indoles from imines.

Reaction Procedure (Scheme 2.16): A reaction tube under a nitrogen atmosphere was charged with XPhos (19.6 mg, 0.04 mmol, 4 mol%), tris(dibenzylideneacetone)dipalladium (0) (18.3 mg, 0.02 mmol, 2 mol%), sodium *tert*-butoxide (268 mg, 2.8 mmol, 2.8 equiv.) and dioxane (3 mL). After 1 min, the dihalide (1 mmol) was added under nitrogen and heated to 110 °C. After 5 min, the imine (1 mmol) was added under nitrogen with stirring until the complete disappearance of the dihalide as determined by GC. The mixture was allowed to cool to room temperature, taken up in hexanes (15 mL), and filtered through Celite. The solvents were evaporated under reduced pressure. Purification by flash chromatography (SiO_2, hex–EtOAc, 20 : 1) afforded pure indoles.

same palladium catalyst. The reaction with 1,2-dibromobenzene shows a wide scope and allows the introduction of aryl, alkyl, and vinyl substituents at different positions on the five-membered ring of the indole. The regioselective synthesis of indoles substituted in the six-membered ring can be carried out by employing *o*-dihalobenzene derivatives with two different halogens, taking advantage of the different reactivities of I, Br, and Cl in oxidative addition reactions. They found that a very high yield of indoles can be achieved by using *o*-chlorononaflates as substrates, which is particularly appropriate for the synthesis of the challenging 6-substituted indoles. Soon after, they found that this transformation can be performed in water with the assistance of microwaves.

Benzimidazoles are an important class of compounds for the pharmaceutical industry; the benzimidazole core structure can be found in many commercial drugs such as Prilosec, Nexium, Protonix, Atacand, Famvir, and Vermox, as well as numerous experimental drug candidates in a wide range of therapeutic areas. Regarding their importance, many procedures have been developed for their preparation. In 2009, Deng and co-workers found that CuI/DMEDA (*N*,*N'*-dimethylethylenediamine) is an efficient catalyst system for the amination of arylhalides with guanidines.[25] Under their conditions, 1-*H*-2-aminobenzimidazoles can be prepared in modest yields in a one-pot manner through tandem aminations of 1,2-dihaloarenes (Scheme 2.17). Various other 2-substituted benzimidazoles can be prepared by this methodology as well. Later on in 2012, Schmitt and co-workers found that the combination of CuOAc and L-proline can act as another powerful system for the synthesis of benzimidazoles from the same types of substrates.[26]

Regarding the limitations with the copper-promoted systems, palladium-catalysts were explored and applied in benzimidazole synthesis as well. In 2011, You and co-workers developed a palladium-catalyzed synthesis of 1,2-disubstituted (hetero)aryl fused imidazoles from 1,2-dihaloarenes.[27] This palladium-catalyzed cascade amination allows regiospecific and modular synthesis of a library of structurally diverse 1,2-disubstituted (hetero)aryl fused imidazoles in good to excellent yields (Scheme 2.18). The prepared

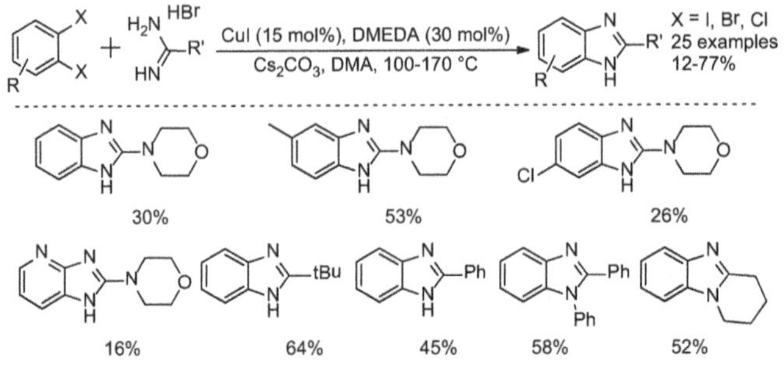

Scheme 2.17 Cu-catalyzed synthesis of benzimidazoles.

Reaction Procedure (Scheme 2.17): A Schlenk tube equipped with a strong magnetic stir bar was charged with CuI (57 mg, 0.3 mmol, 0.15 equiv.), morpholine-4-carboxamidine hydrobromide (588 mg, 2.8 mmol, 1.4 equiv.), and Cs_2CO_3 (2.6 g, 8 mmol, 4 equiv.). The Schlenk tube was evacuated and backfilled with N_2 three times. Under a N_2 atmosphere, DMA or NMP (5 mL), 1,2-diiodobenzene (659 mg, 2 mmol, 1.0 equiv.) and *N,N′*-dimethylethylenediamine (53 mg, 0.6 mmol, 0.3 equiv.) were added sequentially *via* syringe. The reaction mixture was stirred in a preheated oil bath at 150 °C for 24 h and then cooled to room temperature. EtOAc (25 mL) was added, and the suspension was stirred for 30 min. The inorganic salt was filtered off and washed with EtOAc. The volatile EtOAc was removed under reduced pressure, and the remaining DMA or NMP solution was directly loaded on a preparatory HPLC for purification to afford the product.

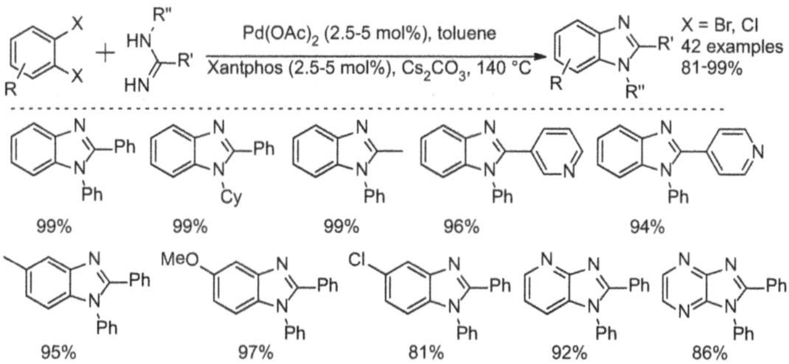

Scheme 2.18 Pd-catalyzed synthesis of benzimidazoles.

Reaction Procedure (Scheme 2.18): A flame-dried Schlenk test tube with a magnetic stir bar was charged with $Pd(dba)_2$ or $Pd(OAc)_2$ (2.5–5.0 mol%), Xantphos (2.5–5.0 mol%), *t*BuONa (96 mg, 1.0 mmol) or Cs_2CO_3 (324 mg, 1.0 mmol), amidine (0.30 mmol), *o*-dihaloarene (0.25 mmol) and toluene (1 mL) in the presence of 4 Å sieves (100 mg) under N_2. A rubber septum was replaced with a glass stopper, and the system was then evacuated three times and backfilled with N_2. The reaction mixture was stirred for 10 min at room temperature, and then heated at 140 °C for 24 h. The reaction mixture was then cooled to ambient temperature, diluted with 10–15 mL of CH_2Cl_2, filtered through a plug of silica gel, and washed with 10–20 mL of CH_2Cl_2. The combined organic extracts were concentrated and the resulting residue was purified by column chromatography on silica gel to provide the desired product.

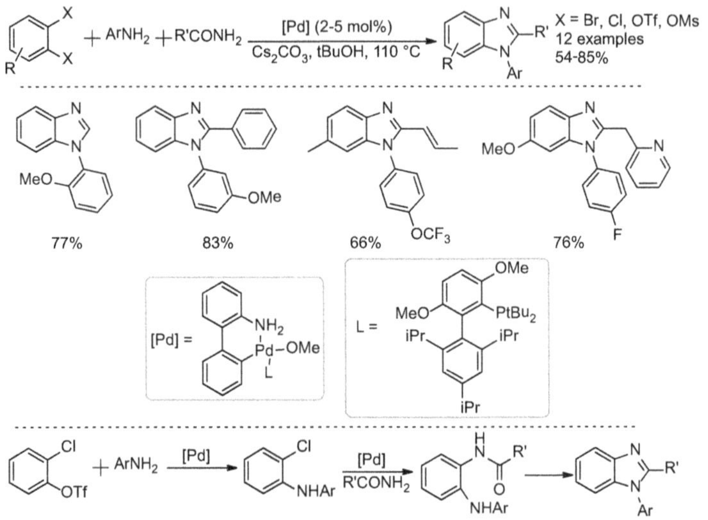

Scheme 2.19 Pd-catalyzed synthesis of benzimidazoles from amines and amides.

> **Reaction Procedure** (Scheme 2.19): An oven-dried test tube equipped with a magnetic stir bar and a Teflon screw-cap was charged with *t*BuBrettPhos precatalyst (2–5 mol%), arylamine (1.0 mmol, if solid), amide (1.3 mmol, if solid), 2-chloroaryl triflate or 2-chloroaryl bromide (1.0 mmol, if solid), and Cs_2CO_3 (2.4 mmol). The tube was evacuated and backfilled with nitrogen using standard Schlenk techniques and this process was conducted three times. The arylamine (1.0 mmol, if liquid), amide (1.3 mmol, if liquid) and 2-chloroaryl triflate or 2-chloroaryl bromide (1.0 mmol, if liquid) were added followed by *t*-BuOH (1.5 mL) *via* syringe and the punctured cap was replaced with a new one under a stream of nitrogen. The test tube was placed in a preheated oil bath at 110 °C and stirred for the indicated time. After cooling to room temperature, the mixture was diluted (3 mL CH_2Cl_2), filtered through Celite (CH_2Cl_2 rinse), and concentrated using a rotary evaporator. The residue was purified by silica gel chromatography using a Biotage Isolera 4 or by neutral alumina column chromatography to provide the pure product.

products exhibited various interesting fluorescent properties and elucidated the correlation between chemical structures and fluorescent characteristics.

In 2013, Buchwald and Jui developed a novel approach to palladium-catalyzed regioselective *N*-arylbenzimidazole synthesis.[28] This methodology involves cascade intermolecular amination and amidation reactions of 2-chloroaryl sulfonates (or halides); the desired benzimidazoles were isolated in good yields in general (Scheme 2.19). In their procedure, a single catalyst based on *t*BuBrettPhos was able to selectively perform both catalytic

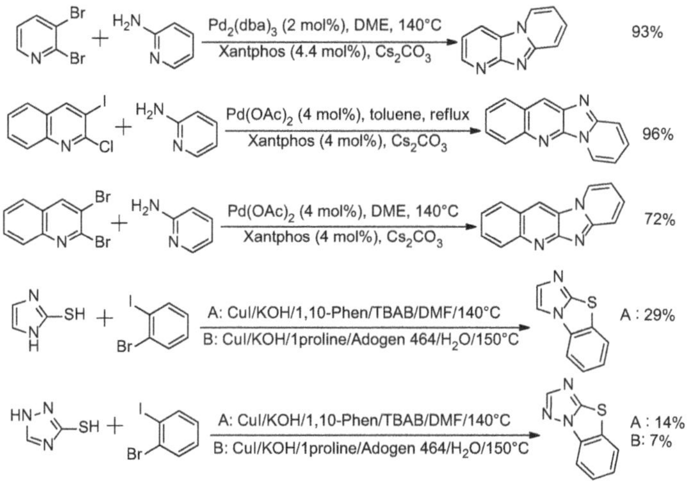

Scheme 2.20 Synthesis of dipyrido[1,2-*a*:2′,3′-*d*]imidazole and imidazothiazoles.

Reaction Procedure (Scheme 2.20): A round-bottomed flask of 50 mL was charged with $Pd_2(dba)_3$ (0.030 mmol, 0.028 g, 2.0 mol%), Xantphos (0.066 mmol, 0.038 g, 4.4 mol%), and DME (1,2-dimethoxyethane; 5 mL). The obtained mixture was flushed with N_2 for 10 min under magnetic stirring. Meanwhile, a pressure tube of 80 mL was charged with 2,3-dibromopyridine (1.5 mmol, 0.355 g), amidine (1.8 mmol), and caesium carbonate (6.0 mmol, 1.955 g). To this mixture was added the preformed Pd catalyst under a N_2 flow. The flask of 50 mL was subsequently rinsed with 2×5 mL of DME. Then, the resulting mixture was flushed with N_2 for 5 min, sealed, and heated (internal temperature: 140 °C, oil bath temperature: 160 °C) under vigorous magnetic stirring for 24 h. After cooling the reaction mixture to room temperature, 20 mL of dichloromethane was added and the suspension was filtered over a glass filter. The filtrate was evaporated under reduced pressure, and the residue was purified by flash column chromatography on silica gel using dichloromethane–methanol (97 : 3) as the eluent.

elements of this process for a broad range of arylamines (or heteroarylamines), amides, and bifunctional electrophile substrates; the corresponding benzimidazole products with complete regioselectivity were afforded. Additionally, they demonstrated that different 2-chloroaryl sulfonates (triflate *versus* mesylate), which are derived from the same chlorophenols, can be reacted under very similar reaction conditions to exclusively afford the opposite regioisomeric heterocycles.

Maes and co-workers developed a regioselective palladium-catalyzed synthesis of dipyrido[1,2-*a*:2′,3′-*d*]imidazole in 2006.[29] Their methodology either involves consecutive palladium-catalyzed intermolecular amination and Cu-catalyzed intramolecular amination or a simple alteration of the reaction temperature with a palladium catalyst. The desired products were formed in good yields by both procedures (Scheme 2.20). This method gives access to unknown polycyclic aza heteroaromatic skeletons with potential antitumor properties in a simple and straightforward way. Meanwhile, they applied their methodology in the synthesis of pyrido[2′,1′:2,3]imidazo[4,5-*b*]quinoline and pyrido[1′,2′:1,2]imidazo[4,5-*b*]quinoline by using 2,3-dibromoquinoline as the coupling partner.[30] By using 3-mercaptotriazole and 1-bromo-2-iodobenzene as substrates, [1,2,4]triazolo[5,1-*b*][1,3]benzothiazole can be produced with copper as the catalyst.[31] A copper catalyst was explored in the cyclization of enamino esters with 1,2-dihaloarenes to indoles and with alkynes to pyridines as well. The enamino esters were generated *in situ* from the reaction between zinc, nitriles and $BrCH_2CO_2Et$.[32,33]

In addition to this reaction, 3-methyleneisoindolin-1-ones can be produced from 1,2-dihaloarenes as well. In 2008, Alper and co-workers reported a palladium-catalyzed synthesis of indolinones through a Sonogashira coupling–carbonylation–hydroamination reaction sequence (Scheme 2.21).[34] They found that the $PdCl_2(PPh_3)_2$/CuI/DBU catalyst system could be successfully employed for the one-pot synthesis of a range of 3-methyleneisolindolin-1-ones from 1,2-dihaloarenes, amines and terminal alkynes in phosphonium salt-based ionic liquids $[(C_6H_{13})_3P^+(CH_2)_{13})CH_3][Br^-]$. All processes were highly stereoselective, and gave the *Z*-isomer as the main product. When different arene dihalides were employed, the

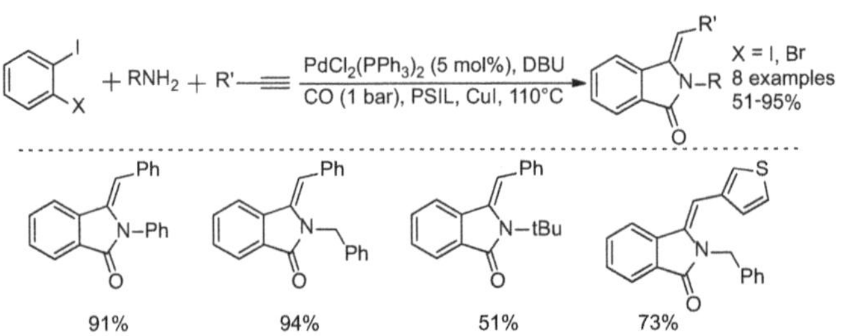

Scheme 2.21 Pd-catalyzed synthesis of indolinones.

Reaction Procedure (Scheme 2.21): A mixture of 1-bromo-2-iodobenzene (1.0 mmol), amine (3.0 mmol), alkyne, $PdCl_2(PPh_3)_2$ (0.05 mmol), CuI (0.1 mmol), DBU (2 equiv.) and trihexyl(tetradecyl)phosphonium bromide (2.0 g), was added to an autoclave. The autoclave was closed, purged three times with carbon monoxide, pressurized with 1 bar of CO, and then heated at 110 °C for 36 h. The autoclave was removed from the heat and cooled to room temperature before excess CO was discharged. The mixture was purified by flash chromatography on silica gel (gradient from hexane to hexane–EtOAc 12.5 : 1) to give the pure product.

1,2-diiodobenzene efficiently provided the isoindolin-1-one derivative in 85% yield, whereas with the 1-chloro-2-iodobenzene the reaction stopped at the first step and gave the Sonogashira coupling product as the only isolated product.

2.2 2-Halophenol Derivatives

2-Halophenol derivatives are promising substrates for the synthesis of oxygen-containing heterocyclic compounds. As early as 1989, the carbonylation of 2-iodophenol with phenyl acetylene or norbornadiene was realized.[35] Coumarin and aurone can be prepared under the same conditions from different substrates (Scheme 2.22).

The carbonylative cyclization of 2-iodophenol with phenyl acetylene will provide 3-phenyl-1-(2-hydroxyphenyl)-2-propyn-1-one as the intermediate, and flavone and aurone are the two theoretically possible products. In 1991, Ortar and co-workers performed a systematic study on the reaction selectivity between these two possible products.[36] In their carbonylative reaction between 2-iodophenols and terminal alkynes, the nature and position of substituents significantly affected the overall yield as well as the ratio between flavone and aurone. For the regiochemical outcome of this carbonylative coupling procedure, they did a study on the cyclization of 3-phenyl-1-(2-hydroxyphenyl)-2-propyn-1-one (Table 2.1).

In 2005, a procedure based on the synthesis of isoaurones, which is a metabolite of the fungus *Thermomonospora alba*, was developed by Merlini and co-workers.[37] A series of isoaurostatins were synthesized by Heck type intramolecular cyclization of cinnamic esters of 2-iodophenols (Scheme 2.23). In antitumor activity tests, the cytotoxic activity of these isoaurones was lower than that of the structurally very similar 4-arylcoumarins.

Benzo-1,3-oxathiole derivatives have been reported with remarkable pharmaceutical and biochemical activities. In 2008, Bao and co-workers described a novel and efficient copper(I)-catalyzed one-pot cascade process for the formation of 2-iminobenzo-1,3-oxathioles from readily available

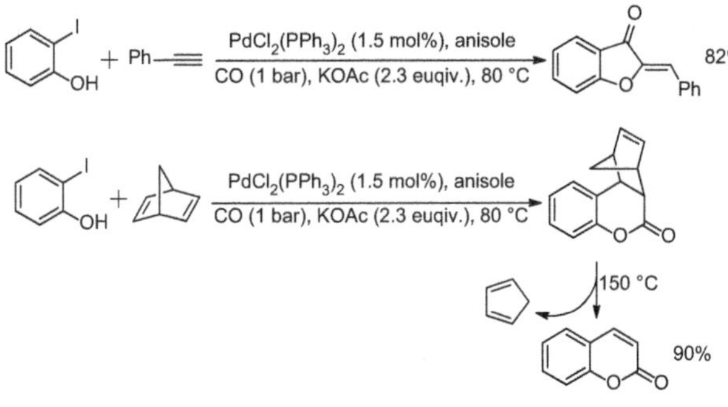

Scheme 2.22 Pd-catalyzed carbonylative synthesis of coumarin and aurone.

Table 2.1 Cyclization of 3-phenyl-1-(2-hydroxyphenyl)-2-propyn-1-one.[a]

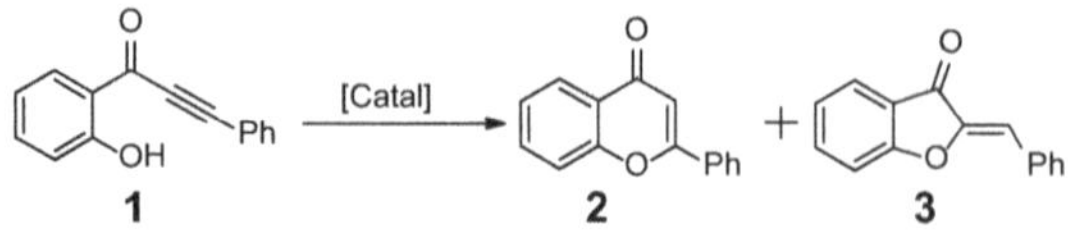

Entry	Reagents	Solvent	Unchanged **1** (%)[b]	**2** (%)[b]	**3** (%)[b]
1	$Pd(PPh_3)_4$, AcOK, CO		25	—	75
2	$Pd(PPh_3)_4$, CO		24	—	76
3	$Pd(PPh_3)_4$	anisole	32	—	68
4	$Pd(OAc)_2(DPPF)_2$, CO		—	—	100
5	AcOK		100	—	—
6	AcOK, 18-crown-6		—	63	37
7	$Pd(PPh_3)_4$, DBU, CO		—	69	31
8	DBU		—	78	22
9	$Pd(OAc)_2(DPPF)_2$, CO, DBU		—	92	8
10	$Pd(OAc)_2(DPPF)_2$, CO	DMF	—	68	32
11	$Pd(PPh_3)_4$		—	55	45
12	DBU		—	91	9

[a]Reactions were carried out for 4 h at 80 °C (in anisole) or 60 °C (in DMF) under N_2 or, when employed, 1 bar CO using 0.5 mmol of substrate, 0.01 mmol of $Pd(PPh_3)_4$, 0.015 mmol of $Pd(OAc)_2$, 0.03 mmol of DPPF, 1.25 mmol of AcOK or DBU, 0.12 mmol of 18-crown-6, and 2 mL of anisole or DMF.
[b]Relative yields determined by ^1H NMR analysis of the reaction residue.

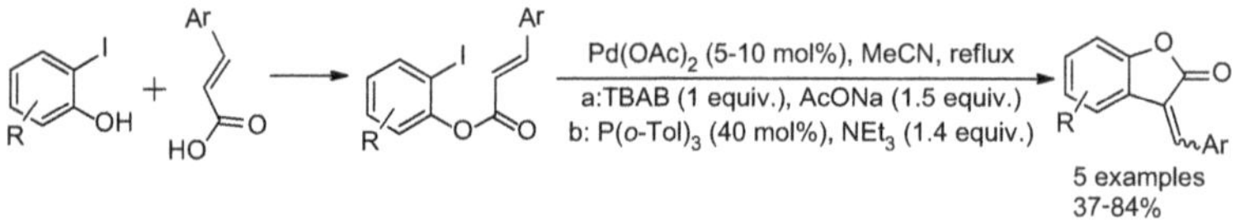

Scheme 2.23 Pd-catalyzed synthesis of isoaurones.

Reaction Procedure (Scheme 2.23): To a solution of the cinnamic ester (1 equiv.) in CH_3CN (about 0.1 M), $Pd(OAc)_2$ (5–10 mol%) and *n*-Bu_4NBr (1 equiv.), AcONa (1.5 equiv.) (method A) or P(*o*-Tol)$_3$ (40 mol%) and Et_3N (1.4 equiv.) (method B) were added. The mixture was stirred and refluxed in an argon atmosphere until the complete disappearance of the ester (hexane–EtOAc 7 : 3). After evaporation of the solvent, the crude product was taken up with EtOAc, washed with water (3×10 mL), dried over Na_2SO_4, and purified by flash column chromatography on silica gel (hexane–EtOAc, 8 : 2 or 9 : 1).

precursors.[38] By using 2-iodophenols and isothiocyanates as substrates, various 2-iminobenzo-1,3-oxathioles were conveniently synthesized in good to excellent yields (Scheme 2.24). Later on, a copper-catalyzed procedure with water as the solvent was established.[39] By using $CuCl_2 \cdot H_2O$/1,10-phen as the catalytic system and DABCO as the base, good yields of the desired products

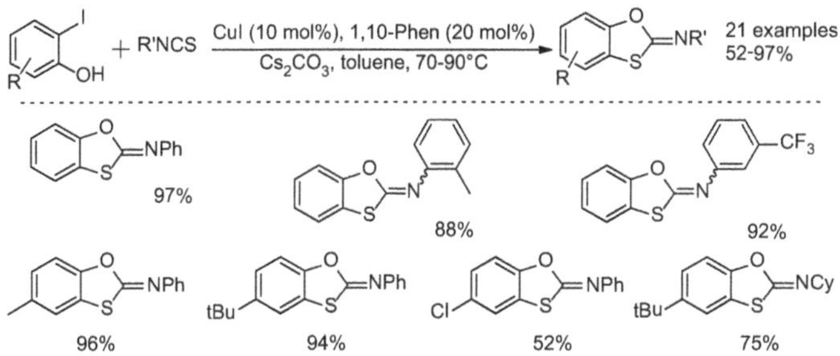

Scheme 2.24 Cu-catalyzed synthesis of benzo-1,3-oxathioles.

Reaction Procedure (Scheme 2.24): An oven-dried Schlenk tube equipped with a Teflon valve was charged with a magnetic stir bar, Cs_2CO_3 (652 mg, 2.0 mmol), CuI (19 mg, 0.10 mmol, 10 mol%), 1,10-phen·H_2O (40 mg, 0.20 mmol, 20 mol%) and *ortho*-iodophenol (1.0 mmol). The tube was evacuated and backfilled with N_2 (this procedure was repeated 3 times). Under a counter flow of N_2, toluene (1.0 mL) was added by syringe and the mixture was pre-stirred for about 0.5 h at room temperature. Then a solution of isothiocyanate (1.0–1.1 mmol) in toluene (1.0 mL) was added *via* syringe under a counter flow of N_2. The tube was sealed and the mixture was allowed to stir at 70–90 °C. The reaction was monitored by TLC. After the starting material was consumed completely and the bottom dot was unchanged, the reaction was stopped and cooled to room temperature. The reaction mixture was directly passed through Celite. After being rinsed with 30 mL of Et_2O, the combined filtrate was concentrated by rotatory evaporation. The residue was purified by column chromatography on silica gel to give the pure product.

can be isolated from water. Recently, ionic liquids have been applied as reaction media as well.[40] With CuI/1,10-phen/Cs_2CO_3/[Bmim][PF_6] as the reaction conditions, the corresponding benzo-1,3-oxathioles were formed in excellent yields. As an advantage, the catalyst can be recycled and reused six times without loss in activity.

C–H activation is at the centre of the research interests of synthetic organic chemists. As early as 1983, Ames and Opalko reported a palladium-catalyzed intramolecular cyclization of 2-bromodiphenyl ethers to dibenzofurans.[41] In the presence of a palladium catalyst, good yields of the desired dibenzofurans were isolated (Scheme 2.25). Here, the substrates applied were prepared by the reaction of substituted fluorobenzene with sodium 2-bromophenoxide at room temperature. Soo after, they succeeded to extend their methodology to 2-iodobenzophenones, 3-benzamido-2-chloropyridine and phenyl 2-bromobenzoate.[42] Thereafter, this novel procedure was studied further by different groups and the reaction conditions could be milder.[43–46] Larock and Liu found that 2-iododiphenyl ethers can be prepared by the

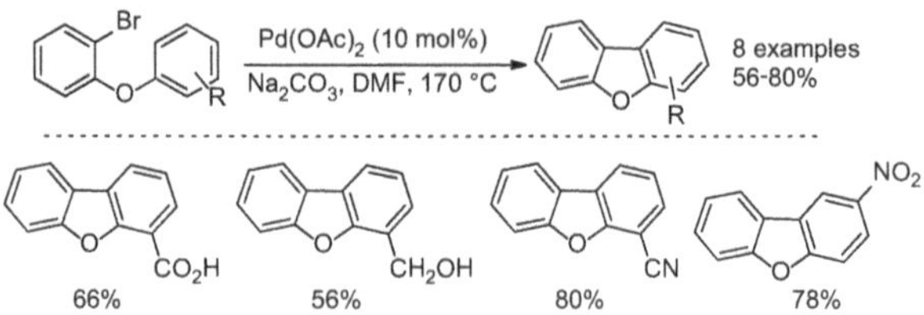

Scheme 2.25 Pd-catalyzed synthesis of dibenzofurans.

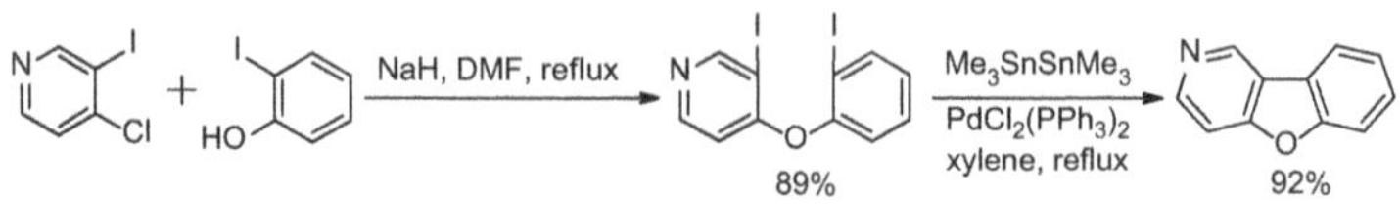

Scheme 2.26 Pd-catalyzed synthesis of benzo[4,5]furopyridines.

reaction of 2-iodophenols with silylaryl triflates in the presence of CsF and then catalyzed by palladium to give the target products. 2-Iodoanilines and 2-iodobenzyl amines were applied as substrates as well.[47] Fagnou and co-workers studied $Pd(OH)_2/C$ (Pearlman's catalyst) as a catalyst for this interesting transformation.[48] At the end, they found that a homogeneous catalyst species was generated from $Pd(OH)_2/C$ under the reaction conditions and that this species was responsible for the observed catalysis.

Alternatively, Li and Yue studied the possibilities for benzo[4,5]furopyridine synthesis.[49] Palladium catalysts were applied in the process and various desired products were formed in good yields (Scheme 2.26).

In 2011, Yoshikai and Liu developed a Pd(II)-catalyzed oxidative cyclization reaction of a 2-arylphenol to dibenzofuran derivatives respectively (Scheme 2.27).[50] The catalytic system features a simple combination of a palladium(II) salt and a pyridine ligand and the use of peroxybenzoate as an inexpensive and convenient oxidant. In their mechanistic experiments, they showed that the reaction involves Pd(II)-mediated C–H bond cleavage as the rate-limiting step, which is followed by oxidation to a high-oxidation-state palladium species and subsequent C–O bond-forming reductive elimination.

Catellani and co-workers studied the cyclization of arylphenols in 2013 (Scheme 2.28).[51] In this system, the formation of a palladacycle with norbornene or norbornadiene *via* aromatic C–H activation was involved. The use of *o*-bromophenols as partners of iodobenzenes has led to ring formation after norbornene deinsertion, or to arylphenols, containing a pendant norbornenyl or norbornadienyl group or to vinylarylphenols. The availability of a reductive elimination step drives the reaction course toward a preferential product.

Additionally, a rhodium-catalyzed silicon-assisted activation of a C–CN and C–Cl bond was reported in 2008.[52] Under these catalytic conditions, carbon–cyano bonds in aryl, alkenyl, allyl, and benzyl cyanides bearing a variety of functional groups can be silylated. The observation of an enamine side product in the silylation of benzyl cyanides and related stoichiometric

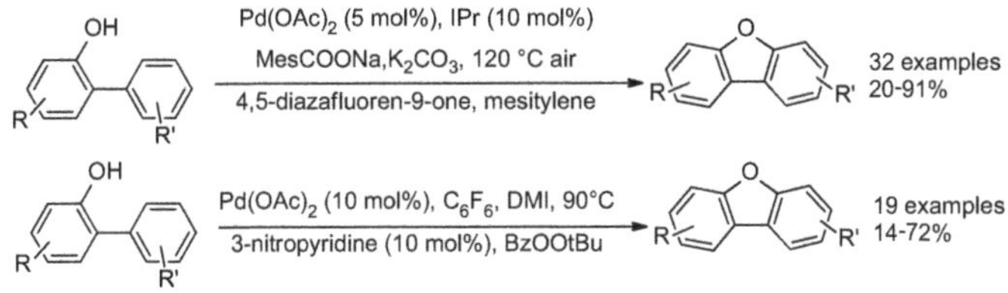

Scheme 2.27 Pd-catalyzed synthesis of dibenzofurans from 2-arylphenol.

Reaction Procedure (Scheme 2.27): A 10 mL Schlenk tube equipped with a stir bar was charged with 2-arylphenol (0.20 mmol), Pd(OAc)$_2$ (4.5 mg, 0.020 mmol, 10 mol%) and 3-nitropyridine (2.5 mg, 0.020 mmol, 10 mol%). The Schlenk tube was evacuated and refilled with N$_2$ three times, followed by sequential addition of C$_6$F$_6$ (0.3 mL), DMI (*N*,*N*′-dimethylimidazolidinone; 0.2 mL) and PhCO$_2$-O*t*Bu (76 μL, 0.40 mmol, 2.0 equiv.). The Schlenk tube was sealed with a Teflon screw cap and then the reaction mixture was stirred at 90 °C for 4 h. Upon cooling to room temperature, the reaction mixture was directly subjected to flash chromatography on silica gel to afford the dibenzofuran product.

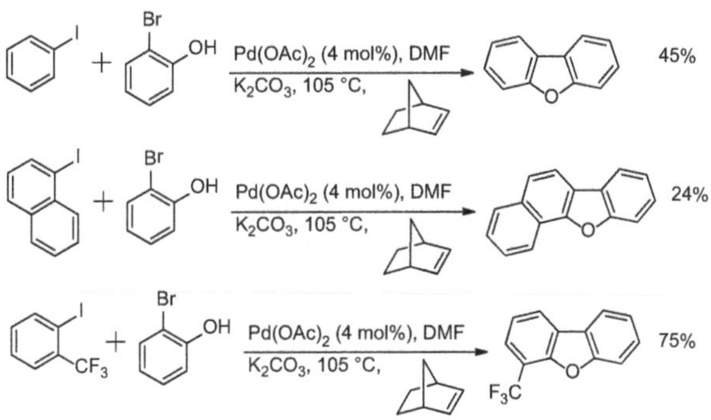

Scheme 2.28 Pd-catalyzed synthesis of dibenzofurans from arylphenols.

studies indicate that the carbon–cyano bond cleavage proceeds through the deinsertion of silyl isocyanide from an η^2-iminoacyl complex. Based on these discoveries, they developed a new intramolecular biaryl coupling reaction through the activation of aryl cyanides and aryl chlorides (Scheme 2.29).

More recently, a palladium-catalyzed intramolecular C–H arylation with mesylated compounds was reported by Kalyani and co-workers.[53] A sequential mesylation–arylation protocol using phenols as substrates was described as well. This method allows for the synthesis of diverse heterocyclic motifs, including dibenzofurans, carbazoles, and indoles, in good yields (Scheme 2.30). The intermolecular arylations were efficient for the coupling of azoles with electronically diverse mesylates.

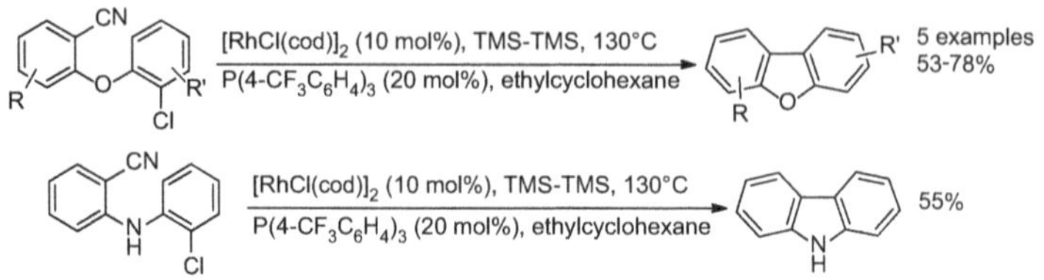

Scheme 2.29 Rh-catalyzed synthesis of dibenzofurans *via* C–CN activation.

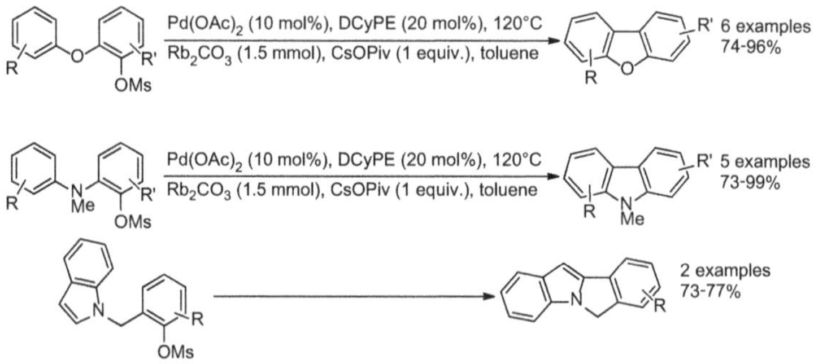

Scheme 2.30 Pd-catalyzed synthesis of dibenzofurans *via* C–OMs activation.

Reaction Procedure (Scheme 2.30): $Pd(OAc)_2$ was weighed into a 20 mL scintillation vial. The vial was taken into a glove box and DCyPE, Rb_2CO_3, and CsOPiv were added. A solution of substrate in toluene (or xylene) was added to this mixture. The vial was sealed with a Teflon lined cap, taken out of the glove box and the reaction mixture was allowed to stir at 120 °C. The reaction mixture was cooled to room temperature and filtered through a 1.5 inch plug of silica gel, eluting with Et_2O (100 mL). The product was isolated after chromatography.

Grigg and co-workers developed a palladium-catalyzed cyclization–carbonylation of *ortho*-vinyl ether substituted aryl iodides in 1994.[54] Under a carbon monoxide atmosphere, nBu$_3$SnR or NaBPh$_4$ can be captured by the palladium intermediate effectively (Scheme 2.31).

Hydroxamic acids, which have been extensively studied as metalloenzyme inhibitors and have been shown to be potent inhibitors of matrix metalloproteinases, are a family of zinc-dependent enzymes associated with diseases such as cancer, arthritis and multiple sclerosis.

In 1999, the same group applied their cyclization–carbonylation methodology in the synthesis of hydroxamic acids in both solution and solid phase.[55] By using hydroxamates as coupling partners, the corresponding products were formed in good to excellent yields (Scheme 2.32).

Due to their interest in this topic, they explored the coupling partners of this palladium-catalyzed cyclization–carbonylation further.[56–60] In addition

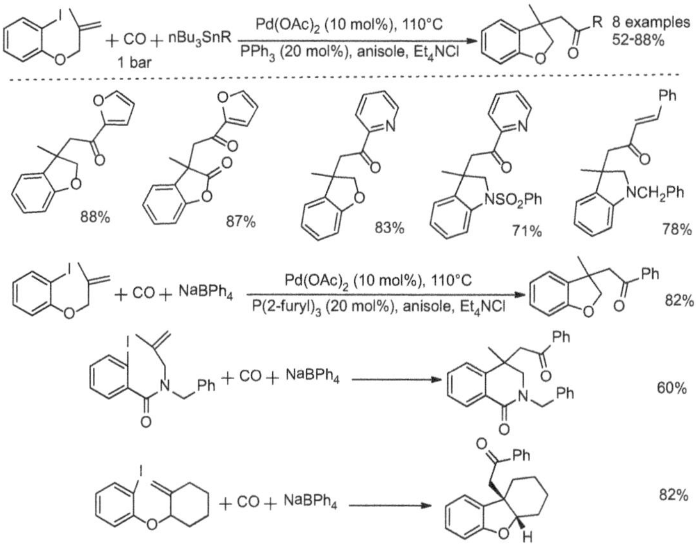

Scheme 2.31 Pd-catalyzed synthesis of heterocycles *via* cyclization–carbonylation.

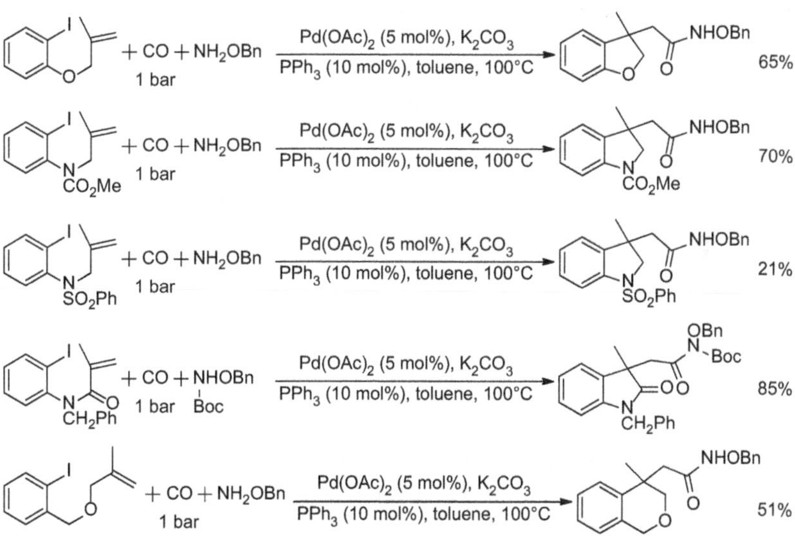

Scheme 2.32 Pd-catalyzed synthesis of hydroxamic acids.

to the nucleophiles discussed previously, arylboronic acids, vinylstannane reagents and arylstannanes can be applied as substrates under carbonylation or non-carbonylation conditions as well. Moreover, in 2002, they described a palladium-catalyzed intermolecular queuing process for the synthesis of various heterocycles.[61] This methodology involving oxidative addition to aryl or vinyl halides followed by cyclisation onto a proximate

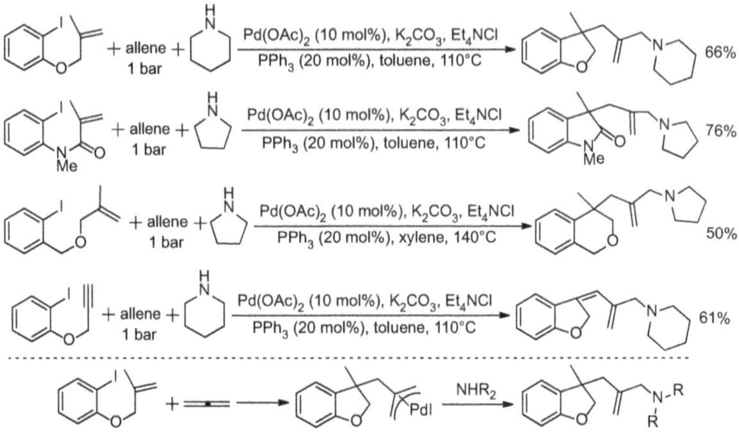

Scheme 2.33 Pd-catalyzed cyclization with allene and amines.

> **Reaction Procedure** (Scheme 2.33): A mixture of substrate (0.5 mmol), secondary amine (1 mmol), $Pd(OAc)_2$ (0.05 mmol), PPh_3 (0.1 mmol), potassium carbonate (1 mmol) and Et_4NCl (0.5 mmol) in dry toluene (10 ml) was degassed using two freeze, pump, thaw cycles. Allene (1 bar) was incorporated into the Schlenk tube and the solution was heated at 110 °C in an oil bath for 20 h. The solution was then concentrated and the residue purified by column chromatography to afford the product.

alkyne or alkene, allene (1 atm) insertion and capture of the resulting π-allyl palladium(II) species by secondary amines affords benzo-fused 5–8-membered rings in good yields (Scheme 2.33). Tetraalkylammonium chlorides are commonly used to accelerate the rates of Heck reactions. In this procedure, the authors proposed the following effects of Et_4NCl: (i) it favours the conversion of RPdI species to RPdCl species in which the palladium centre is more electrophilic, thereby promoting coordination of the olefin to the metal and promoting cyclisation; (ii) it results in the formation of polychloro palladium(0) anionic species which both stabilises the palladium(0) species and promotes oxidative addition; (iii) in common with other quaternary ammonium salts, it acts as a phase transfer catalyst.

In 2007, Liu and co-workers reported a ligand-free palladium-catalyzed synthesis of heterocycles through reductive Heck cyclization.[62] Five and six membered heterocycles, including indolines, 2,3-dihydrobenzofurans, chromans, isochromans, 1,2,3,4-tetrahydroquinolines, and 1,2,3,4-tetrahydroisoquinolines, were produced in 70–99% yields under mild conditions (Scheme 2.34). Water was found to be essential for these reactions. This transformation was investigated further by Lamaty and co-workers in 2007.[63] 3,3-Disubstituted-2,3-dihydrobenzofurans were prepared in good yields.

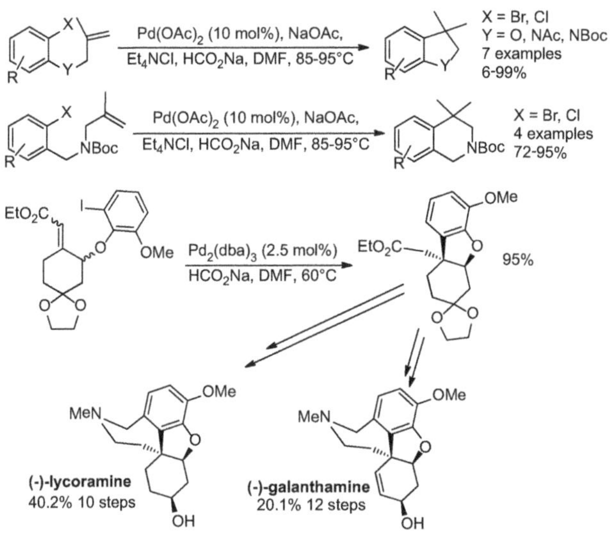

Scheme 2.34 Pd-catalyzed reductive Heck cyclization and application in total synthesis.

Reaction Procedure (Scheme 2.34): A mixture of 3-bromo-4-(3-methylbutenyloxy)benzene (0.5 g, 1.9 mmol), NaOAc (0.43 g, 5.0 mmol), HCO_2Na (0.15 g, 2.4 mmol), $Et_4NCl \cdot H_2O$ (0.41 g, 2.4 mmol), and $Pd(OAc)_2$ (4.7 mg, 0.02 mmol) in DMF (5 mL) was degassed and refilled with nitrogen three times. The mixture was heated to 85–95 °C (oil bath) for 1–4 h. Completion of the reductive cyclization was monitored by HPLC and GC. After completion, the reaction mixture was cooled to room temperature. Water (15 mL) was added and the mixture filtered through a Celite bed and washed with *tert*butyl methyl ether (25 mL). The organic phase was separated and concentrated to give an oily residue that was filtered through silica gel with eluent (EtOAc–hexane, 1 : 20) to give the desired product.

Later on, Zhou and co-workers studied this methodology and applied it in the total synthesis of (−)-galanthamine and (−)-lycoramine, which was isolated from the bulbs of different species of the Amaryllidaceae family, and is a selective, reversible, and competitive acetylcholinesterase inhibitor and has been used in the early treatment of Alzheimer's disease.[64]

Fagnou and co-workers succeeded to combine Heck cyclization with direct arylation in 2009.[65] This interesting domino palladium-catalyzed Heck-intermolecular direct arylation reaction gave new access to a variety of dihydrobenzofurans, indolines, and oxindoles. A variety of sulfur-containing heterocycles such as thiazoles, thiophenes, and benzothiophene could be employed as the direct arylation coupling partner as well and resulted in yields up to 99% (Scheme 2.35). As the authors demonstrated, in addition to

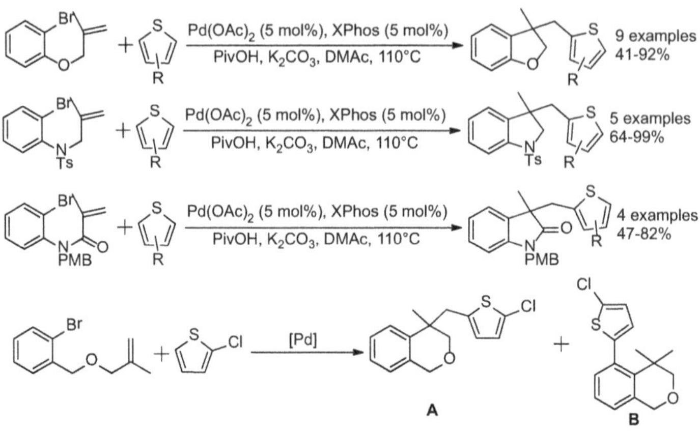

Scheme 2.35 Pd-catalyzed Heck cyclization and arylation.

> **Reaction Procedure** (Scheme 2.35): K_2CO_3 (111 mg, 0.8 mmol, 2.0 equiv.), XPhos (2-dicyclohexylphosphino-2′,4′,6′-triisopropylbiphenyl) (9.5 mg, 0.02 mmol, 5 mol%) and $Pd(OAc)_2$ (4.7 mg, 0.02 mmol, 5 mol%) were weighed to air in a screw cap vial equipped with a magnetic stir bar. The vial was purged with argon for 10 minutes. In a separate vial, the aryl bromide (0.4 mmol, 1.0 equiv.), the heterocycle (1.6 mmol, 4.0 equiv.) and pivalic acid (12.3 mg, 0.12 mmol, 30 mol%) were dissolved in DMAc (1.3 mL) and the solution was added to the reaction mixture by syringe. The reaction mixture was stirred at 110 °C for 16 hours. Volatiles were then removed by horizontal distillation under reduced pressure and the residual solid was purified by silica gel column chromatography without prior removal of insoluble salts to afford the corresponding product.

the synthetic utility, the demonstrated ability to combine direct arylation into this class of tandem process should raise confidence in the viability of other domino processes involving alkylpalladium(II) reactive intermediates. The formation of isochroman heterocycles was also investigated by the homologation of the phenyl ether to benzyl ether. In this case, the formation of the two regioisomers **A** and **B** in a 1.7 : 1 ratio was obtained in 65% isolated yield for the mixture. This outcome is most likely the result of palladium migration from the alkyl group to the adjacent aromatic ring. Such migratory processes can occur when the alkylpalladium intermediate and the adjacent C–H bond on the aromatic ring are in a favourable geometry.

In 2012, an intramolecular version was developed by Zhu's group.[66] In the presence of a catalytic amount of palladium acetate and XPhos with potassium carbonate as a base at 100 °C in DMAc, the desired dihydroquinolin-2-ones spiro-fused to dihydrofuranyl, indolinyl, and indanyl were formed in good to excellent yields (Scheme 2.36). The reaction went through a domino sequence involving a 5-*exo*-trig Heck cyclization followed by an

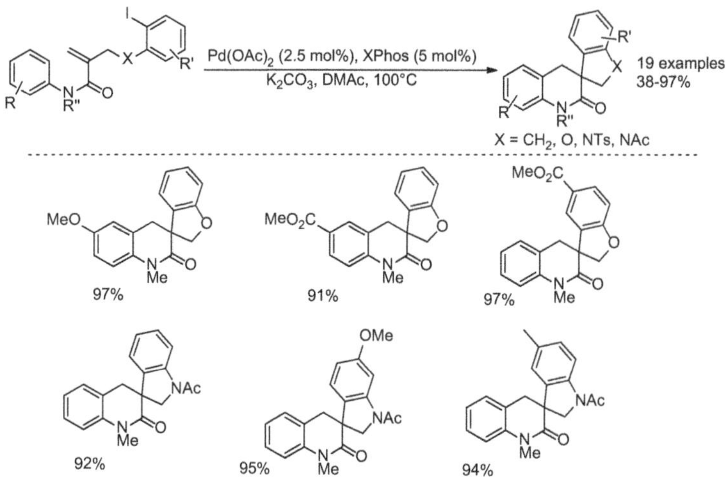

Scheme 2.36 Pd-catalyzed intramolecular Heck cyclization and arylation.

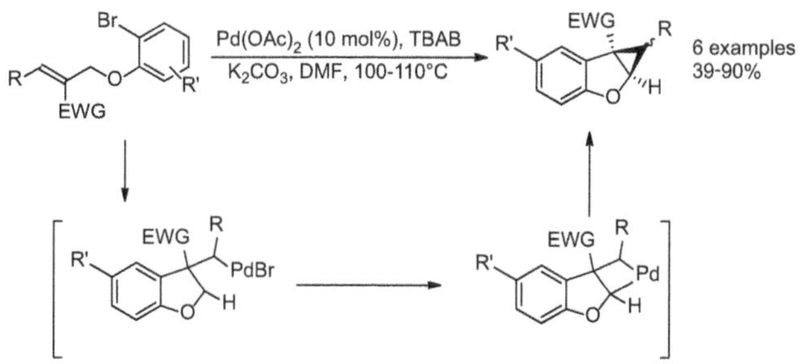

Scheme 2.37 Pd-catalyzed cyclopropaindene synthesis.

intramolecular direct C–H functionalization. In their mechanistic study, they have proven that C–H functionalization might not be the rate-determining step for this domino process.

Interestingly, Kim and co-workers discovered a procedure for the preparation of 1-phenyl-1,6*a*-dihydro-6-oxacyclopropa[*a*]indene-1*a*-carboxylic acid derivatives in 2008.[67] They started from Baylis–Hillman adducts and the reaction went through a palladium-mediated domino carbopalladation involving activation of a C(sp^3)–H bond. Good yields of the desired products were produced (Scheme 2.37).

In 2011, Lautens and co-workers reported a palladium-catalyzed intramolecular carbon–carbon bond-forming reaction between aryl iodides and alkenes.[68] This new cross-coupling reaction forms two new bonds and all of the atoms in the starting materials are incorporated into the product. The use of a palladium catalyst with bulky phosphine ligands was found to be crucial for reactivity as the bulky phosphine ligands favour the reductive elimination of Pd(II) complexes. Good to excellent yields of the desired products were isolated (Scheme 2.38a). In the same year, they found that aryl

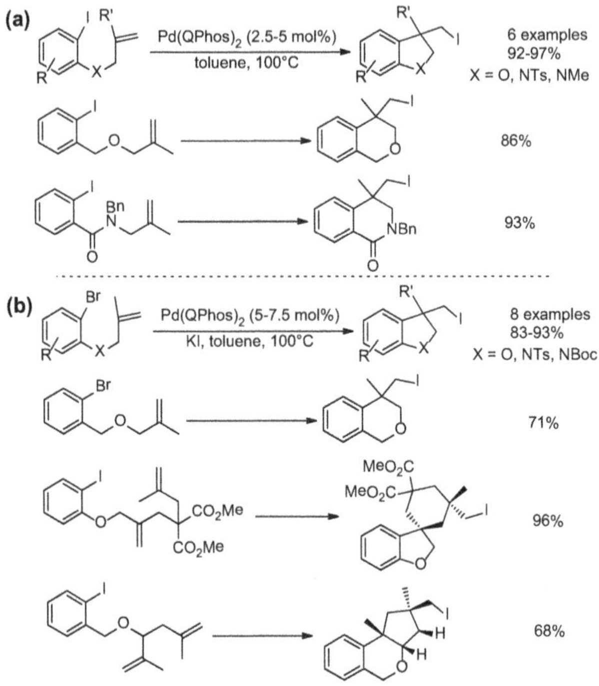

Scheme 2.38 Pd-catalyzed carboiodination of alkenes.

Reaction Procedure (Scheme 2.38a): To a 0.5–2 mL biotage microwave vial equipped with a magnetic stir bar was added the aryl iodide (1 equiv.). The vial was purged with argon for 5 minutes, after which Pd(Q-phos)$_2$ (2.5–5 mol%) and toluene (0.2 M) were added. The vial was capped and added to an oil bath pre-heated to 100 °C. After stirring for 4–16 hours, the vial was cooled and the contents loaded directly on to a silica gel column and purified by flash chromatography.
Reaction Procedure (Scheme 2.38b): To a 0.5–2 mL biotage microwave vial equipped with a magnetic stir bar was added the aryl bromide (1 equiv.). The vial was purged with argon for 5 minutes, after which Pd(Q-Phos)$_2$ (5 or 7.5 mol%), potassium iodide (2 equiv.), and toluene (0.05 M) were added. The vial was capped and added to an oil bath pre-heated to 100 °C. After stirring for 16 hours, the vial was cooled and the contents filtered over a pad of silica gel, washing with ether. The crude material was loaded on to a silica gel column and purified by flash chromatography.

bromides can be used as substrates instead of aryl iodides and applied in the synthesis of a variety of nitrogen- and oxygen-containing heterocycles with a carbon–iodine bond in the products.[69] This palladium-catalyzed carbohalogenation requires potassium iodide for the reaction to proceed in high yields (Scheme 2.38b). Complex products with multiple rings and stereogenic centers are generated in excellent yields with moderate to excellent diastereoselectivities. In addition, polyunsaturated aryl iodide substrates are amenable to domino carbohalogenation reactions yielding complex bicyclic alkyl iodides containing multiple stereogenic centers in high yields with good diastereoselectivities. These concepts can be merged to allow polyunsaturated aryl bromide substrates to participate in domino chemistry, generating bicyclic alkyl iodide products.

In addition to the palladium catalysts mentioned, other transitional metal catalysts were explored and applied in this transformation as well. In 1995, Duñach and Olivero studied the nickel-catalyzed electrochemical reductive deprotection of allyl ethers.[70] Among the various substrates, allyl-*o*-halogenophenols were tested as well. 2-Chlorophenol was produced from allyl-*o*-chlorophenol; phenol was produced from allyl-*o*-bromophenol; while 3-methyl-2,3-dihydrobenzofuran (33%) was produced from allyl-*o*-iodophenol together with phenol (52%).

More recently, a study on intramolecular carbonickelation of alkenes was published.[71] The authors show that the $NiBr_2$bipy complex can be used to catalyze an intramolecular Heck-type reaction in the absence of any additional base and without the need of a glove box. By using 20 mol% of $NiBr_2$bipy and Mn (2 equiv.) in DMF at 50 °C, the desired dihydrobenzofurans or indoles were formed in moderate yields from allylic ethers and allylic amines *via* 5-*exo*-trig cyclization (Scheme 2.39).

Some of the advantages of copper catalysts are that they are inexpensive, abundant *etc.* In 2013, Fu, Peters and their co-workers reported a copper-catalyzed coupling of aryl halides with thiols *via* a single-electron transfer pathway.[72] In order to prove the reaction mechanism, they performed the reaction of allyl-*o*-iodophenol with thiophenol as well. No cyclized product could be produced when the reactions were performed in the dark, while 57% of dihydrobenzofuran was formed when *hν* was introduced (Scheme 2.40).

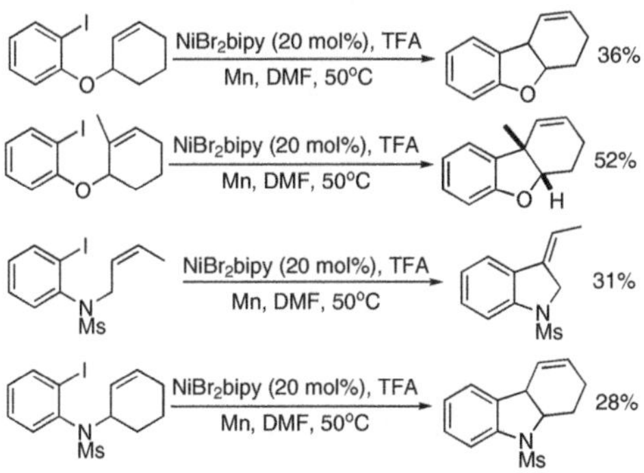

Scheme 2.39 Ni-catalyzed carboiodination of alkenes.

Reaction Procedure (Scheme 2.39): To a solution of aryliodide (0.5–1 mmol, 1 equiv.) in anhydrous DMF (5 mL) under an argon atmosphere at 50 °C was added manganese (2 equiv.) followed by $NiBr_2$bipy (0.2 equiv.) then rapidly TFA (20 μL). The medium was stirred vigorously at 50 °C, and the disappearance of the starting material was monitored by gas chromatography. The mixture was hydrolyzed with water (10 mL), diluted with Et_2O (10 mL), and then filtered through Celite. The aqueous layer was extracted with Et_2O (2×10 mL), and then the combined organic layers were washed with water (3×10 mL) and brine (2×10 mL), dried over anhydrous $MgSO_4$, and concentrated. The crude product was purified by flash chromatography.

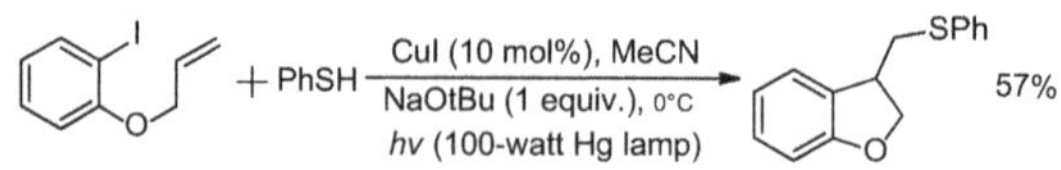

Scheme 2.40 Cu-catalyzed carboiodination of alkenes.

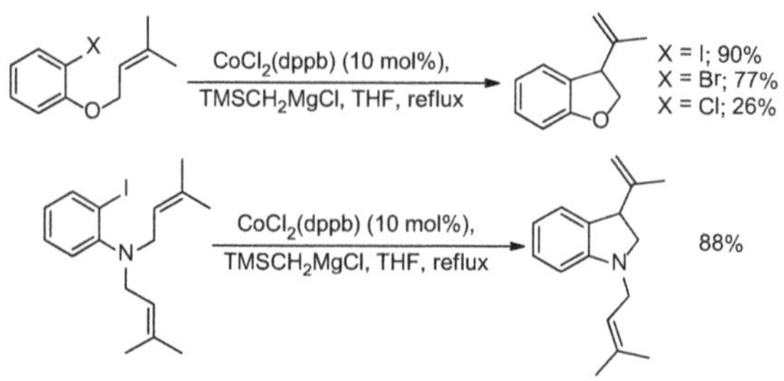

Scheme 2.41 Co-catalyzed carboiodination of alkenes.

Cobalt catalysts were studied by Oshima and co-workers in cyclization reactions in 2002.[73] By using the cobalt complex [$CoCl_2$(dppb)] as the catalyst and assisted by Me_3SiCH_2MgCl, the intramolecular Heck-type cyclization reactions of 6-halo-1-hexenes proceeded *via* a radical process to provide the corresponding tetrahydrofurans in good yields. Allyl-*o*-halogenophenols were tested under identical conditions and gave the desired dihydrobenzofurans in low to excellent yields (Scheme 2.41).

Reactions with other radical initiators, such as AIBN, were reported as well; the desired dihydrobenzofurans could be produced selectively.[74]

Besides the cyclization of alkenes mentioned previously, intramolecular cyclization of allenes and alkynes was reported as well. In 1995, Grigg and co-workers developed a palladium-catalyzed cyclization–amination of allenes.[75] They demonstrated that the effect of the base was obvious—different regioselectivities were observed by using K_2CO_3 and Ag_2CO_3 as bases. The corresponding heterocyclic compounds were isolated in good yields (Scheme 2.42).

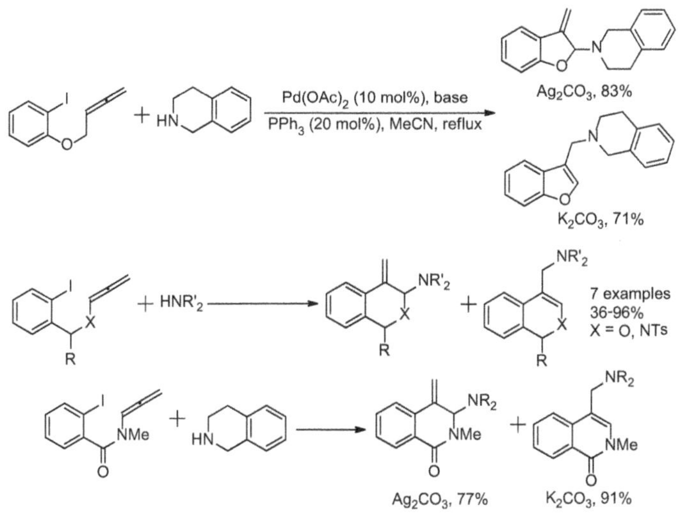

Scheme 2.42 Pd-catalyzed cyclization–amination of allenes.

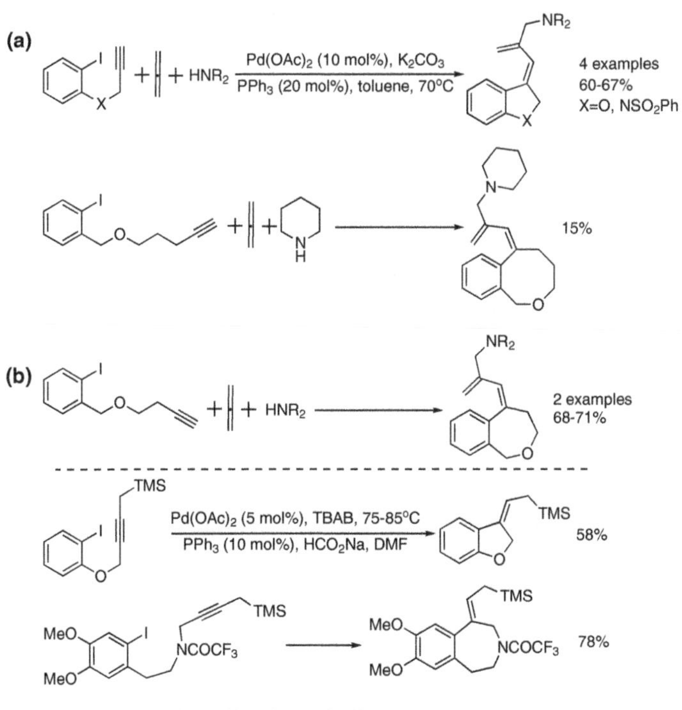

Scheme 2.43 Pd-catalyzed cyclization of alkynes.

The same group described the intramolecular cyclization of alkynes soon after (Scheme 2.43a).[76] In this interesting process, allene and secondary amines were incorporated as well and provided the desired products in good yields.

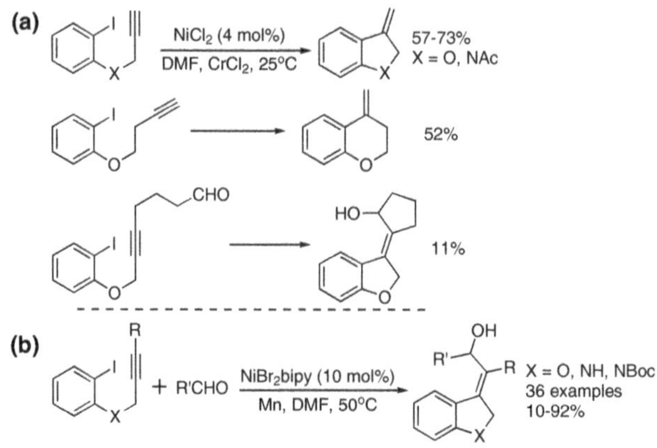

Scheme 2.44 Ni-catalyzed cyclization of alkynes.

The reductive cyclization of alkynes was developed by Tietze's group in 1994 (Scheme 2.43b).[77] By using sodium formate as a hydride source, the corresponding products were isolated in good to excellent yields. Nickel salts were applied in the cyclization of alkynes too. In 1994, Hodgson and Wells reported a chromium(II)-mediated nickel-catalyzed cyclization of alkynes (Scheme 2.44a).[78] As the toxicity of $CrCl_2$, this methodology was further studied in 2011 by Durandetti *et al.* (Scheme 2.44b).[79] In their methodology, no chromium salt was needed. The carbonickelations of alkynes and functionalization of the resulting vinylnickel moiety have been performed efficiently in a nickel-catalyzed domino cyclization–condensation process. This reaction, which does not require the preparation of any other organometallic reagent, proceeds only by *exo*-dig cyclization. This method constitutes a one-pot synthesis of substituted dihydrobenzofurans, chromans, isochromans, indoles, or indanes and the desired products were obtained in good yields and high stereoselectivity.

From a synthetic point of view, intermolecular reactions are more interesting, as a wider variety of substrates is available. In 1990, Larock and co-workers reported a palladium-catalyzed heteroannulation of 1,3-dienes with 2-iodophenols and 2-iodoanilines.[80] Dihydrobenzofurans and indolines were isolated in moderate to good yields (Scheme 2.45). In addition to the 2-iodophenols and 2-iodoanilines, 2-iodobenzyl alcohol and 2-iodobenzyl amine could be applied as substrates as well and gave the corresponding six-membered heterocyclic compounds as the terminal products.

In 2010, Larock and Rozhkov developed an efficient palladium-catalyzed annulation of 1,3-dienes with *o*-iodophenyl acetates and the corresponding dihydrobenzofurans were isolated in good yields (Scheme 2.46).[81] The process was quite general and regio- and stereoselective, and a wide range of electron-deficient and electron-rich *o*-iodoaryl acetates, as well as cyclic, terminal, and internal 1,3-dienes, can be utilized. The presence of the acetyl group greatly improves the yields in this process, particularly with aryl

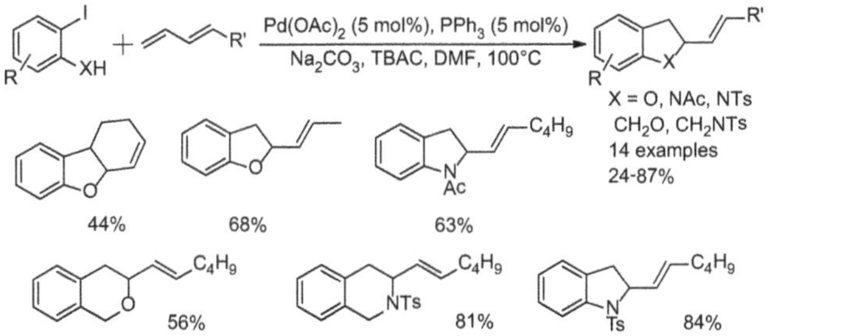

Scheme 2.45 Pd-catalyzed cyclization of 2-iodophenols with 1,3-dienes.

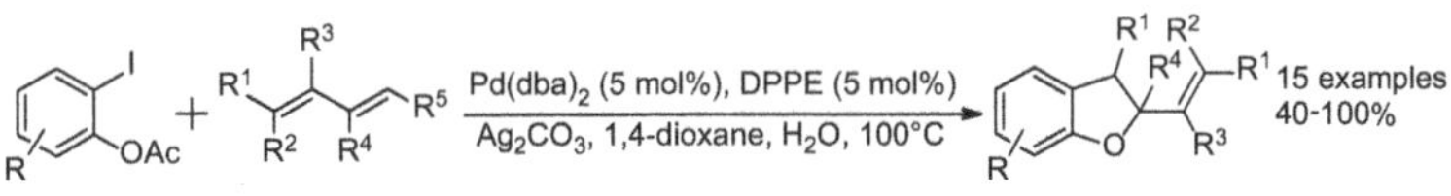

Scheme 2.46 Pd-catalyzed cyclization of 2-iodoaryl acetates with 1,3-dienes.

Reaction Procedure (Scheme 2.46): The *o*-iodoaryl acetate (0.25 mmol), $Pd(dba)_2$ (0.0125 mmol, 5 mol%), dppe (0.0125 mmol, 5 mol%), Ag_2CO_3 (0.5 mmol), and 1,4-dioxane (4 mL) were stirred in a capped vial for 5 min, and then water (1 mL) and the 1,3-diene (1.0 mmol) were added. The resulting reaction mixture was stirred at 100 °C for 24 h, cooled to room temperature, and filtered, then the filtrate was concentrated to give a yellow residue. The resulting residue was purified by column chromatography with silica gel as a solid phase and 4 : 1 hexanes–ethyl acetate as the eluent to afford after solvent removal the final product.

iodides containing electron-donating substituents. However, highly electron-rich *o*-iodoaryl acetates do tend to give somewhat lower yields than electron-deficient aryl iodides and, in the case of internal or aryl-substituted 1,3-dienes, Heck-type products were sometimes observed. Regarding the reaction mechanism, this annulation reaction was believed to proceed *via* (1) oxidative addition of the aryl iodide to Pd(0), (2) *syn*-addition of the resulting arylpalladium complex to the 1,3-diene, (3) intramolecular coordination of the phenolic oxygen to the Pd center, (4) hydrolysis of the acetyl group, and (5) reductive elimination of Pd(0), which regenerates the catalyst.

In addition to the 1,3-dienes, *in situ* formed allenes were applied as substrates and cyclized with 2-iodophenols as well.[82] In 2007, Liang and co-workers reported a novel palladium-catalyzed intermolecular tandem reaction for the synthesis of tetracyclic compounds from propargylic compounds and 2-iodophenols. The corresponding tetracyclic compounds were isolated in moderate to good yields (Scheme 2.47). In respect of the reaction pathway, the possible reaction intermediate was isolated and applied. In detail, this transformation consists of two catalytic cycles. The steps are: (a) initial decarboxylation of propargylic compound by palladium(0) to

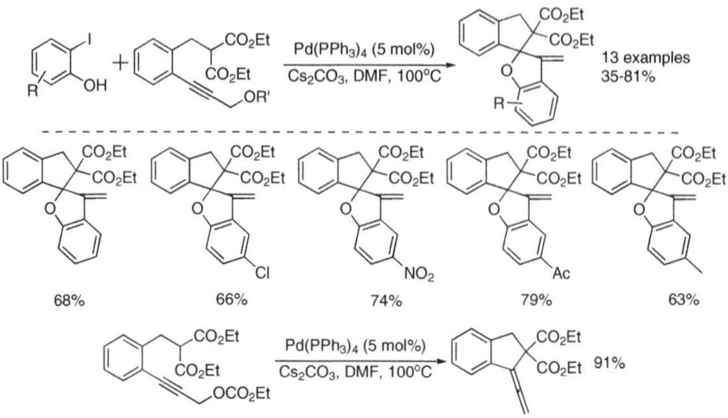

Scheme 2.47 Pd-catalyzed synthesis of tetracyclic compounds.

Reaction Procedure (Scheme 2.47): Cs_2CO_3 (130.4 mg, 0.40 mmol) was added to a solution of 3-(2-(2,2-di(ethoxycarbonyl)ethyl)phenyl)prop-2-ynyl ethyl carbonate (75.2 mg, 0.20 mmol) in DMF (2.0 mL). The mixture was stirred for 5 min and $[Pd(PPh_3)_4]$ (11.5 mg, 0.01 mmol, 5 mol%), and 2-iodophenol (66.0 mg, 0.30 mmol) were added. The resulting mixture was then heated under an argon atmosphere at 100 °C. When the reaction was considered complete as determined by thin-layer chromatography, the reaction mixture was allowed to cool to room temperature and quenched with a saturated aqueous solution of NH_4Cl, and extracted with ethyl acetate. The combined organic extracts were washed with water and saturated brine. The organic layers were dried over Na_2SO_4 and filtered. Solvents were evaporated under reduced pressure. The residue was purified by chromatography on silica gel to afford the pure product.

generate an allenylpalladium complex; (b) regioselective intramolecular nucleophilic attack of the carbanion to form an allene intermediate which could be isolated; (c) oxidative addition of the aryl halide to the palladium(0) catalyst; (d) the addition of arylpalladium compounds to the allenes produces π-allylpalladium compounds; and (e) regioselective intramolecular nucleophilic attack at the more hindered site to afford the terminal products. The selectivity is presumably due to electronic effects at the benzylic position.

Later on, the annulation of allenes with 2-iodophenols and 2-iodoaniline derivatives were further explored. In 2009, Shi and Li developed an efficient method for the coupling reactions of diarylvinylidenecyclopropanes with 2-iodophenols and *N*-(2-iodophenyl)-4-methylbenzenesulfonamide.[83] A variety of cyclopropane-containing 2,2-diaryl-3-(tetramethylcyclopropylidene)-2,3-(dihydro)benzofuran and 2,2-diaryl-3-(tetramethylcyclopropylidene)-1-(toluene-4-sulfonyl)-2,3-dihydro-1*H*-indole derivatives were produced in moderate to good yields with palladium as the catalyst (Scheme 2.48a).

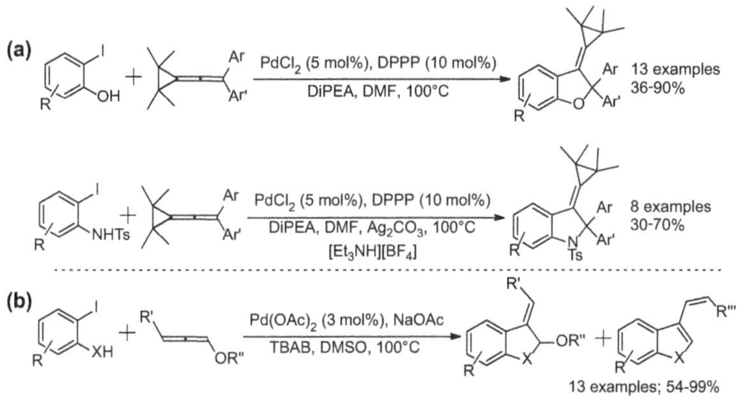

Scheme 2.48 Pd-catalyzed annulation of allenes with 2-iodophenols.

Soon after, Deagostino and co-workers described a phosphine-free annulation reaction for the preparation of substituted 3-alkenylindoles, 2-alkoxy-3-alkylidene-2,3-dihydrobenzofuranes and -indolidines.[84] The desired products were formed in good to excellent yields (Scheme 2.48b). Two different heterocyclic skeletons were obtained, depending on the electron-donating properties of the heteroatom involved in the annulation process.

Alkenes are more widely available then allenes and alkynes. In 2004, Larock and co-worker developed a Pd(0)-catalyzed heteroannulation of cyclic and bicyclic alkenes with *o*-amino- and *o*-hydroxyaryl iodides.[85] By this interesting procedure, indolines and 2,3-dihydrobenzofurans were produced in good yields (Scheme 2.49). Notably, these processes are only successful with cyclic olefins in which the key alkylpalladium intermediate cannot undergo facile palladium β-hydride elimination. An electron-donating group *para* to the hydroxy group of the *o*-iodophenols has been shown to favour formation of the heteroannulation product over the Heck product. The "optimal" phenol process does not work well for all nitrogen-containing aryl halides. However, good to excellent yields of nitrogen heterocycles can be obtained by employing ethylene glycol as the solvent and DMG (dimethylglycine) as an additive. From the point of the reaction mechanism, these reactions appear to involve: (1) oxidative addition of the aryl iodide to the palladium catalyst, (2) arylpalladation of the olefin, (3) possible coordination of the internal nucleophile to the palladium, (4) formation of a six-membered palladacycle, and (5) reductive elimination of the organopalladium intermediate to give the heteroannulation product and regenerate Pd(0).

In 2011, a palladium-catalyzed oxyarylation of dihydronaphthalene and chromenes with *o*-iodophenols and its acetates was developed.[86] The oxyarylation of dihydronaphthalene, two electron-rich and one electron-poor chromenes with different *o*-iodophenols and their acetates has been investigated by using either $Pd(OAc)_2$ (10 mol%) as a precatalyst and Ag_2CO_3 as a base in acetone as a solvent or oxime palladacycle as a precatalyst and dicyclohexylamine as a base in DMA–H_2O. The reactions, which were

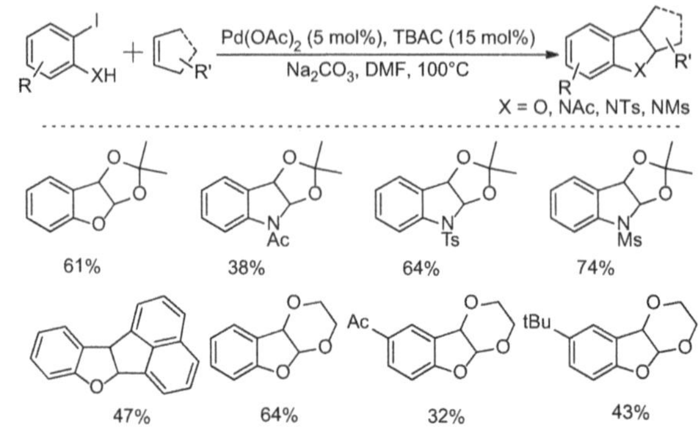

Scheme 2.49 Pd-catalyzed annulation of cyclic alkenes with 2-iodophenols.

Reaction Procedure (Scheme 2.49): The aryl iodide (0.45 mmol), the olefin (0.90 mmol), Na_2CO_3 (48.0 mg, 0.45 mmol), 15 mol% of *n*-Bu_4NCl (19.0 mg, 0.068 mmol), and 5 mol% of Pd(OAc)$_2$ (5.0 mg, 0.023 mmol) were added to a 4-dram vial equipped with a stir bar and Teflon-lined screw cap. One millilitre of DMF was added by syringe. The vial was placed in a mineral oil bath set at 100 °C for 1 day. The reaction mixture was cooled to room temperature then pipetted into 30 mL of diethylether. The ether was washed sequentially with aq. NH_4Cl and brine, dried over $MgSO_4$ and evaporated. The crude product was then purified by silica gel chromatography using 91 : 9 hexane–ethyl acetate.

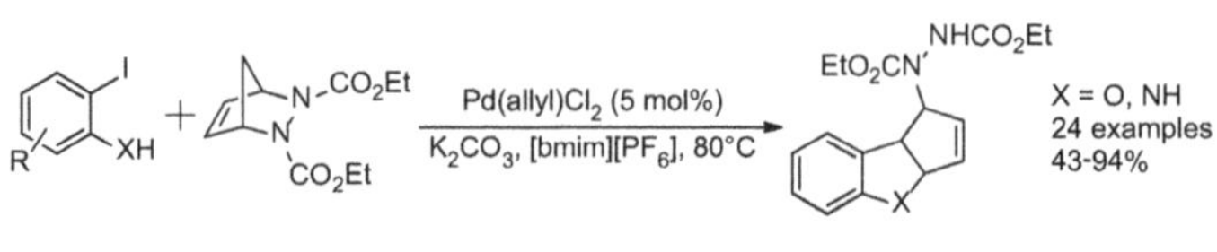

Scheme 2.50 Pd-catalyzed annulation of diazabicyclic alkenes with 2-iodophenols.

strongly accelerated under microwave heating, afforded the best yields when electron-rich chromenes reacted with *o*-iodophenols substituted by electron-withdrawing groups in the presence of Pd(OAc)$_2$ and Ag_2CO_3. However, in the case of electron-poor chromenes the best yield was obtained with the electron-rich *o*-iodophenol. Importantly, this procedure was applied in the synthesis of new pterocarpanquinones as well. These compounds were shown to be active against human leukemic cell lines and human lung cancer cell lines.

More recently, a palladium-catalyzed tandem protocol for the synthesis of cyclopentene fused heterocycles from diazabicyclic alkenes and *ortho*-functionalized aryl iodides was elaborated.[87] A number of cyclopentene fused dihydrobenzofurans and indolines were prepared by this procedure (Scheme 2.50). The reaction can be tuned toward the formation of either

3,4-disubstituted cyclopentenes or cyclopentene fused heterocycles by careful manipulation of the reaction parameters. The reaction was also extended to bicyclic alkenes derived from fulvene, which resulted in the heteroannulation of the azabicyclic system.

Regarding the importance of dihydrobenzofurans, various other procedures were developed as well. In 2007, Fagnou and co-workers reported a palladium-catalyzed intramolecular alkane arylation *via* C–H activation.[88] The corresponding dihydrobenzofurans were prepared in good to excellent yields (Scheme 2.51).

In addition to the dihydrobenzofurans, benzofurans were systematically studied and prepared as well. In 1995, Larock and co-workers described a palladium-catalyzed annulation of internal alkynes with 2-iodoanilines, 2-iodophenols and related derivatives.[89] The corresponding products were isolated in good to excellent yields (Scheme 2.52).

In 1997, a detailed study of the heteroannulation of *o*-iodophenol with acetylenic substrates through palladium-copper catalysis leading to the synthesis of 2-substituted benzofurans was reported by Kundu and co-workers.[90] Using bis(triphenylphosphine)palladium(II) chloride–copper iodide as the catalytic system and triethylamine as the base and dimethylformamide as the solvent, the desired benzofurans were isolated in good yields (Scheme 2.53). An acyclic compound was isolated and proved to be the intermediate in the synthesis of the benzofurans. Some of the benzofurans have been transformed into biologically active compounds. In this

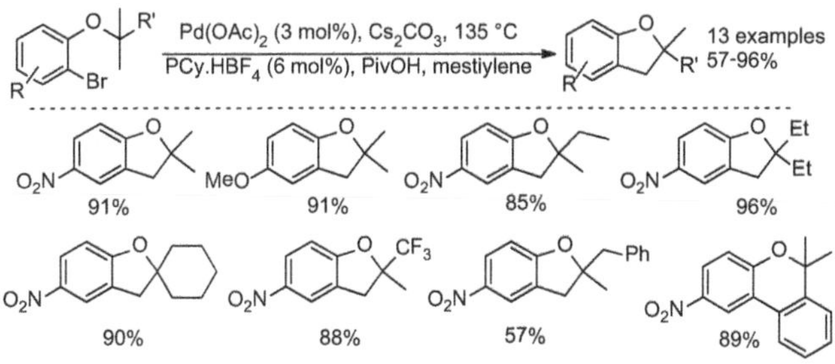

Scheme 2.51 Pd-catalyzed intramolecular alkane arylation.

Reaction Procedure (Scheme 2.51): Cs_2CO_3 (0.77 mmol), $Pd(OAc)_2$ (3 mol%), $PCy_3 \cdot HBF_4$ (6 mol%) and pivalic acid (30 mol%) were weighed in air and placed in a screw capped vial (4 mL) with a magnetic stir bar. The reaction vessel was evacuated and backfilled with argon (×3). The cyclization precursor (0.70 mmol) was then added to the reaction vessel as a solution in mesitylene (3 mL). The reaction was heated to 135 °C for 12 hours. Upon completion, the reaction was cooled to room temperature. The products were loaded directly onto a silica gel packed column chromatography and eluted using ether–hexane mixtures.

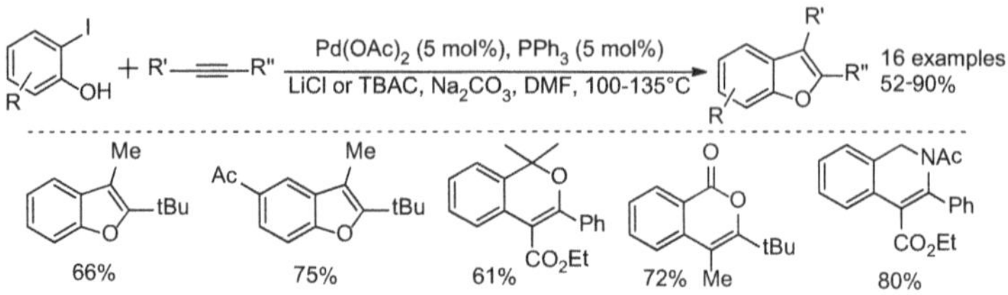

Scheme 2.52 Pd-catalyzed annulation of internal alkynes.

Reaction Procedure (Scheme 2.52): $Pd(OAc)_2$ (0.025 mmol), LiCl (0.50 mmol) or TBAC (0.50 mmol), Na_2CO_3 or K_2CO_3 (0.5–2.5 mmol), the aryl iodide (0.5–1.5 mmol), the alkynes (0.5–2.5 mmol), the solvent (10 or 20 ml) and, where indicated, PPh_3 (0.025 mmol) were added to a 2- or 4-dram vial equipped with a stir bar and heated at the appropriate temperature for the necessary period of time. The reaction mixture was diluted with ether, washed successively with saturated NH_4Cl and water, dried over anhydrous $MgSO_4$, and concentrated. The products were purified by column chromatography.

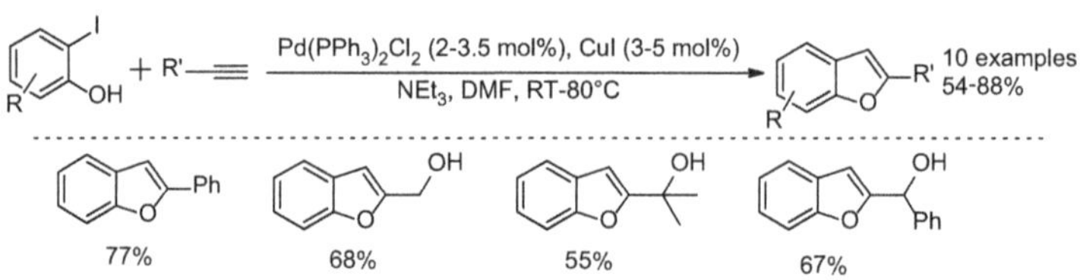

Scheme 2.53 Pd-catalyzed coupling of 2-iodophenols with alkynes.

Reaction Procedure (Scheme 2.53): To a well-stirred mixture of *o*-iodophenol (2 mmol), $Pd(Ph_3P)_2Cl_2$ (2–3.5 mol%), CuI (3–5 mol%) and triethylamine (2 equiv.) in DMF (5 ml), an acetylenic compound was added under an N_2 atmosphere. The mixture was stirred at room temperature for a further period of 1 h. The mixture was then cooled, poured into water (100 ml) and extracted with CH_2Cl_2 (3×50 ml). The combined extracts were washed with 5 mol dm^{-3} aq. NaOH (3×100 ml) and water (3×100 ml), dried (Na_2SO_4) and evaporated. The residue on column chromatography afforded the 2-substituted benzofurans.

methodology, dimethylformamide (DMF) was found to be the best solvent, mainly because of its excellent solubility related to all the starting materials and the catalysts. In a few cases, NaOAc was used as a base with PTC (Bu_4NCl), where the acyclic products could be isolated if the reactions were carried out at room temperature. A similar transformation was previous

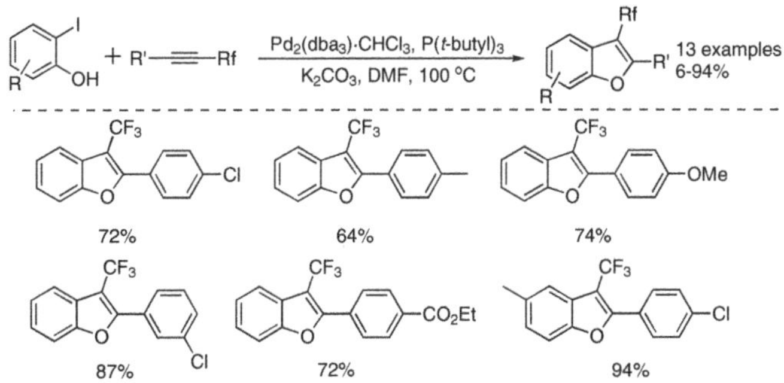

Scheme 2.54 Pd-catalyzed coupling of 2-iodophenols with fluorinated alkynes.

Reaction Procedure (Scheme 2.54): $Pd_2(dba)_3 \cdot CHCl_3$ (0.04 g, 0.048 mmol), P(*t*-butyl)$_3$ (10 wt% in hexane) (0.8 g, 0.395 mmol), K_2CO_3 (0.3 g, 2.44 mmol), the 2-iodophenol (0.214 g, 0.976 mmol), the alkyne (0.488 mmol), and DMF (8.0 mL) were added to a flask equipped with a stir bar. After being heated for 24 h at 100 °C, the mixture was quenched with $NaHCO_3$ aq. and extracted with Et_2O three times. The combined organic layers were dried over anhydrous Na_2SO_4 and concentrated *in vacuo*. The residue was chromatographed on silica gel using hexane to afford the corresponding benzofuran.

reported in 1986 by Arcadi and co-workers with similar catalytic system in piperidine.[91]

Konno and co-workers developed a palladium-catalyzed annulation reaction of a variety of fluorine-containing internal alkynes with 2-iodophenol derivatives in 2004.[92] In this system, the use of P(*t*-Bu)$_3$ as a ligand on palladium was found to be crucial, resulting in the exclusive formation of 3-fluoroalkylated benzofurans in high yields (Scheme 2.54). ^{19}F NMR analysis of the reaction mixture revealed that the addition of phenol to the fluoroalkylated alkynes was followed by an intramolecular Heck reaction, giving the corresponding 3-fluoroalkylated benzofurans.

In 2008, Grigg and co-workers reported the Sonogashira reaction of *N*-alkynylimides with 2-iodophenol or 2-iodo-*N*-tosylaniline.[93] The corresponding 2-(*N*-alkylimino)-benzofurans and indoles were isolated in good yields (Scheme 2.55). Interestingly, selective partial reduction of the product followed by treatment with TsOH generates *N*-acyliminium ions, which cyclise to afford tetra- and pentacyclic lactams in good yields. The formed lactams can be further reduced to the analogous cyclic amines by BH_3.

More recently, a dihydroxyterphenylphosphine bearing cyclohexyl groups on the phosphorus atom (Cy-DHTP) was proved to be a powerful ligand for the palladium-catalyzed one-pot synthesis of 2-substituted benzo[*b*]furans from 2-chlorophenols and terminal alkynes.[94] The Pd-Cy-DHTP catalyst can also be used in the sequential one-pot synthesis of disubstituted benzo[*b*]furans from dichlorophenols, terminal alkynes, and boronic acids. In both cases, the corresponding benzo[*b*]furans were prepared in good yields (Scheme 2.56). In this strategy, the use of XPhos as a second ligand along with an additional base is instrumental in promoting the Suzuki–Miyaura cross-coupling reaction. The use of this catalyst enabled the one-pot formation of chlorobenzo[*b*]furans from dichlorophenols and terminal alkynes and the subsequent Suzuki–Miyaura cross-coupling with boronic acids. The two catalysts were found to work independently, with Pd-Cy-DHTP catalyzing the Sonogashira cross-coupling and Pd-XPhos accelerating the Suzuki–Miyaura cross-coupling. The results of the ESI-MS study provided strong evidence for the formation of heteroaggregates between lithium phenoxides of 2-chlorophenol and Cy-DHTP in which the 2-chloro group is easily accessible to the Pd atom located nearby. A Pd–CuI–PPh_3/KF–Al_2O_3 catalyst system was reported for coupling 2-iodophenols with terminal alkynes as well.[95] The corresponding benzofurans were formed in moderate yields under solvent-free conditions under microwave irradiation. Indoles can be formed under these conditions when the nitrogen is substituted with an activating group. Applying trimethylsilylalkynes and 2-iodophenol as substrates for benzofuran synthesis was described too.[96]

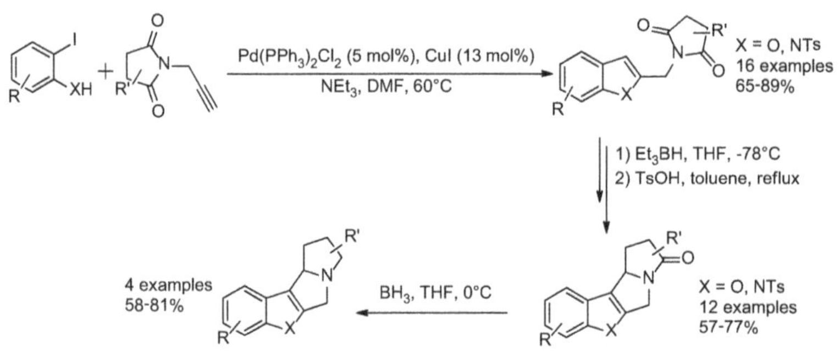

Scheme 2.55 Pd-catalyzed synthesis of lactams.

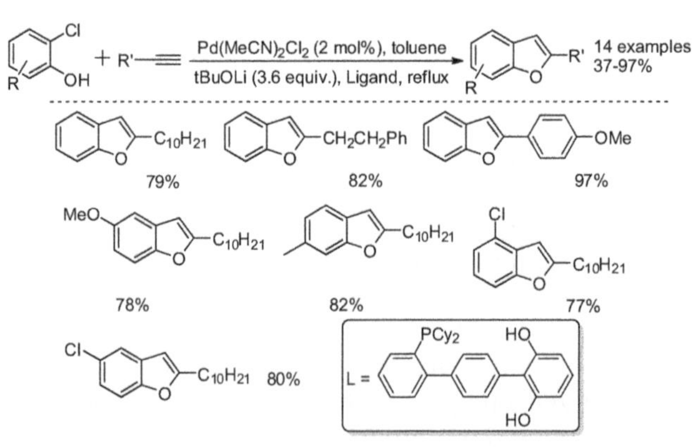

Scheme 2.56 Pd-catalyzed coupling of 2-chlorophenols with alkynes.

Reaction Procedure (Scheme 2.56): Typical experimental procedure for the synthesis of 2-substituted benzo[*b*]furans. Toluene (1.0 mL) was added to $PdCl_2(CH_3CN)_2$ (2.6 mg, 0.01 mmol), Cy-DHTP · HBF_4 (10.6 mg, 0.02 mmol), *t*-BuOLi (144 mg, 1.8 mmol), and 2-chlorophenol (64.3 mg, 0.50 mmol) in a two-neck flask under argon. The reaction mixture was stirred at rt for 30 min, and then 1-dodecyne (112 μL, 0.53 mmol) was added. The reaction mixture was stirred at reflux for 45 min and then cooled. Methanol (1 mL) was added, and the reaction mixture was stirred at reflux for 1 h. The resulting suspension was quenched with aq. NH_4Cl (5 mL) at rt and extracted with ethyl acetate (20 mL×2). The combined organic layers were dried over anhydrous Na_2SO_4 and concentrated, and then the residue was purified by preparative TLC (hexanes) to give 2-decylbenzo[*b*]furan as yellow oil.

Typical experimental procedure for the synthesis of disubstituted benzo[*b*]furans. Toluene (1.0 mL) was added to $PdCl_2(CH_3CN)_2$ (2.6 mg, 0.01 mmol), Cy-DHTP · HBF_4 (10.6 mg, 0.02 mmol), XPhos (9.7 mg, 0.02 mmol), *t*-BuOLi (144 mg, 1.20 mmol), and 2,4-dichlorophenol (64.3 mg, 0.50 mmol) in a two-neck flask under argon. The reaction mixture was stirred at rt for 15 min, and then 1-dodecyne (112 μL, 0.53 mmol) was added. The reaction mixture was stirred at reflux for 45 min and then cooled. Water (1 mL) was added, and the reaction mixture was stirred at reflux for 2 h. After the solution was cooled, 4-methoxyphenylboronic acid (114 mg, 0.75 mmol) and K_3PO_4 (213 mg, 1.0 mmol) were added, and the reaction mixture was stirred at reflux for 6 h. The resulting suspension was quenched with 1 M aq. HCl (7 mL) at rt and extracted with ethyl acetate (20 mL×2). The combined organic layers were dried over anhydrous Na_2SO_4 and concentrated, and then the residue was purified by preparative TLC (SiO_2, 95 : 5 hexanes–ethyl acetate) to give the pure product as a white solid.

Notably, in 2012, Ghosh and co-workers developed a rapid, visible light induced procedure for the synthesis of 2-alkyl/arylbenzofurans by using water as the solvent.[97] Moderate to good yields of the desired benzofurans could be produced from water at room temperature (Scheme 2.57). Compared with previously reported procedures, this new methodology has obvious advantages in terms of efficiency and reaction time. A radical reaction pathway was proposed for this transformation. In detail, in the first step, the *ortho*-iodophenol reacts with Pd(0) under photo irradiation to afford aryl radical (I) and Pd(1)I *via* a one electron transfer that further couples with Pd(1)I to give an aryl palladium(II) intermediate, and gives the coupling product after reductive elimination. Then the benzofurans can be formed after intramolecular cyclization.

As these homogeneous catalyst systems are not easily recyclable, efforts have been made towards recycling as well. In addition to heterogeneous systems, biphasic reaction systems were developed as well. In 2004, a novel

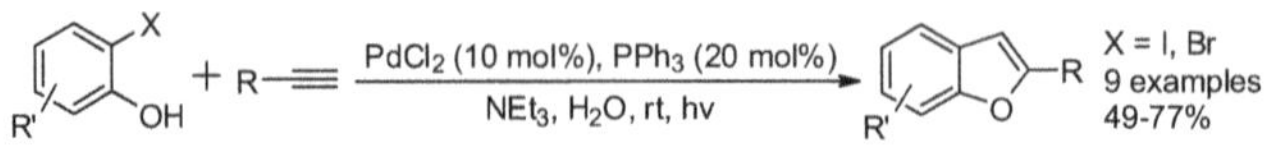

Scheme 2.57 Pd-catalyzed coupling of 2-iodophenols with alkynes stimulated by light.

Reaction Procedure (Scheme 2.57): Palladium chloride (17.7 mg, 10 mol%), PPh_3 (52.4 mg, 20 mol%) and distilled water (25 mL) were placed in an Erlenmeyer flask and heated on a water bath for 10 min followed by cooling at room temperature. Afterwards, 2-halophenols and terminal alkynes were added followed by the addition of triethylamine (400 mg, 4 mmol). The reaction mixture was irradiated with a 150W tungsten lamp (Philips India Ltd) for 1 h and 45 min. After the reaction was complete (monitored by TLC), the product was isolated by Et_2O (20 mL) and purified by column chromatography on silica gel (2% ethyl acetate in light petrol, 60–80 °C) and crystallized further from appropriate solvents.

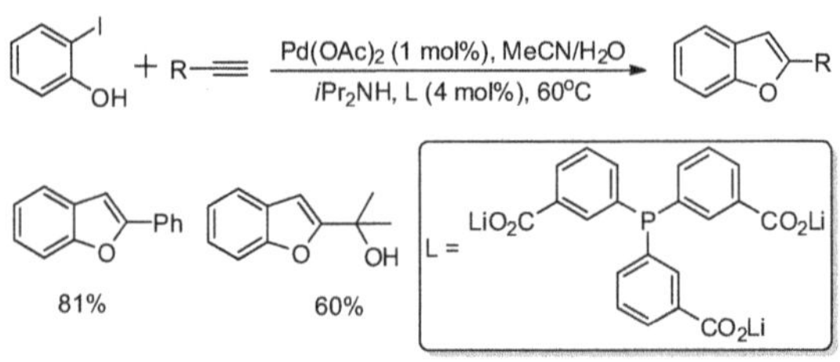

Scheme 2.58 Pd/*m*-TPPTC-catalyzed coupling of 2-iodophenols with alkynes.

water-soluble ligand was developed and applied in palladium- and rhodium-catalyzed coupling reactions.[98] The steric and electronic features of the novel water-soluble phosphane *m*-TPPTC have been determined and compared to the well-known sulfonated analogue TPPTS (3,3′,3″-phosphanetriyl-tris(benzenesulfonic acid) trisodium salt). Through several methods including NMR, IR and UV-VIS spectroscopy, it was found that the *m*-TPPTC phosphane is more basic than the sulfonated TPPTS ligand. This carboxylated phosphane *m*-TPPTC ligand was applied in copper-free palladium-catalyzed Sonogashira cross-couplings on aryl iodides and *ortho*-functionalized aryl iodides. Benzofurans were formed in good yields (Scheme 2.58). This phosphane induced a different selectivity compared to its sulfonated analogue, which may be explained by the difference of basicity between both ligands. Additionally, this ligand was applied in the rhodium-catalyzed arylation of alkynes as well. Remarkably, recycling the Pd/*m*-TPPTC system under mild conditions in the Sonogashira reaction was successful too. No loss in catalytic activity was observed after five cycles.

In 2003, Pal and co-workers described a Pd/C mediated synthesis of 2-substituted benzofurans from 2-iodophenols and terminal alkynes.[99]

This is an efficient procedure for the synthesis of 2-alkyl/aryl substituted benzo[*b*]furans/nitrobenzo[*b*]furans with water as the solvent. In the presence of PPh_3, CuI and prolinol, the corresponding products were formed in good yields (Scheme 2.59). The protocol does not require the use of a phase transfer catalyst or water-soluble phosphine ligands and is free from the use of any organic co-solvent. As the authors demonstrated, this is the first Pd/C mediated synthesis of benzofuran derivatives, but the lack of recycling ability was shown in the manuscript. Temperature proved to be critical; no product could be observed at room temperature.

Felpin and co-workers prepared a heterogeneous bimetallic catalyst and applied it in the synthesis of indoles and benzofurans (Scheme 2.60).[100] The desired products were prepared in water with Pd–Cu/C as the catalyst with 2-iodophenols, 2-iodoanilines and terminal alkynes as the substrates. In this procedure, no NH_2 free substrate succeeded and activating substituents were necessary. In their recycling experiment with indole synthesis as the model system, the yield of indole decreased from 90% in the first reaction to 60% for the second cycle and no reactivity was observed in the third cycle.

Palladium nanoparticles *via in situ* generation were explored and applied in this transformation without exception. In 2010, Ranu and co-workers developed a convenient procedure for the synthesis of 2-substituted benzofuran derivatives.[101] With 2-iodophenols and arylacetylenes as substrates in water in open air, the desired products were isolated in good yields (Scheme 2.61). The reaction does not require any additive, ligand, or co-catalyst. Interestingly, this *in situ* generated palladium nanoparticle gives better reactivity in water then in MeCN. Of the bases tested, organic bases gave much better results than inorganic bases (K_3PO_4, NaOH, K_2CO_3, Cs_2CO_3). This catalyst can be recycled as well. After three cycles, an obvious decrease in reactivity was detected. The loss of activity of the Pd

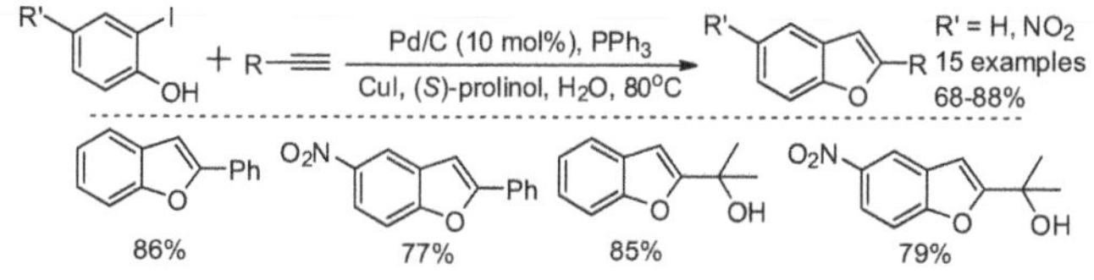

Scheme 2.59 Pd/C-mediated coupling of 2-iodophenols with alkynes.

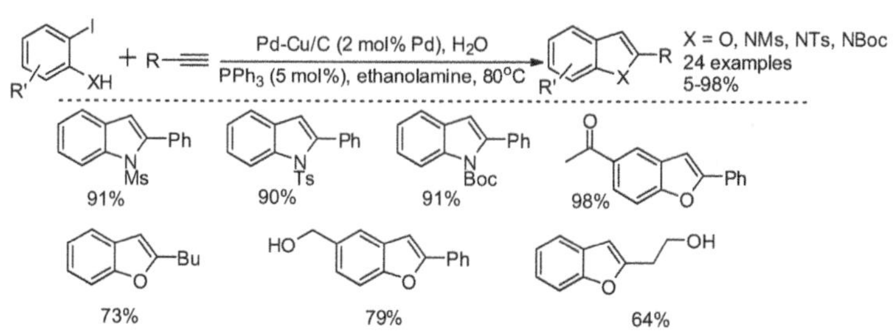

Scheme 2.60 Pd–Cu/C-catalyzed coupling of 2-iodophenols with alkynes.

Reaction Procedure (Scheme 2.60): The Pd–Cu/C was prepared by the following procedure. $Pd(OAc)_2$ (119 mg, 0.5 mmol), $Cu(OAc)_2$ (170 mg, 0.9 mmol) and charcoal (1 g) were dispersed in MeOH (100 mL). Then, hydrogen gas was bubbled through the solution for 5 minutes to remove oxygen. The resulting mixture was stirred for 12 h at 25 °C under H_2 (1 atm, balloon). The catalyst was filtered under a Millipore membrane (filters nylon 0.45 μm, 47 mm), washed with MeOH and dried under vacuum. ICP analyses were performed on the filtrate to verify the final Pd-metal loading on carbon to be 5 wt% and Cu-metal loading on carbon to be 3.6 wt%.

In a sealed tube, aryl iodide (0.5 mmol, 1.0 equiv.), PPh_3 (5 mol%) and catalyst Pd–Cu/C (2 mol%) were suspended in H_2O (3 mL) previously degassed. Then, acetylene derivative (1.0 mmol, 2.0 equiv.) and ethanolamine (1.5 mmol, 3.0 equiv.) were added. The resulting mixture was stirred 20 h at 80 °C under an argon atmosphere. After cooling to room temperature, DCM (10 mL) and H_2O (10 mL) were added and the mixture was filtered over a pad of Celite. The aqueous layer was extracted with DCM (2×10 mL). The collected organic extracts were washed with brine (20 mL), dried over $MgSO_4$, filtered and concentrated under reduced pressure.

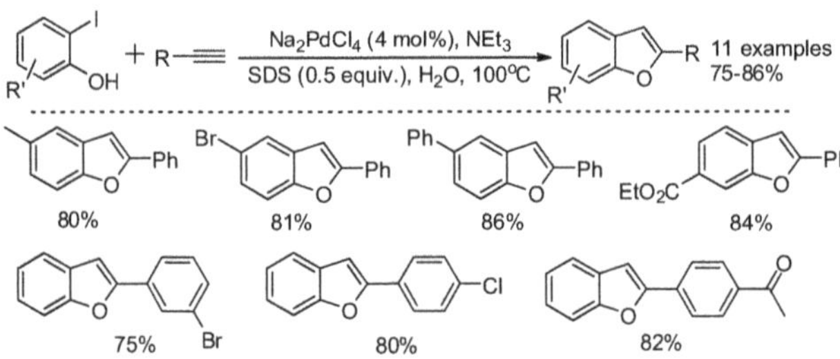

Scheme 2.61 PdNPs-catalyzed coupling of 2-iodophenols with alkynes.

Reaction Procedure (Scheme 2.61): To a mixture of Na_2PdCl_4 (12 mg, 0.0408 mmol) and sodium dodecyl sulfate (144 mg, 0.5 mmol) in water (3 mL) were added 2-iodophenol (220 mg, 1 mmol), phenylacetylene (123 mg, 1.2 mmol), and Et_3N (303 mg, 3 mmol). The mixture was heated to reflux (oil bath) for 14 h (TLC). After cooling, the reaction mixture was extracted with ethyl acetate (3×10 mL). The extract was washed with water and brine and dried with Na_2SO_4. Evaporation of the solvent left the crude product, which was purified by column chromatography over silica gel (hexane–ether, 92 : 8) to afford pure product.

nanoparticles is due to the agglomerization of the nanoparticles after each cycle. The catalytic efficiency of the nanoparticles is very much dependent on the particle size. Usually, an increase in the size of the nanoparticles results in a decrease in the activity.

In 2011, Ohtaka, Nomura and their co-workers stabilized palladium nanoparticles on linear polystyrene and applied them in Suzuki coupling and Sonogashira coupling in water.[102] The stabilized PdO nanoparticles (PS-PdONPs) were prepared in water by thermal decomposition of $Pd(OAc)_2$ in the presence of polystyrene. The immobilization degree of palladium was dependent on the molecular weight of polystyrene, while the size of the Pd nanoparticles was not. Linear polystyrene-stabilized Pd nanoparticles (PS-PdNPs) were also prepared using $NaBH_4$ and phenylboronic acid as reductants. The catalytic activity of PS-PdONPs was slightly higher than that of PS-PdNPs for the Suzuki coupling reaction in water. PS-PdONPs exhibited high catalytic activity for Suzuki and copper-free Sonogashira coupling reactions in water and was recycled without loss of activity. Benzofurans were produced in excellent yields when 2-iodophenols and terminal alkynes were applied as starting materials (Scheme 2.62).

In 2013, Park and co-workers developed a procedure for the synthesis of benzofurans from *o*-iodophenol and various terminal alkynes, by using a Pd catalyst supported on nano-sized carbon balls (NCB) as the catalyst.[103] Interestingly, this procedure works under copper- and ligand free conditions. Additionally, this recyclable catalyst could be reused more than five times with similar reactivity in the same heteroannulation reaction. This procedure tolerates various functional groups and gives the corresponding 2-substituted benzofurans in moderate to good yields (Scheme 2.63).

Instead of using 2-halophenols as substrates, simple phenols can be applied as well *via* C–H activation. In 2011, Li, Wang and their co-workers showed that benzofurans could be produced from commercially available phenols and propiolate through direct oxidative cyclization.[104] In the presence of $Pd(OAc)_2/PPh_3$ and CF_3CO_2Ag, (*E*)-type 3-phenoxyacrylates underwent reactions smoothly to generate the corresponding benzofurans in good yields in benzene at 110 °C under air pressure.

Based on the methodologies developed in this topic, this type of transformation was applied in the synthesis of various biologically active molecules and natural products as well, for example polysubstituted benzofurans as parasitic growth inhibitors; optically active 2- and 3-indolylglycine derivatives; benzofurans with antimicrobial activity and so on.[105–109]

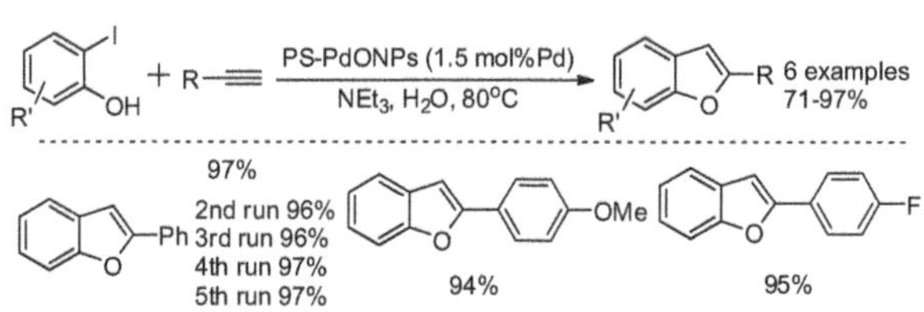

Scheme 2.62 PS-PdONPs-catalyzed coupling of 2-iodophenols with alkynes.

Reaction Procedure (Scheme 2.62): Preparation of PS-PdONPs. To a screw-capped vial with a stir bar were added 9.0 mg of polystyrene (85 μmol of styrene unit), $Pd(OAc)_2$ (5.5 mg, 25 μmol), and 1.5 M aqueous K_2CO_3 solution (3 mL). After stirring at 90 °C for 1 h, the reaction mixture was filtered with hot water. Subsequently, the polystyrene-stabilized Pd nanoparticles were washed with hot water (5×1.0 mL) and acetone (5×1.0 mL).

Typical procedure for the synthesis of benzo[*b*]furan. To a screw-capped vial with a stir bar were added 2-iodophenol (110 mg, 0.5 mmol), phenylacetylene (56.2 mg, 0.55 mmol), PS-PdONPs (2.9 mg, 1.5 mol% of Pd), triethylamine (152 mg, 1.5 mmol), and H_2O (1.0 mL). After stirring at 80 °C for 20 h, the reaction mixture was cooled to room temperature by immediately immersing the vial in water (~20 °C) for about 10 min. Subsequently, the aqueous phases were removed, and the recovered catalyst was washed with H_2O (5×3.0 mL) and diethyl ether (5×3.0 mL), which were then added to the aqueous phase. The aqueous phase was extracted eight times with diethyl ether. The combined organic extracts were dried over $MgSO_4$ and concentrated under reduced pressure. The recovered PS-PdONPs was dried *in vacuo* and reused. The resulting product was purified by PTLC (hexane–CH_2Cl_2).

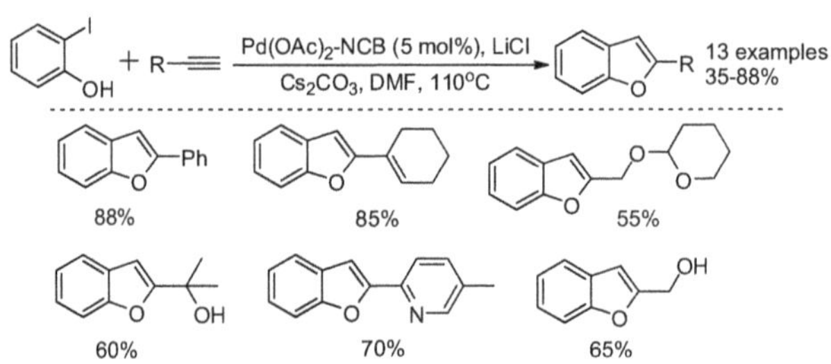

Scheme 2.63 $Pd(OAc)_2$-NCB-catalyzed coupling of 2-iodophenols with alkynes.

Reaction Procedure (Scheme 2.63): Procedure for the preparation of 5 mol% $Pd(OAc)_2$-NCB. To a stirred solution of $Pd(OAc)_2$ (0.56 g, 2.5 mmol) in THF (200 mL) was added NCB (9.5 g). After the mixture was stirred for 1 day at room temperature, $Pd(OAc)_2$-NCB was filtered and washed with THF. The solid was dried at room temperature under vacuum. Mol% Pd contents at $Pd(OAc)_2$-NCB was determined using an inductively coupled plasma (ICP) spectrometer.

General synthetic procedure. A mixture of $Pd(OAc)_2$-NCB (5 mol%), *o*-iodophenol (0.5 mmol), LiCl (0.5 mmol), Cs_2CO_3 (1.0 mmol) and terminal alkyne (1.0 mmol) was dissolved in 10 mL of DMF in a pressure tube. After the resulting solution was stirred for an appropriate time at 110 °C, the reaction mixture was filtered and neutralized with saturated NH_4Cl. The mixture was extracted with ethyl acetate, dried over $MgSO_4$, filtered and concentrated *in vacuo*, then purified by column chromatography using hexane and ethyl acetate as eluents.

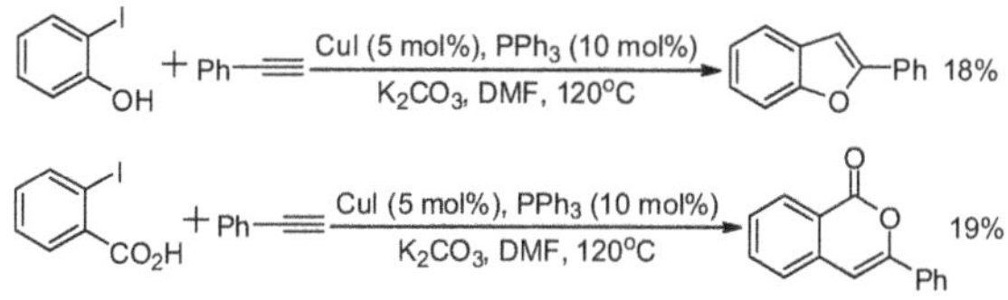

Scheme 2.64 CuI-catalyzed transformations of alkynes.

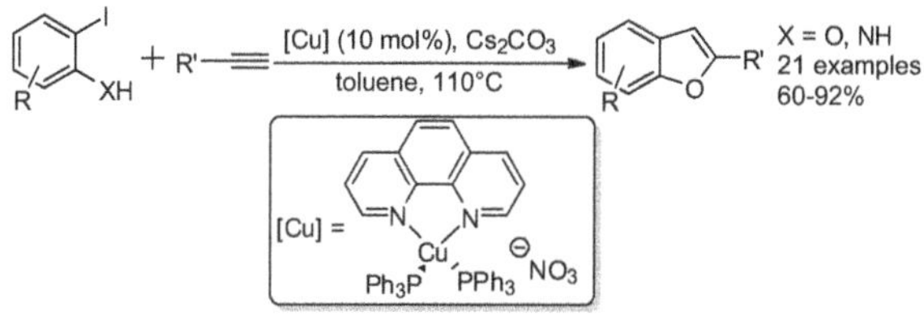

Scheme 2.65 Well-defined Cu(I) complex-catalyzed transformations of alkynes.

Besides these palladium catalysts, copper catalysts were developed and explored as well. As early as 1993, Miura and co-workers reported a copper-catalyzed reaction of aryl iodides with terminal alkynes.[110] With CuI–PPh_3 as the catalyst and K_2CO_3 as the base, excellent yields of the corresponding products were isolated. In the case of using 2-iodophenol and 2-iodobenzoic acid as substrates, the corresponding annulated products could be formed but in low yields (Scheme 2.64). The reaction of copper acetylide with 2-iodophenol in refluxing pyridine was reported previously.[111]

In 2002, Venkataraman and co-workers reported a copper(I)-catalyzed procedure for the synthesis of 2-arylbenzo[*b*]furans.[112] With a well-defined copper complex as the catalyst, a variety of 2-arylbenzo[*b*]furans were prepared in good to excellent yields by this procedure (Scheme 2.65). This method can tolerate a variety of functional groups, does not require the use of expensive additives and palladium catalysts. Later on, they succeeded to extend this procedure to indole synthesis by using 2-iodoanilines as substrates. Interestingly, the authors proved that bases have an important role in the success of this transformation. Cs_2CO_3 gave the best yield, while no product was formed by using NEt_3 as the base.

Later on, Li and co-workers reported an inexpensive and highly efficient CuI/DABCO system for the cross-coupling reactions of aryl halides and vinyl halides with arylboronic acids or terminal alkynes.[113] In the presence of TBAB, CuI-catalyzed Suzuki–Miyaura cross-coupling of vinyl halides and aryl halides with arylboronic acids was conducted smoothly to afford the corresponding diarylethenes and polyaryls in moderate to good yields using DABCO (1,4-diazabicyclo[2.2.2]octane) as the ligand. They also found that the inexpensive CuI/DABCO catalytic system was effective for Sonogashira cross-couplings of aryl halides and vinyl halides. A variety of aryl halides and vinyl halides, including activated aryl chlorides, underwent the coupling with terminal alkynes in moderate to excellent yields. In the examples tested, benzofurans were produced when 2-iodophenol was applied as a substrate (Scheme 2.66).

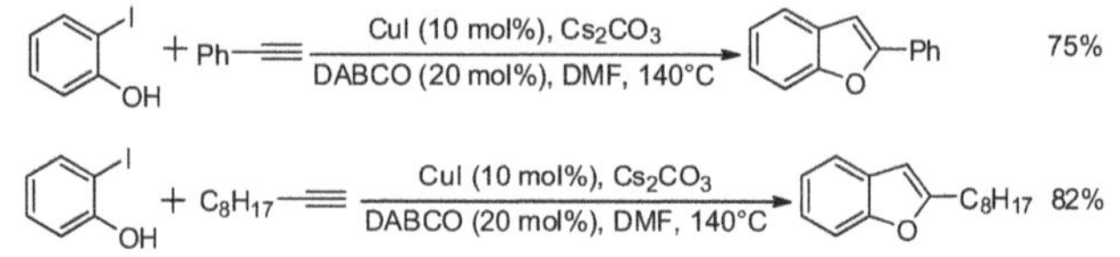

Scheme 2.66 CuI-DABCO-catalyzed synthesis of benzofurans.

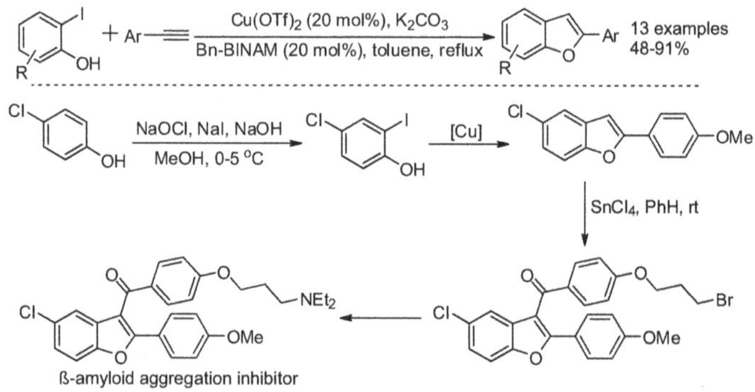

Scheme 2.67 CuI/Bn-BINAM-catalyzed synthesis of benzofurans.

Reaction Procedure (Scheme 2.67): Dibenzyl-BINAM (23.2 mg, 0.05 mmol), $Cu(OTf)_2$ (18.1 mg, 0.05 mmol), K_2CO_3 (103.7 mg, 0.75 mmol), and *o*-iodophenol (55 mg, 0.25 mmol) were placed in a 10 mL reaction tube capped with a septum. The tube was evacuated and backfilled with nitrogen. Toluene (3 mL) was added to the reaction mixture at room temperature. Then the resulting solution was refluxed, followed by phenyl acetylene (33 mL, 0.3 mmol) addition at refluxing temperature. The septum was replaced with a glass stopper and the reaction mixture was refluxed until the complete disappearance of *o*-iodophenol (TLC), then the reaction mixture was allowed to cool to room temperature and the solvent was evaporated. The crude residue was purified by column chromatography on silica gel using hexane as the eluent.

In 2010, Sekar and co-workers developed a copper(II)-catalyzed system for the synthesis of benzofurans from 2-iodophenols and terminal alkynes.[114] A wide range of 2-arylbenzo[*b*]furans were synthesized through copper(II)-catalyzed coupling of 2-iodophenols and aryl terminal acetylenes in moderate to good yields (Scheme 2.67). This method requires neither expensive palladium catalysts nor oxophilic phosphine ligands, and can tolerate different functional groups. The methodology was successfully utilized in the formal synthesis of β-amyloid aggregation inhibitor 5-chloro-3-[4-(3-diethylaminopropoxy)benzoyl]-2-(4-methoxyphenyl) benzofuran.

In a study from Shen's group, they demonstrated a concise and practical copper-catalyzed method for the synthesis of indoles and benzo[*b*]furans.[115] Both heterocycles could be obtained in high yield by the reactions of *N*-substituted 2-iodoanilines or 2-iodophenol with terminal alkynes under mild conditions, namely in the presence of CuI (10 mol%), traces of palladium, and a base in EtOH or 1,4-dioxane without using a phosphine ligand. Notably, simple aliphatic substituted terminal alkynes could be tolerated to smoothly produce indole and benzo[*b*]furan derivatives as well. In their comparison study, they found that palladium contaminants as low as 100 ppb are responsible for these successful couplings. From this point, this study serves to further highlight the importance of testing for trace metal impurities in reagents.

In addition to homogeneous copper catalysts, heterogeneous copper was prepared and applied in benzofuran synthesis as well. In 2012, Ramón and coworkers impregnated copper on a magnetite catalyst and applied it in domino Sonogashira-cyclization processes between 2-iodophenol and different alkynes to give the corresponding substituted benzo[*b*]furans (Scheme 2.68).[116]

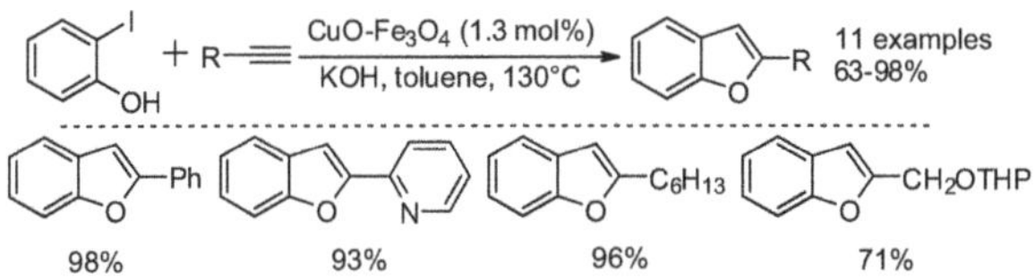

Scheme 2.68 CuO–Fe_3O_4-catalyzed synthesis of benzofurans.

Reaction Procedure (Scheme 2.68): General procedure for the synthesis of catalysts. To a stirred solution of $CuCl_2$ (1 mmol, 130 mg) in deionized water (120 mL) was added commercially available Fe_3O_4 (4 g, 17 mmol, powder <5 mm, BET area: 9.86 m^2 g). After 10 min at room temperature, the mixture was slowly basified with NaOH (1 M) until the pH was around 13. The mixture was stirred over one day at room temperature in air. After that, the catalyst was filtered and washed several times with deionized water (3×10 mL). The solid was dried at 100 °C over 24 h in a standard glassware oven to obtain the expected catalyst. The copper catalyst gave an incorporation of copper of 1.6% according to XRF.

General procedure for the synthesis of benzofurans. To a stirred solution of 2-iodophenol (1 mmol) in toluene (3 mL) under an argon atmosphere were added CuO–Fe_3O_4 (50 mg), KOH (67 mg, 1.2 mmol) and the corresponding alkyne (1.2 mmol). The resulting mixture was stirred at 130 °C until the end of reaction. The catalyst was removed by a magnet and the resulting mixture was quenched with water and extracted with EtOAc. The organic phases were dried over $MgSO_4$, followed by evaporation under reduced pressure to remove the solvent. The product was purified by chromatography on silica gel (hexane–ethyl acetate) to give the corresponding products.

The catalyst could be recovered ten times without losing its activity. This catalyst avoids the use of any type of expensive and difficult to handle organic ligand, showing excellent yields, under mild reaction conditions. The catalyst is very easy to remove from the reaction medium, just by using a simple magnet. However, a relatively high reaction temperature was required.

Additionally, the intramolecular cyclization of *o*-allyl-iodophenols offers another option for the synthesis of benzofuran derivatives. In 1986, a 'salen'cobalt(III)-catalyzed cyclization of *o*-allyl-iodophenols and *o*-but-3-enyl-iodophenol was reported by Pattenden and co-workers.[117] Benzofurans and benzopyrans were produced in low yields. Interestingly, a cobalt(I) complex was produced which can give the desired product. In 1988, a palladium-catalyzed cyclization of aryl allyl ethers was described by Larock and Stinn.[118] A series of benzofurans were isolated in moderate to good yields from the corresponding *o*-iodoaryl allyl ethers (Scheme 2.69). The cyclization of the analogous arylmercurials was discussed as well. An equal equivalent of Li_2PdCl_4 was required in this case. Unfortunately, a relatively long reaction time (2 days) was needed.

In 2004, Pan, She and co-workers studied this transformation further and found that the reaction conditions can be milder by using ionic liquid as the solvent.[119] By using a catalytic amount of $PdCl_2$ as the catalyst, NBu_3 as the base, in ionic liquid ([BMIm]BF_4), the desired products were formed in good yields at 60 °C for 24 hours (Scheme 2.70). Remarkably, the ionic liquid

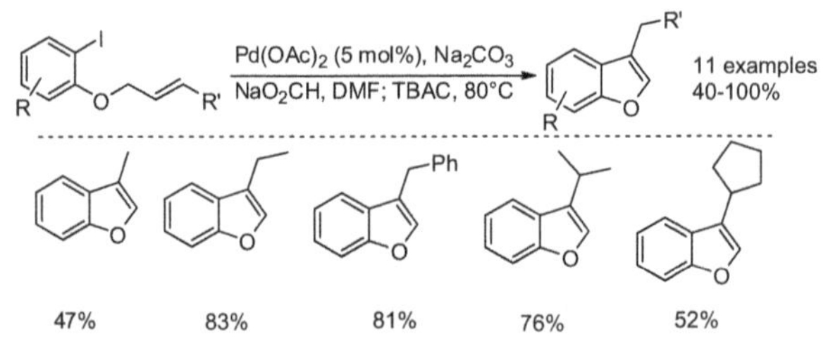

Scheme 2.69 Palladium-catalyzed cyclization of *o*-iodoaryl allyl ethers.

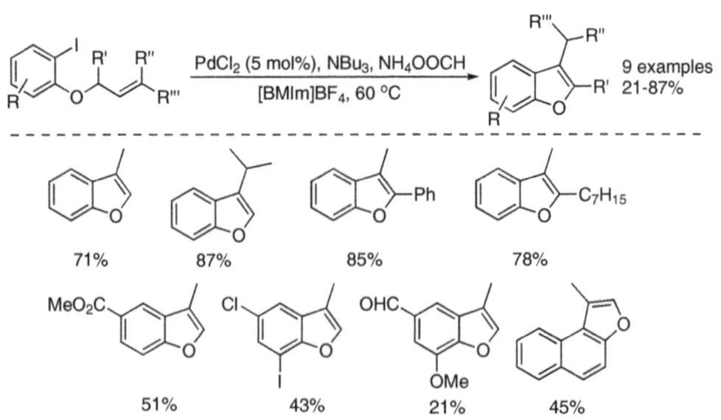

Scheme 2.70 Palladium-catalyzed cyclization of *o*-iodoaryl allyl ethers in ionic liquid.

Reaction Procedure (Scheme 2.70): *Ortho*-iodo aryl allyl ether (1 mmol), $(n\text{-Bu})_3N$ (1.5 mmol), NH_4OOCH (1 mmol) and $PdCl_2$ (0.05 mmol) were added to [BMIm]BF_4 (1.5 mL). The mixture was heated to 60 °C and maintained for 24 h. When the mixture was cooled to room temperature, it was extracted by diethyl ether (3×15 mL). The Pd catalyst was kept and suspended in the ionic liquid layer. The diethyl ether layers were collected and concentrated in vacuum. Then the crude mixture was purified by column chromatography on silica gel to give the pure benzofurans.

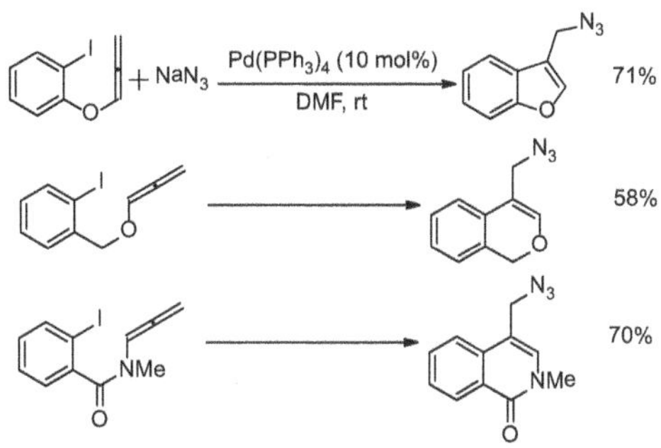

Scheme 2.71 Palladium-catalyzed cyclization of 2-iodophenoxyallenes.

containing palladium catalyst can be used four times with little loss of activity (1st 71%; 2ed 66%; 3rd 60%; 4th 57%).

Furthermore, 2-iodophenoxyallene was applied as a substrate in palladium-catalyzed benzofuran synthesis as well.[120,121] In the presence of palladium catalyst and another nucleophile, 3-substituted benzofuran can be formed in good yields (Scheme 2.71).

In 2006, Cheng and co-workers developed a new palladium-catalyzed multistep reaction involving ring closure of 2-iodophenoxy-, 2-iodobenzyloxy-, and 2-iodobenzylaminoallenes and ring opening of bicyclic alkenes.[122] This method provides the construction of two different and new C–C bonds in one pot and allows an efficient synthesis of various benzo[*b*]furan, ^{1}*H*-isochromenyl, and 1,2-dihydroisoquinoline-substituted 1,2-dihydro-1-naphthalenols in good yields (Scheme 2.72). On the basis of known palladium-catalyzed allene chemistry and the mechanisms for the catalytic reactions involving π-allylpalladium complexes as key intermediates, a mechanism was proposed by the authors. The reduction of $PdCl_2(PPh_3)_2$ to a Pd(0) by zinc metal likely initiates the catalytic reaction. Oxidative addition of 2-iodophenoxyallene to Pd(0) gives organopalladium species. Coordination of the allenyl group in organopalladium to the palladium center followed by insertion into the palladium–carbon bond affords the π-allyl palladium complex. Then, this is followed by the coordination of oxabenzonorbornadiene *via* the *exo* face of the carbon–carbon double bond to the π-allyl palladium complex and then the insertion of the double bond of

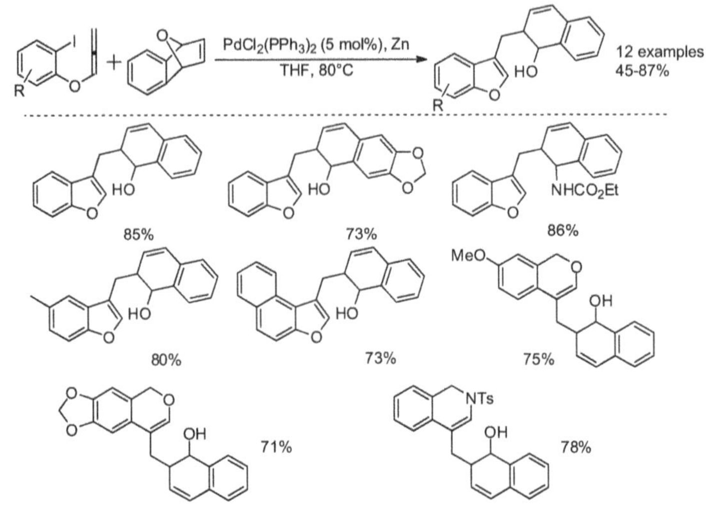

Scheme 2.72 Palladium-catalyzed cyclization of 2-iodophenoxyallenes with bicyclic alkenes.

> **Reaction Procedure** (Scheme 2.72): A 25 mL round-bottomed side-arm flask containing $PdCl_2(PPh_3)_2$ (0.0500 mmol), and Zn powder (1.5 mmol) was evacuated and purged with nitrogen gas three times. To the flask were then added iodoallenes (1.00 mmol), bicyclic alkenes (1.20 mmol) and THF (3.0 mL) *via* syringes. The reaction mixture was allowed to stir at 80 °C for 16 h. At the end of the reaction, the reaction mixture was diluted with CH_2Cl_2, filtered through Celite and silica gel, and the filtrate was concentrated. The crude residue was purified through a silica gel column using hexane and ethyl acetate as eluent to give a pure ring opening product.

oxabenzonorbornadiene into the allylpalladium bond to give the desired alkoxypalladium intermediate. Subsequent transmetalation with zinc halide leads to the corresponding Zn(OR)X and palladium(II) halide. The latter is then reduced by zinc metal powder to regenerate the Pd(0) catalyst. The Zn(OR)X is hydrolyzed after workup to give the final product.

Allenes and alkynes are considered to be analogue compounds in organic synthesis. In 2009, a nickel-catalyzed cyclization of iodoarylethers and direct trapping by various electrophiles was reported by Durandetti, Maddaluno and co-workers.[123] The desired cyclized products were formed in low to good yields (Scheme 2.73a). Later on, Wang, Zhang and their co-workers developed a palladium-catalyzed cyclization and carbine migratory insertion reaction for the synthesis of 3-vinylindoles and 3-vinylbenzofurans.[124] This reaction merged the alkyne-based Pd-catalyzed cyclization and palladium carbene migratory insertion in a single catalytic cycle, generating a C–C single bond and a C=C double bond in one operation. The desired products were formed in good yields (Scheme 2.73b).

Werz and co-workers developed a novel domino reaction with diynyl-substituted bromoarenes as substrates.[125] Benzofuran, pyridinofuran, indole, and isochromene systems were successfully obtained by this method in yields of 50–80% (Scheme 2.74). Regarding the reaction mechanism, the involvement of a Pd(IV) species was assumed.

From a synthetic point of view, the intermolecular reaction is more interesting than the intramolecular reaction as a wider variety of substrates can be used in the previous case. In 2005, Ma and co-workers reported a procedure for benzofuran preparation that was based on the copper-catalyzed coupling of vinyl halides with phenols, followed by palladium-catalyzed cyclization.[126] The CuI-catalyzed coupling reaction of vinyl halides and phenols occurs at 60–90 °C with *N,N*-dimethylglycine hydrochloride as the additive, giving vinyl aryl ethers in good yields. Then the palladium-catalyzed cyclization results in benzofurans at 100 °C in DMF in modest to good yields (Scheme 2.75). This procedure was applied later by Costa's group in the synthesis of 5-deoxypterocarpens, pterocarpens, and coumestans in good yields.[127]

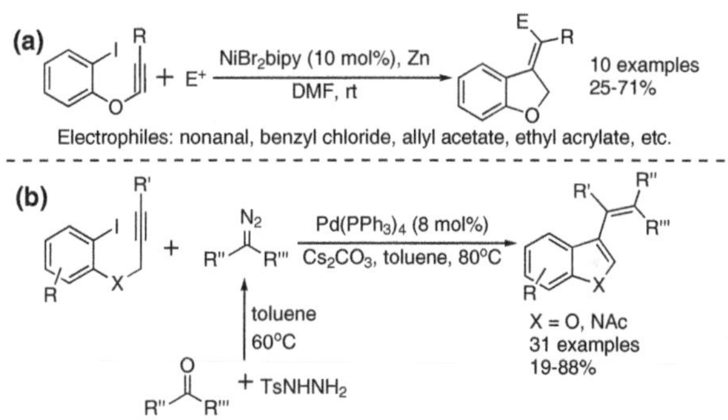

Scheme 2.73 Cyclization of alkynes.

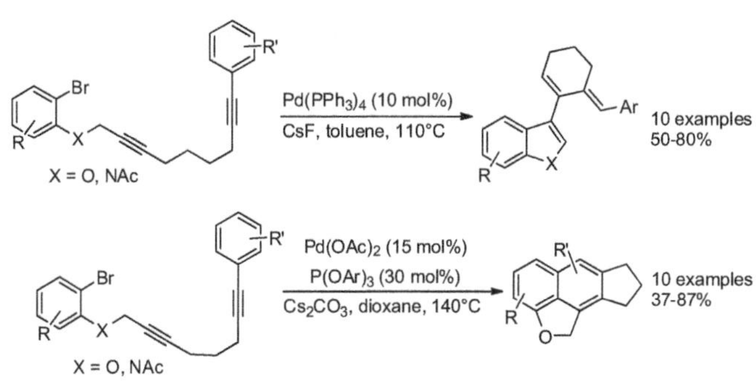

Scheme 2.74 Palladium-catalyzed cyclization of diynyl-substituted bromoarenes.

Reaction Procedure (Scheme 2.74): The alkylated bromophenol (0.10 mmol, 1.0 equiv.) was dissolved in toluene (3.0 mL). $Pd(PPh_3)_4$ (0.01 mmol, 10 mol%) and CsF (1.00 mmol, 10.0 equiv.) were added to the reaction mixture. The mixture was degased for 5–10 minutes with Argon. The reaction was stirred for at least 3 h at 110 °C. After cooling, the reaction mixture was filtered through a pad of Celite and washed with CH_2Cl_2. The solvent was removed by rotary evaporation and the residue was purified by silica gel column chromatography (pentane–EtOAc) to afford the desired compound.

The alkylated bromophenol (0.10 mmol, 1.0 equiv.) was dissolved in dioxane (3.0 mL). $Pd(OAc)_2$ (0.015 mmol, 15 mol%), tris(2,4-di*tert*butylphenyl)phosphite (as ligand) (0.03 mmol, 30 mol%) and Cs_2CO_3 (0.50 mmol, 5.0 equiv.) were added to the reaction mixture. The mixture was degased for 5–10 minutes with Argon. The reaction was stirred overnight at 140 °C. After cooling, the reaction mixture was filtered through a pad of Celite and washed with CH_2Cl_2. The solvent was removed by rotary evaporation and the residue was purified by silica gel column chromatography (pentane–EtOAc) to afford the desired compound.

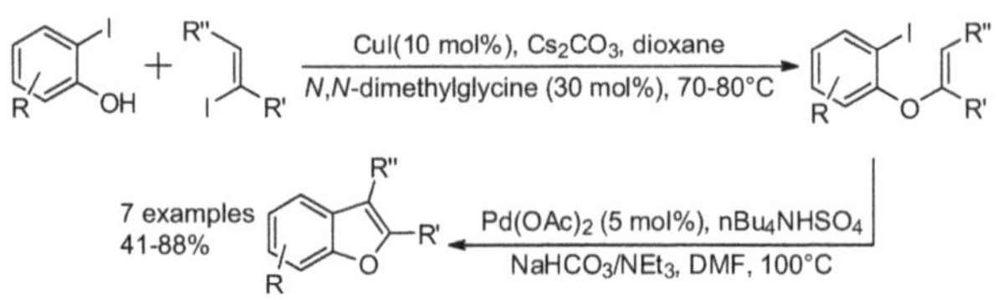

Scheme 2.75 Pd/Cu-catalyzed synthesis of benzofurans from vinyl halides and phenols.

Reaction Procedure (Scheme 2.75): A dried Schlenk tube equipped with a magnetic stir bar and a rubber stopper was charged sequentially with aryl vinyl ether (1.0 mmol), $NaHCO_3$ (2.5 mmol, if indicated), *n*-Bu_4NHSO_4 (1.0 mmol), 4 Å molecular sieves (400 mg, crushed into a powder before use), and $Pd(OAc)_2$ (0.05 mmol). The tube was evacuated and backfilled with argon (4 times). Et_3N (2.5 mmol, if indicated) was injected, followed by anhyd. DMF (1.0 mL). The tube was heated to 100 °C until the starting material had disappeared (monitored by TLC). The cooled solution was partitioned between ethyl acetate and saturated aqueous NH_4Cl. The organic layer was washed with brine, dried, filtered, and concentrated on a rotary evaporator. The material thus obtained was purified by silica gel chromatography to give the corresponding benzofuran.

For vinyl ether preparation. A mixture of vinyl halide (1.0 mmol), phenol (1.5 mmol), Cs_2CO_3 (2.1 mmol), CuI (0.1 mmol), *N*,*N*-dimethylglycine hydrochloride (0.3 mmol) in 1,4-dioxane (2 mL) was heated at 60–90 °C under a nitrogen atmosphere. The cooled solution was partitioned between water and ethyl acetate. The aqueous layer was extracted with ethyl acetate. The combined organic layers were washed with brine, dried over Na_2SO_4, and concentrated. The residue was chromatographed to afford the corresponding vinyl aryl ether.

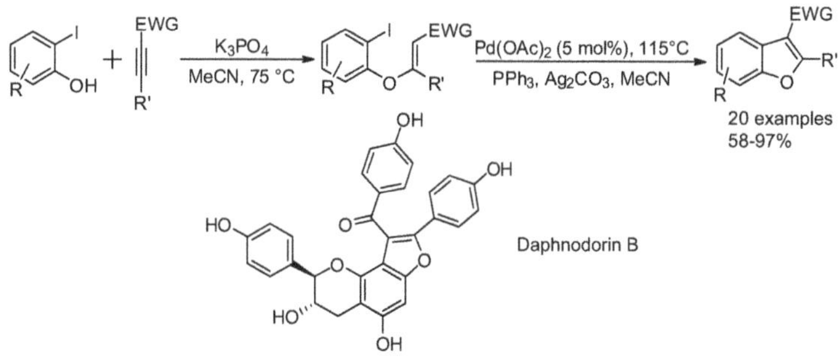

Scheme 2.76 Pd-catalyzed synthesis of Daphnodorin B.

Reaction Procedure (Scheme 2.76): To a solution of *o*-iodophenol (0.5 mmol) and activated alkynes (0.5 mmol) in CH_3CN (3.0 mL) was added K_3PO_4 (0.5 mmol), and then the reaction mixture was heated until the reaction reached completion. After cooling, the reaction mixture was diluted with water (15 mL), and extracted with EtOAc (2×15 mL). The combined organic extracts were washed with brine, dried ($MgSO_4$), and the solvent was removed *in vacuo*. The crude product was put into the next step without purification. A mixture of conjugate addition intermediates (0.5 mmol), $Pd(OAc)_2$ (5.7 mg, 0.025 mmol), PPh_3 (13.1 mg, 0.05 mmol), Ag_2CO_3 (110 mg, 0.5 mmol), and CH_3CN (5.0 mL) was stirred at 115 °C for 15 h under an Ar atmosphere. The reaction mixture was cooled to rt, diluted with water (15 mL), and extracted with EtOAc (3×15 mL). The combined organic extracts were washed with brine, dried ($MgSO_4$), and the solvent was removed *in vacuo*. The crude product was chromatographed on silica gel (EtOAc–petroleum ether, 1 : 20) to give the title compound.

In 2013, Zhang and co-workers developed a procedure for the syntheses of various 2-substituted-3-functionalized benzofurans (Scheme 2.76).[128] They started from *o*-iodophenols and activated alkynes, and the reaction went through addition and an intramolecular Heck reaction sequence. The interesting feature of this methodology was further illustrated by an enantioselective total synthesis of natural product Daphnodorin B.

More recently, Kapdi and co-workers developed an efficient and practical synthetic route for the palladium-catalyzed intramolecular C–H bond functionalization of coumaryl ethers to obtain benzofurocoumarins.[129]

Interestingly, the catalyst they applied is a well-defined and characterized palladacyclic complex, which proved to be an efficient catalyst for the intramolecular C–H bond functionalization of coumaryl and pyronyl ethers leading to the synthesis of tri, tetra and pentacyclic compounds in good yields (Scheme 2.77). A microwave-assisted one-pot procedure for obtaining the tetracyclic rings from 4-chlorocoumarin has also been carried out. Later on, they prepared another palladium complex, $[Pd(PPh_3)_2(saccharinate)_2]$, which was found to be active for the C–H bond functionalization of coumaryl and pyronyl ethers too.[130]

The activation of the C–H bond on aromatic rings was explored as expected. As early as 1982, palladium-catalyzed intramolecular cyclization of diaryl ethers and diaryl amines were mentioned.[131] In the presence of a palladium catalyst, low yields of the desired indole and benzofuran derivatives could be observed (Scheme 2.78).

Yin and co-workers described a novel entry to functionalized benzofurans and indoles starting from furan derivatives in 2012.[132] The protocol involves palladium(0)-catalyzed dearomatizing C2 arylation of the furan ring, the

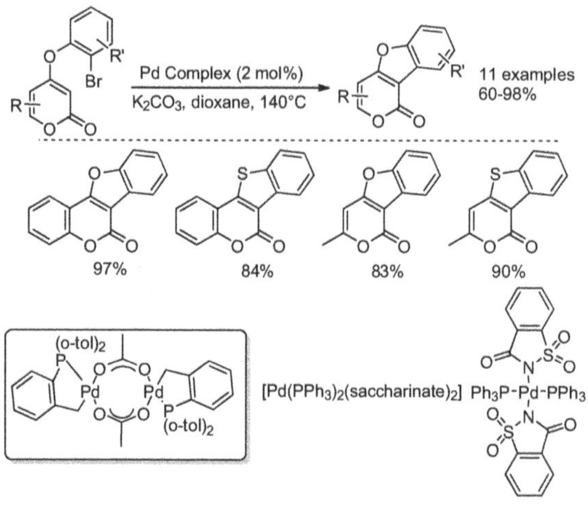

Scheme 2.77 Pd-complex-catalyzed synthesis of benzofurocoumarins.

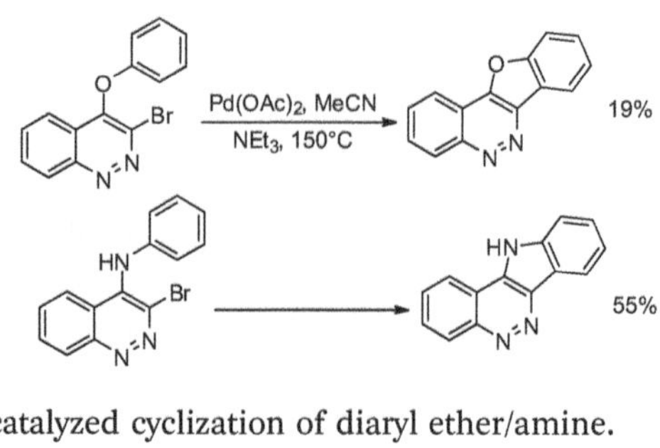

Scheme 2.78 Pd-catalyzed cyclization of diaryl ether/amine.

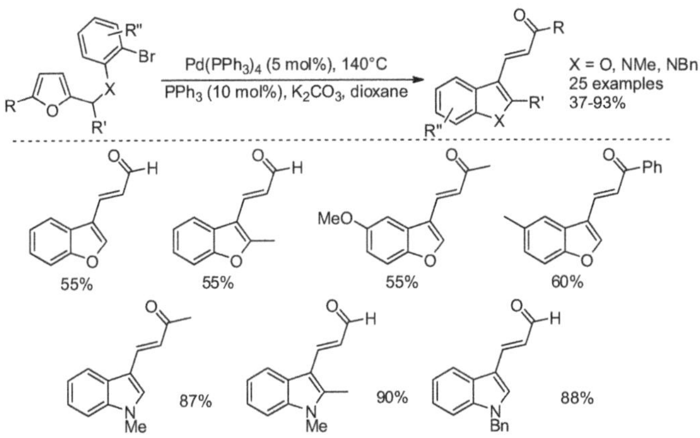

Scheme 2.79 Pd-catalyzed ring opening of furans.

formation of a π-allylic palladium complex, furan ring opening, and a β-hydride elimination. The corresponding benzofurans were isolated in low yields and good yields of the desired indole derivatives can be produced (Scheme 2.79).

Besides benzofurans, benzofuranones could be prepared using 2-halophenols as substrates as well. In 1986, Setsune and co-workers developed a copper-mediated coupling of 2-bromophenols with ethyl cyanoacetate and diethyl malonate.[133] In the presence of an equal amount of CuBr, the desired benzofuranones were formed in satisfied yields.

2.3 2-Haloaniline Derivatives

Applying 2-haloaniline derivatives in organic synthesis provides a promising pathway for the synthesis of nitrogen-containing heterocyclic compounds. Among the various five-membered nitrogen containing compounds, indole derivatives are representative examples which have been reported with broad applications in different areas.

In 1983, Dieck and co-workers studied the reactions between aryl and vinyl halides with 1,3-dienes and amines.[134] The catalytic formation of π-allylic palladium complexes *via* addition reactions to 1,3-dienes was involved, then the complexes formed reacted with amines and gave the final products. In order to explore the potential value of this methodology in organic synthesis, they tested 2-iodoaniline as a substrate as well. By using isoprene and 1,3-cyclohexadiene as the coupling partner, the desired cyclized products were formed in good yields (Scheme 2.80).

Later on, Larock and co-workers performed a systematic study on this transformation.[80] By adding TBAC (nBu_4NCl) as the additive, the corresponding compounds were formed in good yields (Scheme 2.81). Mechanistically, as previous proposed, heteroannulation proceeds *via* aryl- and π-allylpalladium intermediates.

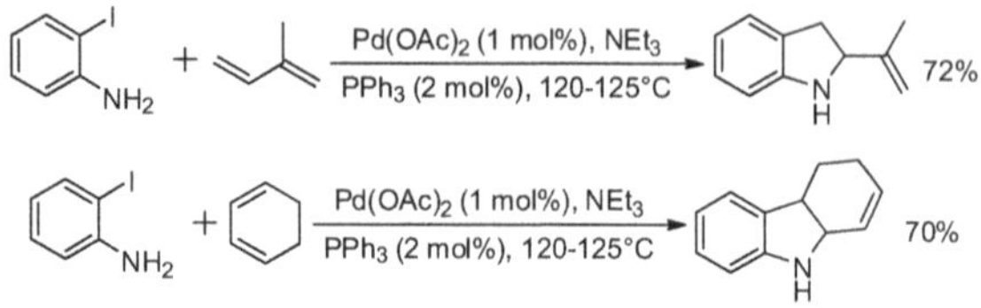

Scheme 2.80 Pd-catalyzed synthesis of indolines from 1,3-dienes and 2-iodoaniline.

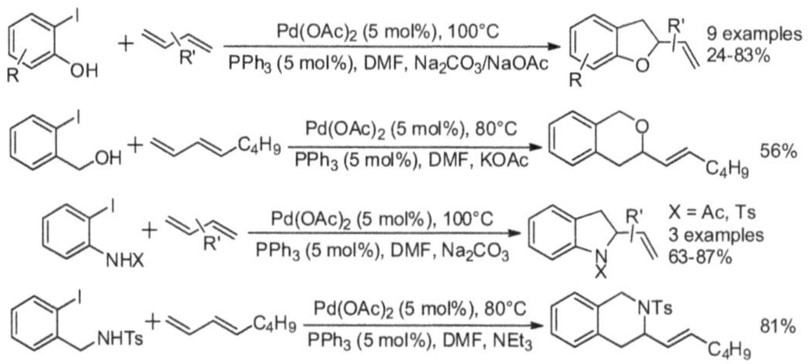

Scheme 2.81 Palladium-catalyzed annulation of 1,3-dienes.

As the work of Larock's group continued, they reported a palladium-catalyzed annulation of vinylic cyclopropanes and cyclobutanes in 1996.[135] This new palladium-catalyzed hetero- and carboannulation of unsaturated cyclopropanes and cyclobutanes by aryl halides bearing functionality in the *ortho* position provided a novel and efficient process for the synthesis of a wide variety of five- and six-membered ring heterocycles and carbocycles (Scheme 2.82a). This process is reasonably general with regard to the types of functional groups on the arene that can be employed and the substitution pattern allowable in the unsaturated cyclopropane or cyclobutane. Different bases, with or without PPh_3, have proven optimal for each different functional group present on the arene. Temperatures of 80–100 °C and reaction times of several days are usually necessary to obtain the best yields.

While analogous products are obtained by the Pd-catalyzed annulation of 1,3-dienes, this procedure may have advantages when the unsaturated cyclopropanes or cyclobutanes are more readily prepared. However, a large excess of alkenes was required. In 2004, they succeeded to extend their methodology to functionalized alkenes.[85] Under the same reaction conditions, the corresponding annulated products were produced in moderate to good yields (Scheme 2.82b).

In 2009, a palladium-catalyzed tandem ring opening–ring closing reaction of diazabicyclic alkenes was developed.[136] This procedure represents a facile one-pot strategy for cyclopentannulation of heterocycles. Using this methodology, the group could tune the reaction to the formation of either 3,4-disubstituted cyclopentenes or cyclopentene fused heterocycles by careful

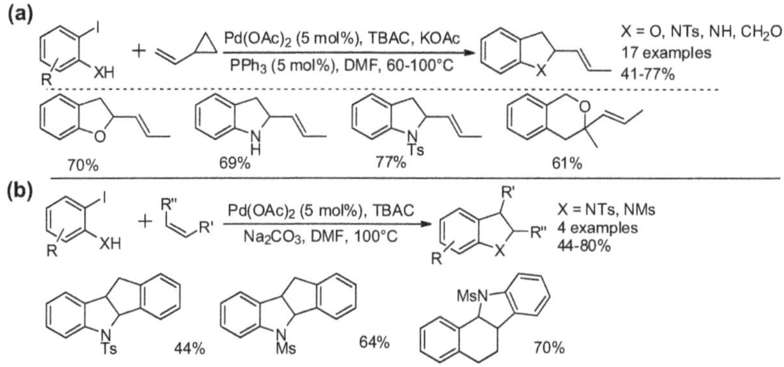

Scheme 2.82 Palladium-catalyzed annulation of vinylic cyclopropanes and alkenes.

Reaction Procedure (Scheme 2.82a): Palladium acetate (0.0125 mmol), n-Bu_4NC1 (0.25 mmol), the appropriate base (1.0 mmol), the aryl iodide (0.25 mmol), the alkene (1.25 mmol), DMF (1 ml) and, where indicated, PPh_3 (0.0125 mmol) were added to a 1-dram vial equipped with a stir bar and teflon-lined screw cap. After heating for the appropriate time, the reaction mixture was diluted with ether (20 ml), washed with satd NH_4CI (3×20 ml) and dried over $MgSO_4$. The reaction mixture was filtered, concentrated and purified by flash column chromatography using hexane–ethyl acetate. The following compounds were obtained using the above general procedure.

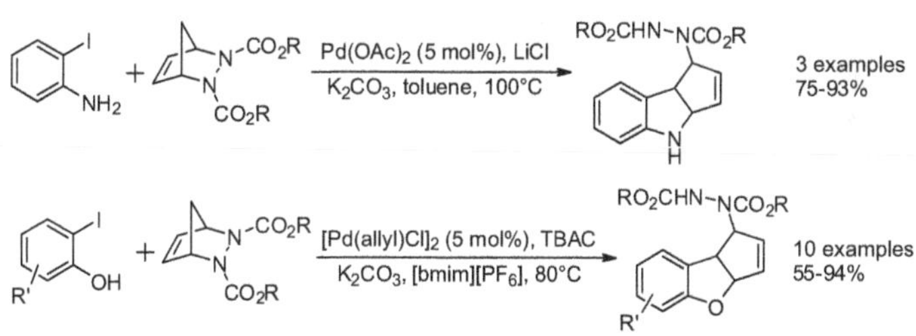

Scheme 2.83 Palladium-catalyzed annulation of diazabicyclic alkenes.

manipulation of the reaction parameters. The desired dihydroindoles and dihydrobenzofurans were isolated in good yields in general (Scheme 2.83).

Additionally, palladium catalysts were applied in the intramolecular cyclization of allyl moieties for dihydroindole synthesis as well. As early as 1994, Grigg and co-workers performed this type of work and applied it in the synthesis of heterocycles (Scheme 2.84a).[137] In 2005, Lautens's group reported a new rapid synthetic approach to a variety of *trans*-2,4-disubstituted 1,2,3,4-tetrahydroquinolines with excellent diastereoselectivities as well as to various five- to seven-membered carbo- and heterocycles *via* conventional thermal and microwave-assisted conditions (Scheme 2.84b).[138]

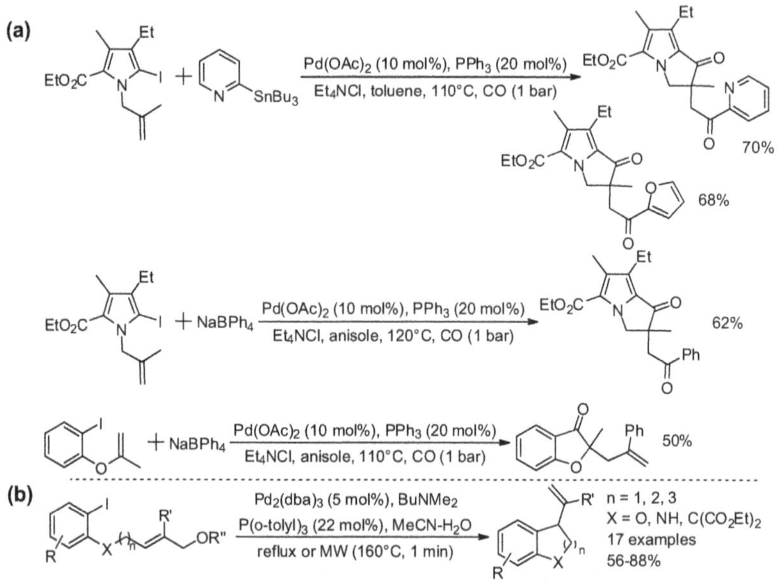

Scheme 2.84 Palladium-catalyzed cyclization of allyl moieties.

> **Reaction Procedure** (Scheme 2.84b): To a mixture of substrate (1.0 equiv.), (o-tolyl)$_3$P (22 mol%) and n-BuNMe$_2$ (2.0 equiv.) in $CH_3CN–H_2O$ (10 : 1, 0.1 M) was added $Pd_2(dba)_3 \cdot CHCl_3$ (5 mol%). The mixture was degassed and filled with N_2 gas quickly and refluxed for 6 h under a N_2 atmosphere. The resulting mixture was cooled to room temperature and treated with saturated NH_4Cl. Extractive workup and purification of the residue by column chromatography on silica gel (hexane–Et_2O, 40 : 1 to 20 : 1 as eluent) gave 2,4-disubustituted-1,2,3,4-tetrahydroquinoline.

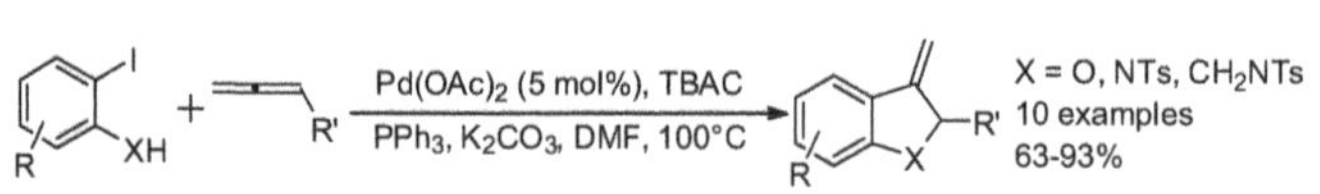

Scheme 2.85 Palladium-catalyzed annulation of allenes.

Allenes as an interesting class of unsaturated compounds have been explored in cyclizations with 2-iodoanilines and 2-iodophenols as well. In 1991, Larock and co-workers reported the cyclization of allenes with double functionalized arenes.[139] Various five- and six-membered heterocycles were formed in high yields (Scheme 2.85). But the main problem is the long reaction time; one or two days are usually needed.

Ma and co-workers performed extensive studies on allenes; various interesting synthetic procedures have been developed. In 2009, they reported the enantioselective cyclization of 2-iodoanilines with allenes.[140] With the newly developed spiro-BOX ligand, a palladium catalyst can catalyse

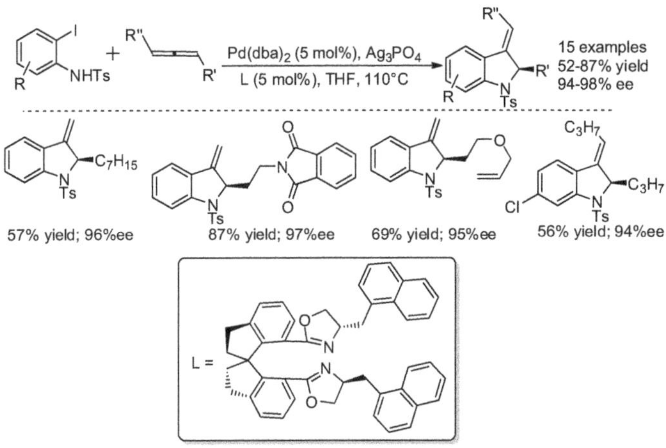

Scheme 2.86 Palladium-catalyzed enantioselective annulation of allenes.

Reaction Procedure (Scheme 2.86): To a Schlenk tube with a screw cap were added $Pd(dba)_2$ (6 mg, 0.011 mmol), (R_a,*S*,*S*)-L7 (7 mg, 0.011 mmol), and 1 mL of THF. The resulting mixture was stirred for 2 h at room temperature, which was followed by sequential introduction of Ag_3PO_4 (34 mg, 0.081 mmol), 2-iodoaniline (0.20 mmol), allene (0.81 mmol), and 1 mL of THF at room temperature. The resulting solution was stirred at 110 °C. When the reaction was completed as monitored by TLC, the solvent was evaporated under vacuum, and the residue was purified by chromatography on silica gel (eluent: petroleum ether–ethyl acetate, 70 : 1) to afford the product as an oil; the ee was determined by HPLC analysis [Chiralcel AD-H, hexane–*i*-PrOH = 85 : 15, 0.7 mL min^{-1}, 230 nm].

asymmetric allylic annulation of readily available 2-iodoanilines with allenes affording the potentially useful 3-alkylideneindolines in good yields with high to excellent enantiomeric excesses (94–98% ee; Scheme 2.86). The configuration of the double bond was confirmed by an NOE study.

More recently in 2013, Ma and Zhu have developed a one-pot sequential cyclization reaction of 2-iodoaniline derivatives and propargylic bromides.[141] The reaction goes through allene as the intermediate, and offers a diverse and efficient procedure for the synthesis of polysubstituted indoles starting from the readily available substrates. This tandem coupling cyclization reaction is believed to proceed *via* palladium(0)-catalyzed carbon–carbon bond formation, involving organoindium reagents generated *in situ*, and cycloisomerization. Here, NaI may help to generate the indium reagent and increase the group-transfer ability of the indium reagent generated *in situ*. This protocol was also applied to the gram scale synthesis of indomethacin—a clinically used non-steroid anti-inflammatory drug (NSAID) to reduce fever, pain, stiffness, and swelling (Scheme 2.87).

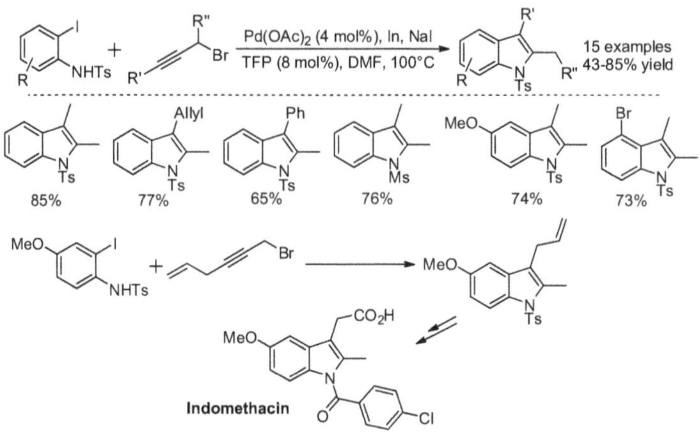

Scheme 2.87 Palladium-catalyzed annulation of allenes generated *in situ*.

> **Reaction Procedure** (Scheme 2.87): To a flame dried Schlenk tube were added indium powder (230.0 mg, 2.0 mmol), NaI (450.2 mg, 3.0 mmol)/DMF (2 mL), and 1-bromobut-2-yne (3.0 mmol)/DMF (2 mL) sequentially under an Ar atmosphere. After the mixture was stirred at room temperature for 15 min, *N*-tosyl 2-iodoaniline (1.0 mmol), $Pd(OAc)_2$ (9.0 mg, 0.040 mmol), TFP (18.8 mg, 0.081 mmol), and DMF (2 mL) were added to the reaction mixture sequentially. The resulting mixture was stirred at 100 °C and after 12 h the reaction was over as monitored by TLC. The resulting mixture was poured into H_2O (50 mL) and extracted with Et_2O (3×50 mL). The combined organic layer was dried over anhydrous Na_2SO_4, filtered, evaporated, and purified *via* column chromatography on silica gel (eluent: petroleum ether–ethyl acetate, 40 : 1) to afford the desired product.

In the same year (2013), Doyle and co-workers reported an intramolecular version of the cyclization of the allenes.[142] They combined allene chemistry with fluorination, and their report represents the first example of tandem C–C and C–F bond formation for the palladium-catalyzed carbofluorination of allenes. The intramolecular Heck-fluorination cascade provides monofluoromethylated heteroarenes, an important class of products in medicinal chemistry, in good yields (Scheme 2.88). The intermolecular variants for the three-component coupling of allenes, aryl iodides, and AgF were reported as well. Mechanistic studies indicate that C–F bond formation occurs by an outer sphere attack of fluoride on an allylpalladium fluoride intermediate.

A copper-catalyzed intermolecular cyclization of 2-iodoanilines with allenes was developed by Huang and co-workers in 2012.[143] This new methodology allows for the synthesis of 2-methyl-1*H*-indole-3-carboxylates in moderate to good yields from readily available substrates (Scheme 2.89).

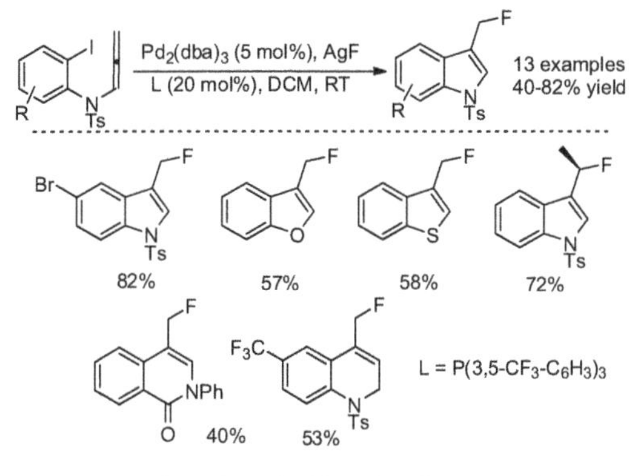

Scheme 2.88 Palladium-catalyzed carbofluorination of allenes.

Reaction Procedure (Scheme 2.88): After cooling to room temperature in a desiccator, an oven-dried flask was charged with the iodoallene (1.0 equiv.), silver(I) fluoride (1.5 equiv.), $P(3,5\text{-}CF_3\text{-}C_6H_3)_3$ (20 mol%), and tris(dibenzylideneacetone)dipalladium (5 mol%). Three times, the flask was purged under a high vacuum and filled with nitrogen, before charging with dichloromethane (0.1 M) and removing the nitrogen line. The reaction mixture was allowed to stir at room temperature, covered in aluminum foil, for 24 h. The reaction mixture was then filtered through a pad of Celite, eluting with dichloromethane. After concentrating the filtrate under reduced pressure, the residue was purified by manual column chromatography to afford the desired product.

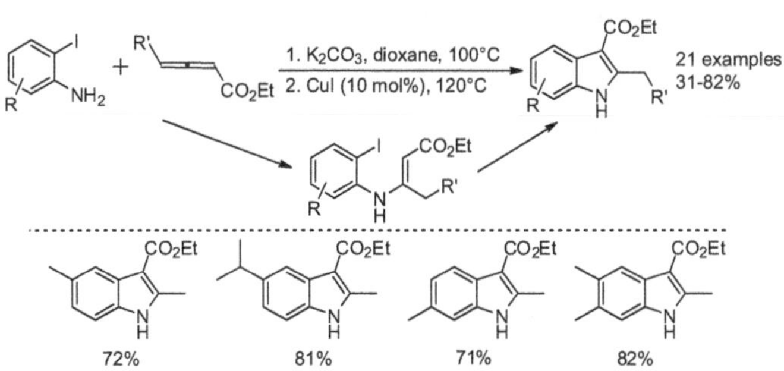

Scheme 2.89 Copper-catalyzed annulation of 2-iodoanilines and allenes.

Reaction Procedure (Scheme 2.89): A mixture of 2-iodoaniline (0.2 mmol), K_2CO_3 (0.3 mmol), and allene (0.24 mmol) in 1,4-dioxane (0.9 mL) was heated at 100 °C for 8 h in a sealed reaction tube. After cooling to rt, CuI (3.8 mg, 0.02 mmol) was added and the mixture was heated at 120 °C for 7 h. After removal of the solvent under reduced pressure, the crude product was purified by column chromatography on silica gel (EtOAc–PE, 1 : 6) to give the pure product.

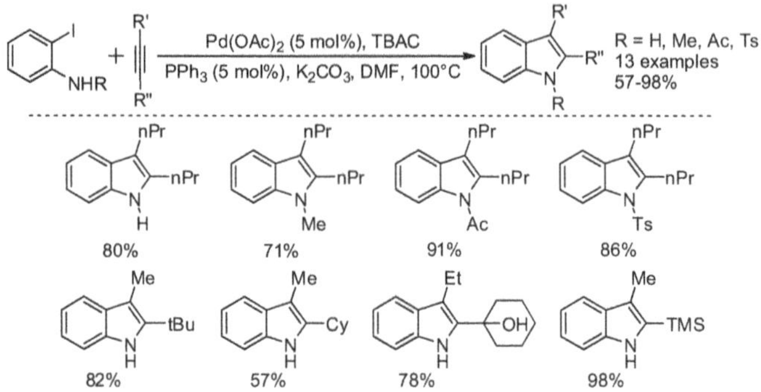

Scheme 2.90 Palladium-catalyzed annulation of 2-iodoanilines with alkynes.

From the point of the reaction mechanism, this reaction consisted of two parts: a Michael addition reaction and an intramolecular Heck type coupling reaction.

In 1991, Larock and Yum reported the heteroannulation of internal alkynes with 2-iodoanilines, which is now known as Larock indole synthesis.[144] In the presence of a palladium catalyst, the desired indoles were isolated in moderate to good yields under mild reaction conditions (Scheme 2.90). However, a large excess of alkynes was necessary in some cases (5 equiv.).

This procedure was further extended to *ortho*-iodo-heteroaniline derivatives for preparing natural amino acids.[145] Later on, a $Pd(OAc)_2$/TPPTS catalyst system was reported and the reaction was realized in aqueous media.[146] In addition to indoles, benzofurans were also produced in good yields under milder conditions (RT–65 °C, NEt_3, MeCN–H_2O). Using $Pd(PPh_3)_2Cl_2$ as a catalyst with CuI as an additive, the cyclization of 2-iodoanilines and even 2-bromoanilines with terminal alkynes can be realized at room temperature in DMF;[147] 68–82% of the corresponding indole derivatives were produced. In 2006, ultrasound irradiation was applied in the annulation of 2-iodoanilines with terminal alkynes as well.[148] With the assistance of ultrasound irradiation, the reaction can proceed much faster and without the addition of a copper ligand. Interestingly, as the author demonstrated, Bu_4NOAc was used as a base in this methodology with acetonitrile as the solvent.

In 2006, Lu and co-workers developed a practical and economical one-pot process for the synthesis of 2,3-disubstituted indoles.[149] This palladium-catalyzed domino indolization procedure allows rapid access to a variety of 2,3-disubstituted indoles regiospecifically under fairly mild conditions (Scheme 2.91). The reaction consisted of a palladium-catalyzed Sonogashira reaction and organopalladium-mediated amidopalladation.

In addition to homogeneous palladium catalysts, heterogeneous palladium catalysts were applied in the annulation of 2-iodoanilines with alkynes as well. In 2004, Yum and co-workers reported a Pd(II)–NaY

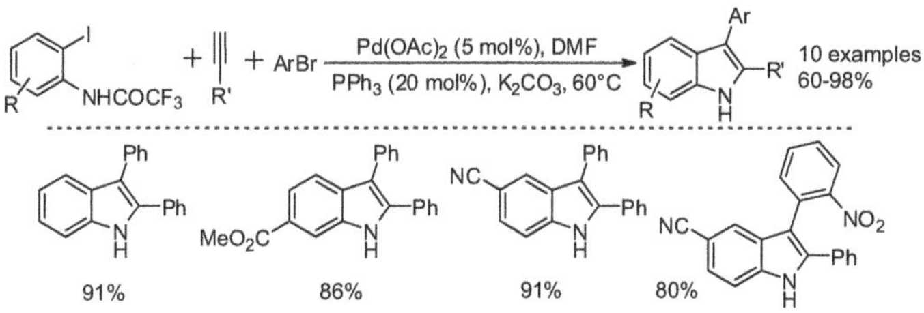

Scheme 2.91 Palladium-catalyzed three-component synthesis of indoles.

Reaction Procedure (Scheme 2.91): A 10 mL, 3-neck flask, equipped with a magnetic stir bar and thermocouple, and Argon inlet, were charged 2-trifluoroacetylamido-iodobenzene (0.5 g, 1.54 mmol, 1.0 equiv.), $Pd(OAc)_2$ (17.32 mg, 0.08 mmol, 0.05 equiv.), triphenylphosphine (80.94 mg, 0.154 mmol, 0.2 equiv.), and potassium carbonate (0.851 g, 6.16 mmol, 4 equiv.), followed by the addition of 5 mL of anhydrous DMF. With stirring at room temperature, were added into the reaction mixture aryl acetylene (1.85 mmol, 1.2 equiv.) and aryl bromide (1.85 mmol, 1.2 equiv.). The reaction was heated then at 60 °C until the reaction was complete monitored by HPLC. The mixture was quenched with water, and the aqueous solution was extracted with AcOEt three times. The organic solution was washed with brine, and dried over Na_2SO_4. The product was purified by column chromatography.

zeolite-catalyzed heteroannulation of *o*-haloanilines and terminal alkynes to indoles.[150] This method provides a convenient new route to various heteroannulation products in a one-pot reaction. Various 2-substituted indoles were prepared in a one-pot reaction in good yields (Scheme 2.92). The product formation largely depended on the solvent, base, and reaction temperature. No product was formed in 1,4-dioxane and only a trace of product was detected at 120 °C. The recycled catalyst showed good reusability in this heteroannulation reaction as well; good yields resulted even after six runs. In 2006, other heterogeneous palladium catalysts ($[Pd(NH_3)_4]^{2+}$/NaY and [Pd]/SBA-15) were applied to the synthesis of 2-functionalised indoles, giving generally high conversions and selectivities (>89% yield) using only 1 mol% [Pd]-catalyst under standard reaction conditions (NMP, 80 °C).[151] For the synthesis of 2,3-functionalised indoles by cross-coupling arylation, the [Pd]/SBA-15 catalyst was found to be particularly interesting, producing the expected compound with 35% yield after 12 days of reaction, which is comparable to the homogeneous catalyst, $Pd(OAc)_2$ (48% yield). However, the catalyst was not reused, which is known as the main purpose of applying heterogeneous catalysts.

Pd/C as a commonly applied and commercially available heterogeneous catalyst was reported in this transformation as well. In 2010, Sajiki and coworkers reported a Pd/C (10 mol%) catalyzed cyclization of 2-iodoanilines

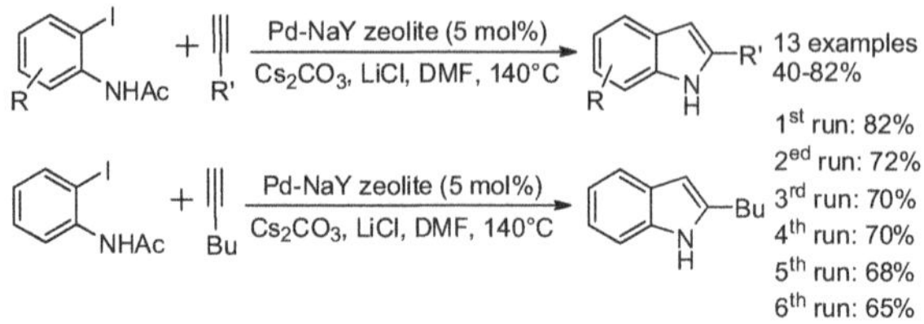

Scheme 2.92 Pd–NaY zeolite-catalyzed synthesis of indoles.

Reaction Procedure (Scheme 2.92): The Pd-loaded zeolite (0.125 mmol), lithium chloride (0.5 mmol), caesium carbonate (1 mmol), *N*-acetyl 2-iodoaniline (0.5 mmol), phenyl acetylene (1.0 mmol), and DMF (10 ml) were added to a sealed tube. The reaction mixture was stirred for 6 h at 140 °C. The reaction mixture was diluted with saturated aqueous ammonium chloride. The product was isolated with ethyl acetate. The organic layer was dried over anhydrous magnesium sulfate. The reaction mixture was filtered and concentrated. The product was purified by silica gel column chromatography using hexane–ethyl acetate. 2-Phenyl indole was obtained as a pure product.

with both terminal and internal alkynes. Good yields of the desired indoles were formed in NMP at 120 °C with NaOAc as base and without additional additive.[152] In 2013, another Pd/C (10 mol%) catalyzed cyclization of 2-iodoanilines with terminal alkynes was reported by Ray's group.[153]

By using $ZnCl_2$ (10 mol%) and PPh_3 as an additive, NEt_3 as the base, in DMF at 110 °C, the corresponding indoles were isolated in excellent yields (11 examples; 90–97%). But the recyclability of the Pd/C catalyst was not demonstrated in both cases. In the same year, a bimetallic catalyst Pd–Cu/C was prepared and applied in the synthesis of indoles and benzofurans *via* annulation of 2-iodoanilines and 2-iodophenols with terminal alkynes.[98] Good yields of the products could be achieved, but again no recyclability data was given.

Regarding the high price of palladium catalysts, copper salts were considered and explored. In 2003, Cacchi and co-workers developed a general method for the synthesis of 2-aryl and 2-heteroaryl indoles from aryl iodides and 1-alkynes through a domino copper-catalyzed process (Scheme 2.93a).[154] The best results were obtained with $[Cu(phen)(PPh_3)_2]NO_3$ in the presence of K_3PO_4 in toluene or dioxane at 110 °C. 2-Aryl and 2-heteroaryl indoles can also be isolated in good yields by using catalysts derived from CuI and PPh_3 or CuI and (±)-1,2-*trans*-cyclohexanediamine in dioxane at 110 °C. In 2008, Fu and co-workers reported an efficient method for the copper-mediated Sonogashira cross-coupling of *o*-iodoacetanilide derivatives with alkynes at room temperature.[155] By using copper(I) iodide (30 mol%)/*N*-methylpyrrolidine-2-carboxamide (100 mol%) as the catalyst, the corresponding coupling products were obtained in good to excellent yields (Scheme 2.93b).

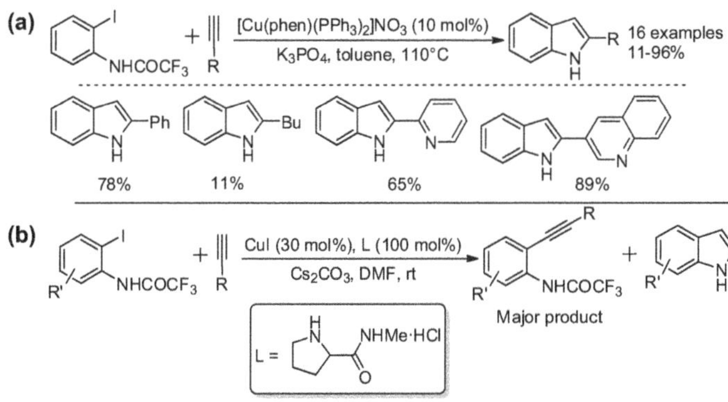

Scheme 2.93 Copper-catalyzed synthesis of indoles from 2-iodoanilines.

Reaction Procedure (Scheme 2.93a): In a Carousel Tube Reactor (Radleys Discovery Technologies), to a solution of *o*-iodotrifluoroacetanilide (0.50 mmol) and 1-chloro-4-ethynyl-benzene (0.50 mmol) in 2.0 mL of toluene, [Cu(phen)(PPh_3)]NO_3 (0.050 mmol), and K_3PO_4 (1.00 mmol) were added. The mixture was stirred for 6 h at 110 °C under an argon atmosphere. After cooling, the reaction mixture was diluted with ethyl acetate, washed with water, dried over Na_2SO_4, and concentrated under reduced pressure. The residue was purified by chromatography (*n*-hexane–EtOAc, 85 : 15 v/v).

(Scheme 2.93b): In a Carousel Tube Reactor (Radleys Discovery Technologies), to a solution of *o*-(heptyn-1-yl)trifluoroacetanilide (0.50 mmol) in 2.0 mL of 1,4-dioxane, CuI (0.025 mmol), (±)-1,2-*trans*-cyclohexanediamine (0.050 mmol), and K_3PO_4 (1.00 mmol) were added. The mixture was stirred for 1.5 h at 110 °C under an argon atmosphere. After cooling, the reaction mixture was diluted with ethyl acetate, washed with water, dried over Na_2SO_4, and concentrated under reduced pressure. The residue was purified by chromatography.

This catalyst system has a high tolerance towards various functional groups in the substrates. This represents the lowest reaction temperatures for copper-catalyzed Sonogashira cross-coupling thus far.

Later on in 2007, a comprehensive study on the copper-catalyzed domino coupling–cyclization process of 1-alkynes and 2-bromotrifluoroacetanilides under relatively mild conditions was reported by Ma and Liu.[156] This process allows the preparation of indoles bearing substituents at the 2,5-, 2,6-, or 2,7-positions by applying the appropriate 2-bromotrifluoroacetanilides as substrates. A wide range of functional groups were tolerated under these reaction conditions; all the desired products were isolated in good yields (Scheme 2.94). Additionally, a Sonogashira-type coupling reaction between vinyl iodides and 1-alkynes was developed as well.

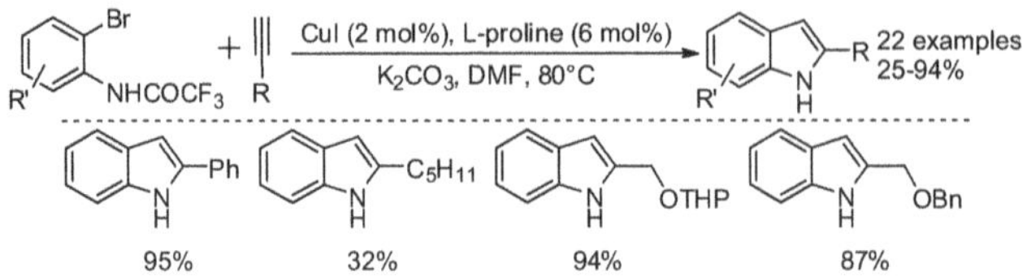

Scheme 2.94 CuI/L-proline-catalyzed synthesis of indoles.

Reaction Procedure (Scheme 2.94): A mixture of bromide (1.1 mmol), alkyne (1.0 mmol), CuI (0.02 mmol), L-proline (0.06 mmol), and K_2CO_3 (2 mmol) in 2 mL of DMF was heated in a sealed tube to 80 °C under argon. After reaction completion, monitored by TLC, the cooled mixture was partitioned between ethyl acetate and water. The organic layer was separated, and the aqueous layer was extracted with ethyl acetate twice. The combined organic layers were washed with brine, dried over Na_2SO_4, and concentrated *in vacuo*. The residual oil was loaded onto a silica gel column and eluted with 10 : 1 to 5 : 1 petroleum ether–ethyl acetate to afford the corresponding indole.

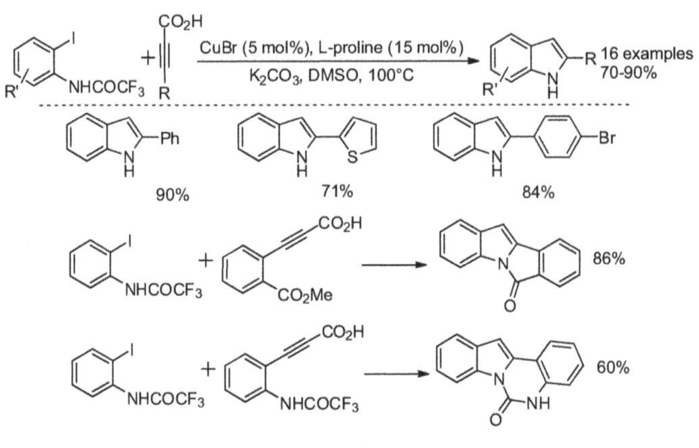

Scheme 2.95 CuI/L-proline-catalyzed synthesis of indoles from propiolic acids.

Recently, a methodology for the synthesis of a 2-aryl indole-related skeleton starting from propiolic acids and 2-iodoanilines was reported.[157] This conversion tolerates a wide range of functional groups which can be useful for further synthetic transformation. Additionally, the synthesis of 6*H*-isoindolo[2,1-*a*]indol-6-one and [1,2-*c*]quinazolin-6(5*H*)-one were achieved under the same conditions. The desired products were formed in good to excellent yields from readily available starting materials (Scheme 2.95). In this procedure, from our understanding, the decarboxylation to generate alkynylcopper happened at the first stage.

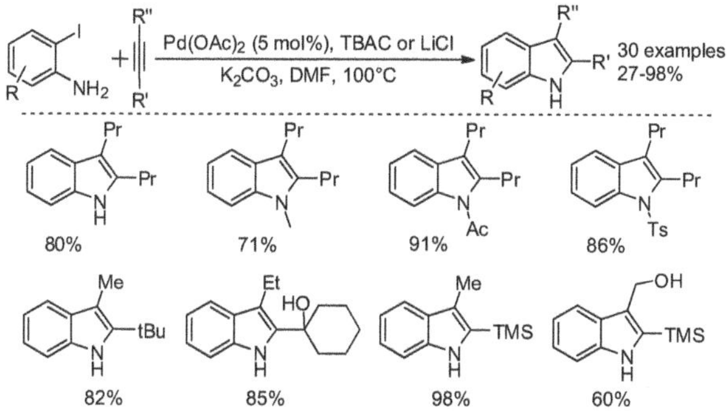

Scheme 2.96 Palladium-catalyzed synthesis of indoles from internal alkynes.

Reaction Procedure (Scheme 2.96): $Pd(OAc)_2$ (0.0125 mmol), LiCl (0.5 mmol) or n-Bu_4NCl (0.5 mmol), the appropriate base (2.50 mmol), the aryl iodide (0.50 mmol), the alkyne (0.50–2.50 mmol), DMF (10 mL), and where indicated PPh_3 (0.025 mmol) were added to a 4-dram vial equipped with a stir bar and Teflon-lined screw cap. After being heated for the appropriate time at 100 °C, the reaction mixture was diluted with ether and washed with saturated aqueous NH_4Cl and H_2O. The organic layer was dried over $MgSO_4$. The reaction mixture was filtered and concentrated, and the product was purified by flash column chromatography using hexanes–ethyl acetate.

In 1998, Larock and co-workers reported a palladium-catalyzed annulation of 2-iodoanilines with internal alkynes.[158] Various 2,3-disubstituted indoles were formed in good to excellent yields *via* this palladium-catalyzed coupling of 2-iodoaniline and the corresponding *N*-methyl, -acetyl, and -tosyl derivatives with a wide variety of internal alkynes (Scheme 2.96). The best results were obtained by employing an excess of the alkyne and a sodium or potassium acetate or carbonate base plus 1 equiv. of either LiCl or n-Bu_4NCl, occasionally adding 5 mol% PPh_3. The yields with LiCl appeared to be higher than those obtained with n-Bu_4NCl. The process is quite general regarding the substituents on the nitrogen of the aniline and the two ends of the alkyne triple bond. The reaction is quite regioselective, placing the aryl group of the aniline on the less sterically hindered end of the triple bond and the nitrogen moiety on the more sterically hindered end. This methodology readily affords 2-silylindoles, which can be easily protodesilylated, halogenated, or reacted with alkenes and $Pd(OAc)_2$ to produce 3-substituted indoles, 2-haloindoles, or 2-(1-alkenyl)-indoles, respectively. The presence of alcohol groups in the alkyne seems to have a particularly strong directing effect, perhaps due to coordination with palladium. This catalytic process apparently involves arylpalladium formation, regioselective addition to the C–C triple bond of the alkyne, and subsequent intramolecular palladium displacement.

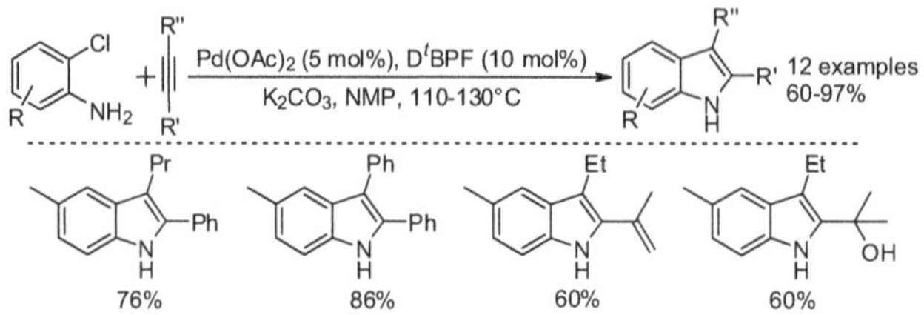

Scheme 2.97 Palladium-catalyzed synthesis of indoles from 2-chloroanilines and internal alkynes.

Reaction Procedure (Scheme 2.97): $Pd(OAc)_2$ (0.05 mmol), 1,1′-bis(di-*tert*-butylphosphino)ferrocene (0.1 mmol), and potassium carbonate (346 mg, 2.5 mmol) were charged to an oven-dried reaction vial. The vial was purged with argon. 2-Chloro-4-methylaniline (1 mmol), 1-pentyl-1-pentyne (1.2 mmol), and NMP (2 mL) were added *via* syringe. The reaction was heated to 130 °C, while stirring and monitoring by HPLC. The reaction was complete after 4 h. The ratio of the regioisomers was 91 : 9. The mixture was filtered through a pad of Celite. The Celite was washed with ethyl acetate. The organic phase was washed with water and brine, dried over $MgSO_4$, filtered, and concentrated to give a brown residue. The product was purified *via* column chromatography.

Later on, the use of 2-bromo- or 2-chloroanilines and internal alkynes was developed.[159] By the proper choice of ligand, base, solvent, and concentration, the corresponding indole derivatives were formed in good yields (Scheme 2.97). The dimerization of 2-chloroanilines could be detected during the optimization process (Table 2.2). It is interesting to note that under the optimal conditions except using Cy_3P as the ligand, the reaction of 4-methyl-2-chloroaniline and diphenylacetylene afforded the desired indole product in 86% isolated yield. Thus, this ligand can safely be employed if regiochemistry issues are not a concern.[160] A similar reaction with 2-iodoaniline produced no desired indole product and the use of dialkyl acetylenes led to a clean reaction. Later on, they applied this catalyst system to prepare polycyclic indole skeletons by using 2-chloroanilines bearing tethered acetylenes as substrates through intramolecular heteroannulation.[161]

Recently, a highly efficient one-pot synthesis of 5,6-disubstituted-5*H*-pyrrolo[2,3-*b*]pyrazine-2,3-dicarbonitriles from the reaction between 5-(alkylarylamino)-6-chloropyrazine-2,3-dicarbonitriles and phenylacetylene was reported as well.[162] The reactions were carried out in water in SDS (10 mol%), with $Pd(PPh_3)_2Cl_2$ (5 mol%) and CuI (10 mol%) as the catalyst and K_2CO_3 as the base, at 70 °C; good to high yields (68–96%) of the desired products were isolated. Under the best reaction conditions, only Sonogashira product was formed in the absence of CuI or PPh_3. Interestingly, 20%

Table 2.2 Effect of ligand on the annulation of alkynes with 4-methyl-2-chloroaniline.[a]

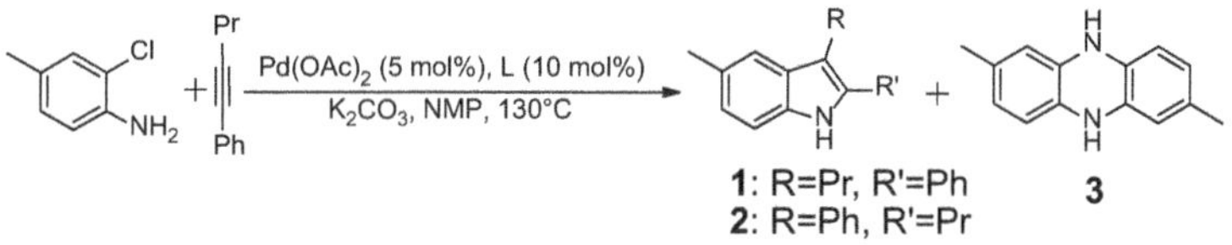

Entry	Ligand	**1**:**2**[b]	Yield of **3**[c] (%)
1	PCy_3	67:33	0
2	P^tBu_3	76:14	79
3	DPPF	80:20	20
4	D^iPPF	79:21	10
5	D^tPPF	91:9	19
6	D^tPPF	91:9	8[d]
7	Cy-JohnPhos	71:29	21
8	JohnPhos	80:20	55
9	DavePhos	72:28	29
10	^{t}Bu-DavePhos	84:16	62

[a]4-Methyl-2-chloroaniline (1 mmol), alkyne (1.2 mmol).
[b]Ratios were determined by HPLC at $\lambda = 228$ nm.
[c]Yields were determined by HPLC at $\lambda = 248$ nm.
[d]NMP (10 mL).

of indole derivate could be produced by using Pd/C as the catalyst instead of $Pd(PPh_3)_2Cl_2$.

In 2004, Konno and co-workers reported the annulation of 2-iodoanilines with fluorine-containing terminal alkynes.[163] 2-Fluoroalkylated indoles were formed in high yields with the assistance of $Pd(PPh_3)_4$ in DMF at 80 °C for 8 h. The use of $P(o\text{-Tol})_3$ instead of PPh_3 as a ligand led to the preferential formation of 3-fluoroalkylated indoles in high yields. Interestingly, the reaction of trifluoromethylated alkynes bearing a benzylic substituent afforded 2- or 3-trifluoroethylated indole derivatives in good yields. In the latter case, the carbopalladation reaction of the fluoroalkylated allenes generated *in situ* was proposed.

Due to the importance of indole and pyridine moieties in natural products and pharmaceutical compounds, the cyclization between 2-iodoanilines with 2-alkynylpyridines was developed in 2008.[164] 3-Substituted-2-pyridin-2-ylindoles were produced in good yields in DMP at 140 °C with $Pd(OAc)_2$ (5 mol%) as the catalyst, DPPF (7.5 mol%) as the ligand and KOAc as the base. However, the regioselectivity was rationalized by a combination of steric, coordinative and electronic effects.

Afterwards, several parameters, such as ligand and palladium precursor, were optimized. In 2008, phenylurea was found to be an effective ligand for the palladium-catalyzed annulation of 2-bromoanilines with internal alkynes.[165] Good yields of 2,3-disubstituted indoles were prepared by this catalyst system [$Pd(OAc)_2$ (1 mol%), phenylurea (4 mol%), K_2CO_3, DMF, 130 °C].

Well-defined palladium complexes are interesting from an academic point of view. In 2013, Shi, Cao and their co-workers prepared a well-defined ferrocenyl functionalized NHC–palladium complex and applied it as an efficient catalyst for Larock heteroannulation.[166] 2,3-Disubstituted indoles were isolated in good yields (69–90%) with high regioselectivity from the reactions between 2-iodoanilines or 2-bromoanilines and their derivatives with various internal alkynes (Scheme 2.98). The reactions occur in a broad scope and with a high tolerance of functional groups. NHC–palladium complexes could be excellent candidates to replace expensive palladium–phosphine complexes for Larock indole catalysis.

Rao and co-workers prepared a phosphine-free thiopseudourea–Pd(II) complex and applied it in the heteroannulation of internal alkynes with 2-haloamines and *N*-tosyl substituted 2-bromoanilines.[167] The corresponding products were afforded in good to high yields under a low catalyst loading of 0.1 mol% (Scheme 2.99a). However, in the case of non-asymmetric diaryl substituted alkynes, no regioselectivity could be observed. Lam, Lo and their co-workers developed a fluorous, oxime-based palladacycle, which proved to be very efficient for the annulation of internal alkynes

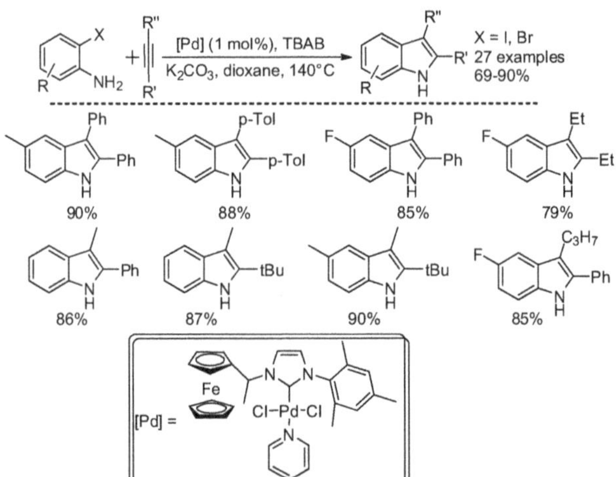

Scheme 2.98 NHC–Pd complex-catalyzed synthesis of indoles from internal alkynes.

Reaction Procedure (Scheme 2.98): Aniline (1 mmol), alkyne (2 mmol), Pd catalyst (1 mol%), TBAB (1 mmol), and K_2CO_3 (3 mmol) were dissolved in dioxane (2 mL) in a 5 mL vial in air and heated at 140 °C for 24 h. After the reaction was completed, the mixture was diluted with ethyl acetate (10 mL), filtered through a pad of Celite, and washed multiple times with ethyl acetate. The combined organic layer was dried over anhydrous Na_2SO_4, and evaporated under reduced pressure. The residue was purified by flash chromatography on silica gel to afford the desired product.

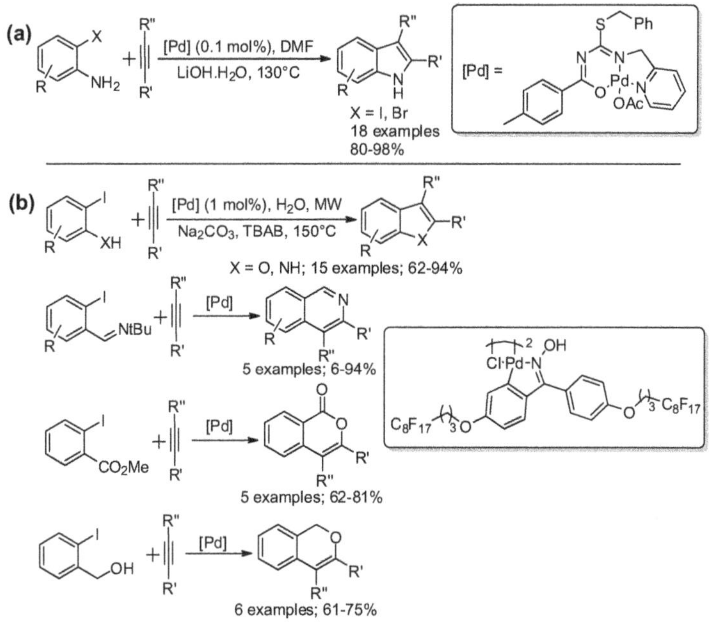

Scheme 2.99 Well-defined Pd(II) complex and synthesis of indoles.

Reaction Procedure (Scheme 2.99a): A 25 mL round-bottomed flask was charged with 2-haloamines (1 mmol), diphenylacetylene (1.5 mmol), $LiOH \cdot H_2O$ (4 mmol) and catalyst (0.001 mol% of complex in 2 mL *N,N*-dimethylformamide). The reaction mixture was stirred at 130 °C for 20 h. The reaction mixture was cooled to room temperature, diluted with ethyl acetate (20 mL) and washed with brine water. The combined organic phase was dried over anhydrous Na_2SO_4. After removal of the solvent, the residue was subjected to column chromatography on silica gel using ethyl acetate and hexane to afford the indole product in high purity. In the case of 2-bromoanilines, 0.1 mol% of the catalyst complex was applied.

with functionalized aryl iodides in an aqueous medium (Scheme 2.99b).[168] Notably, this catalyst can be reused several times without a significant decrease in reactivity.

Additionally, a copper complex was applied in 2-arylindole synthesis as well. In 2010, Zhou and co-workers prepared a water soluble copper(II) complex and applied it in the coupling of aryl iodides with alkynes.[169] This is a simple, highly efficient, and environmentally friendly protocol by using the sulfonato–Cu(salen) complex as the catalyst in water. This method avoids the use of stringent inert conditions. Both alkyl- and arylsubstituted terminal alkynes were found to react efficiently with aryl iodides with good functional group compatibility. More importantly, this methodology could be

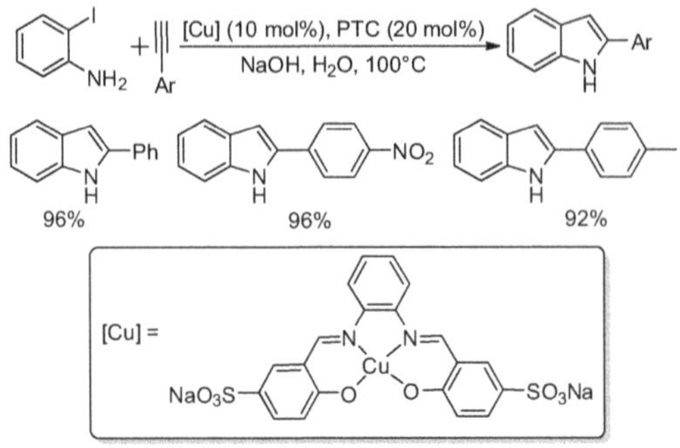

Scheme 2.100 Sulfonato–Cu(salen) complex-catalyzed synthesis of indoles in water.

Reaction Procedure (Scheme 2.100): Catalyst (0.1 mmol), aryl halide (1.0 mmol), NaOH (1.0 mmol), alkyne (1.2 mmol), $(n\text{Bu})_4$NBr (0.2 mmol; PTC), and water (2 mL) were added to a sealed tube. The reaction mixture was stirred at 100 °C for 24 h and then cooled to room temperature and extracted with ethyl acetate. The organic layer was then dried with anhydrous Na_2SO_4, and the solvent was removed under reduced pressure. The product was finally obtained by column chromatography on silica gel.

successfully applied to the domino synthesis of 2-arylindoles with 2-iodoanilines and terminal alkynes in good yields, which contributed to the first example of the copper-catalyzed synthesis of indoles in water (Scheme 2.100). In addition, the catalyst can be easily recovered and reused.

Gold (and its complexes)-catalyzed organic transformations have been a focus of attention in recent years. In 2008, Wang and co-workers reported a novel gold(I)-catalyzed Sonogashira coupling reaction.[170] The cross-coupling reactions of terminal alkynes with aryl iodides and aryl bromides generate the corresponding coupling products in good to excellent yields under the present reaction conditions. Furthermore, aromatic terminal alkynes could also react with 2-iodoaniline to form substituted indoles in the presence of AuI (1 mol%) and DPPF (1 mol%) in toluene in excellent yields through a coupling–cyclization reaction sequence under the present reaction conditions (Scheme 2.101).

In 2012, Sun's group applied the cyclization of 2-iodoaniline with alkynes to the synthesis of 2-arylindoles and evaluation as nitric oxide synthase and NFκB inhibitors.[171]

Iboga alkaloids are pharmacologically important indole alkaloids, consisting of a seven-membered indoloazepine ring fused with a rigid isoquinuclidine ring. In 2012, a mild and efficient synthetic route to the iboga scaffold by employing annulation reactions was described by Sinha and Jana (Figure 2.1).[172] The utility of this process is demonstrated by the direct

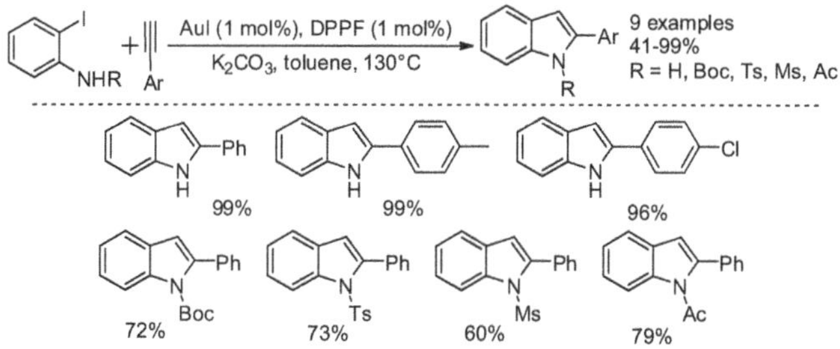

Scheme 2.101 Gold-catalyzed synthesis of indoles.

> **Reaction Procedure** (Scheme 2.101): Phenylacetylene (1.0 mmol), *p*-iodoanisole (1.0 mmol), AuI (3.2 mg, 0.01 mmol), DPPF (5.5 mg, 0.01 mmol) and K_2CO_3 (276 mg, 2.0 mmol) were added to a round-bottomed flask containing toluene (2 mL). The reaction mixture was stirred at 130 °C for 24 h. After completion of the reaction, the mixture was filtered and washed with diethyl ether (2×5 mL). The combined organic phase was dried with Na_2SO_4, filtered, and concentrated, and the residue was purified by flash chromatography on silica gel.

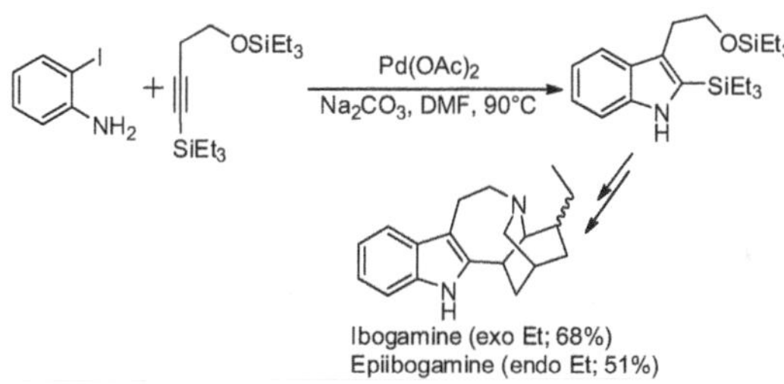

Figure 2.1 Total synthesis of Iboga.

access to the ibogamine, epiibogamine and iboga-analogs. The cyclization precursors were readily obtained from 2-iodoaniline by heteroannulation reaction with suitable alkynes followed by iodination.

Besides the alkynes described previously, more functionalized alkynes were tested and applied in the synthesis of more functionalized indole derivatives. In 1998, Smith and co-workers reported a palladium-catalyzed solid phase synthesis of 2,3-disubstituted indoles.[173] By applying Ellman's THP resin, good yields of the desired indoles were produced by treating with 10% TFA, which was stable with cold acetic acid and 2 M aqueous HCl. The released resin can be recovered. In 2000, a palladium-catalyzed stereoselective synthesis of α-aryl-2-benzofuranmethanamines and α-aryl-1*H*-indole-2-methanamines, which are valuable chiral synthons for the synthesis of biologically active compounds, was reported by Botta, Corelli and their co-workers.[174] Through palladium-mediated annulation of chiral

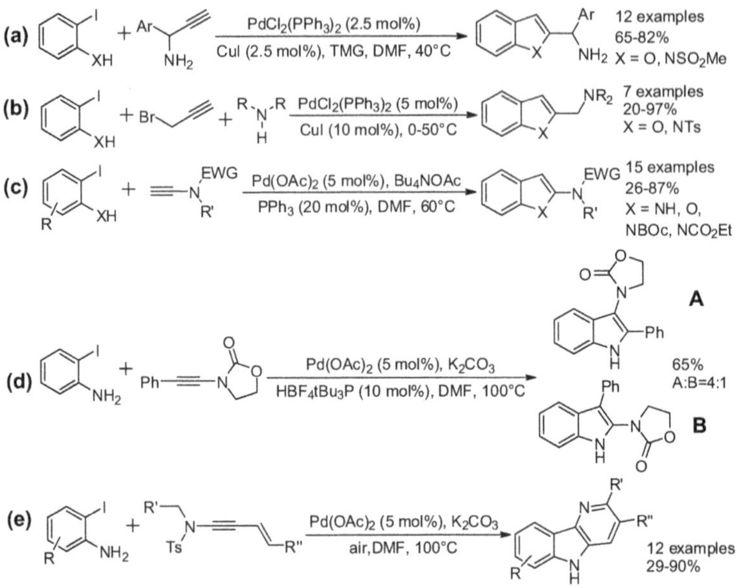

Scheme 2.102 Palladium-catalyzed annulation of functionalized alkynes.

α-arylpropargylamines with 2-iodophenols or 2-iodo-*N*-mesylaniline, the corresponding products were formed in good yields and with high stereoselectivity (Scheme 2.102a). Alami and co-workers described the annulation of the *in situ* generated propargylic amines with 2-iodophenols and 2-iodoanilines in a three-component manner in 2004 (Scheme 2.102b).[175,176] Indole and benzofuran derivatives were produced in high yields through the procedure. Variation is allowed in each of the three components, thus making a wide range of accessible products. This process is not only of interest for the combinatorial synthesis of propargylic amines and heterocycles, but in many cases also offers considerable synthetic advantages in terms of yield, selectivity, and simplicity of the reaction procedure. In 2009, Skrydstrup and co-workers developed the palladium-catalyzed cyclization of ynamides with 2-iodoanilines.[177] 2-Amidoindoles were prepared in good yields through this palladium-catalyzed one-pot two-step procedure from the corresponding ynamides and *o*-iodoanilines (Scheme 2.102c). Here, interestingly, Bu_4NOAc was applied as a base which was found to be superior to Cs_2CO_3, KOAc, K_3PO_4, K_2CO_3 and Na_2CO_3. A key highlight of this sequence is that after the Sonogashira reaction, intramolecular cyclization to the indole occurs spontaneously without activation of the alkyne. The same group succeeded to extend their methodology to internal ynamides by using a more basic ligand (Scheme 2.102d).[178] Cao, Lai and their co-workers described an interesting palladium-catalyzed sequential reaction for the synthesis of δ-carbolines from 2-iodoanilines and *N*-tosyl-enynamines.[179] The control experiments demonstrate that the protocol consists of Larock heteroannulation, elimination, electrocyclization, and oxidative aromatization, leading to an efficient method to construct δ-carboline derivatives (Scheme 2.102e). Regarding the reaction mechanism, based the isolated intermediates, the following plausible mechanism was proposed: (i) the Larock heteroannulation reaction of 2-iodoaniline and *N*-tosylenynamine affords the intermediate sulfonamide; (ii) subsequently, elimination of a molecule of 4-methylbenzenesulfinic acid gives the imine; (iii) that reaction is followed by an electrocyclization reaction to form tricyclic intermediate; (iv) finally, oxidative aromatization by air gives the final product δ-carboline.

In 2010, the synthesis of 2-indolylphosphines by palladium-catalyzed annulation of 1-alkynylphosphine sulfides with 2-iodoanilines was developed.[180] Palladium-catalyzed annulation of 1-alkynylphosphine sulfides with 2-iodoanilines followed by desulfidation affords 3-substituted 2-indolylphosphines. This annulation–desulfidation sequential protocol offers a conceptually new approach to bulky heteroarylphosphines. In the scope of 2-iodoaniline derivatives, an alkyl substituent on nitrogen was found to be crucial. Primary and secondary alkyl-substituted anilines underwent the reaction smoothly to yield the corresponding indoles in good yields (Scheme 2.103). However, the reactions of *N*-acetyl-, *N*-*tert*butoxycarbonyl-, *N*-tosyl-2-iodoanilines and 2-iodoaniline did not proceed at all. Interestingly, the use of 1-alkynylphosphine oxides as substrates expanded the scope of accessible 2-indolylphosphines. In detail, annulation of 1,2-bis(diphenylphosphinyl)ethyne occurred smoothly to yield diphosphine dioxide in good

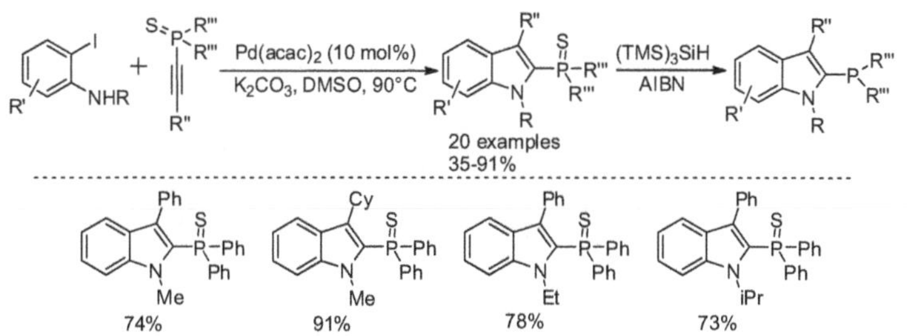

Scheme 2.103 Pd-catalyzed annulation of 1-alkynylphosphine sulfides with 2-iodoanilines.

Reaction Procedure (Scheme 2.103): $Pd(acac)_2$ (7.6 mg, 0.025 mmol) and K_2CO_3 (0.069 g, 0.50 mmol) were placed in a 20 mL reaction flask under argon. DMSO (2.0 mL), alkyne (0.25 mmol), and aniline (0.50 mmol) were sequentially added. The resulting mixture was stirred at 90 °C for 11 h. After the mixture was cooled to room temperature, saturated aqueous NH_4Cl (10 mL) was added, and the product was extracted with ethyl acetate. The combined organic layer was dried over Na_2SO_4 and concentrated *in vacuo*. Chromatographic purification on silica gel gave the pure product.

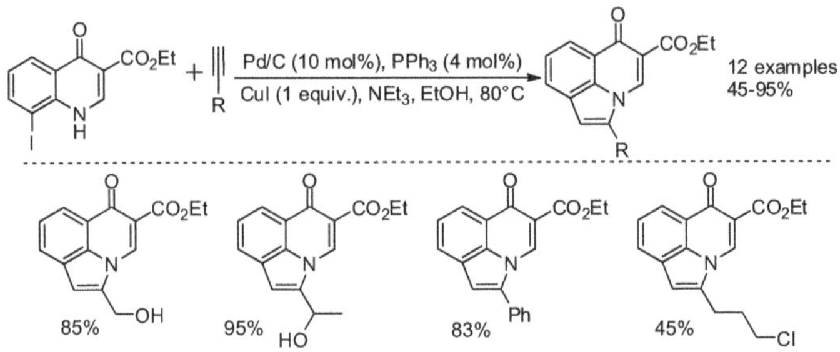

Scheme 2.104 Pd-catalyzed annulation of propargylic alcohols.

Reaction Procedure (Scheme 2.104): A mixture of 8-iodo-4-oxo-1,4-dihydroquinoline-3-carboxylic acid ethyl ester (0.869 mmol), 10% Pd/C (9.21 mg, 0.0087 mmol), PPh_3 (0.034 mmol), CuI (0.087 mmol) and Et_3N (2.17 mmol) in ethanol (3.0 mL) was stirred for 1 h under nitrogen. The acetylenic compound (1.30 mmol) was added and the mixture was stirred at 80 °C for the time required. After completion, the reaction mixture was cooled to room temperature, filtered through a Celite bed and the filtrate was concentrated under vacuum. The crude mass was diluted with chloroform (30 mL) and water (15 mL) and the mixture was extracted with chloroform (3×30 mL). The organic layers were collected, combined, washed with saturated aq NaCl (2×25 mL), dried over anhydrous Na_2SO_4 and concentrated under vacuum. The product was isolated by column chromatography on silica gel using 9:1 hexane–ethyl acetate.

yield, while the reaction of 1,2-bis(diphenylthiophosphinyl)ethyne did not proceed. The corresponding trivalent diphosphine would serve as a bidentate ligand. (1-Alkynyl)di-*tert*-butylphosphine oxide also underwent the annulation reaction in moderate yield although a longer reaction time was required.

A general and practical synthesis of 2-substituted 6-oxopyrrolo[3,2,1-*ij*]quinolines was achieved by a Pd/C-mediated coupling–cyclization strategy in 2009 (Scheme 2.104).[181] The methodology involves the reaction of 8-iodo-4-oxo-1,4-dihydro quinoline-3-carboxylic acid ethyl ester with a variety of terminal alkynes in the presence of 10% Pd/C–PPh_3–CuI as a catalyst system in EtOH. The reaction mechanism and utility of this methodology was discussed as well. However, the recyclability was not demonstrated.

In 2013, Takemoto and co-workers developed a palladium-catalyzed isocyanide insertion and alkyne functionalization for the synthesis of 2-arylindole derivatives, including tetracyclic carbazoles (Scheme 2.105).[182] In this process, the formation of two C–C bonds *via* isocyanide insertion was achieved, and isocyanide was effectively used by incorporating both the C and N atoms as components of the indole skeleton. Regarding the reaction mechanism, the reaction started with the oxidative addition of Pd(0) to aryl

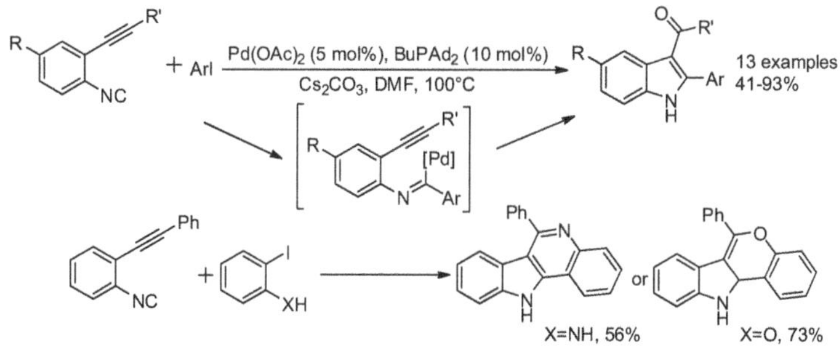

Scheme 2.105 Pd-catalyzed cyclization with isocyanides.

Reaction Procedure (Scheme 2.105): To a stirred solution of iodobenzene (0.100 mmol) in DMF (1 mL) were added $Pd(OAc)_2$ (0.0049 mmol), *n*$BuPAd_2$ (0.0101 mmol) and Cs_2CO_3 (97.7 mg, 0.300 mmol) at room temperature and the reaction mixture was heated to 100 °C. After stirring for 10 min, freshly prepared 1-isocyano-2-(phenylethynyl)benzene in DMF (2 mL) was added for 3 h and stirred at 100 °C.

After stirring for 1 h, the reaction mixture was neutralized by a saturated aqueous solution of NH_4Cl and extracted with EtOAc. The combined extracts were washed with brine, dried over Na_2SO_4 and concentrated under reduced pressure. The obtained residue was purified by silica gel column chromatography (hexane–EtOAc, 9 : 1).

iodide followed by isocyanide insertion to generate the imidoylpalladium intermediate, which is the key intermediate for subsequent steps.

Additionally, the intramolecular cyclization of 2-halo-*N*-allylanilines provides optional procedures for indole derivative synthesis. Many transition metal catalysts have been developed in this field, such as cobalt,[183] nickel[184] and palladium,[185] but stoichiometric amounts of catalysts are needed. In 1987, a catalytic version was developed by Larock and Babu.[186] In the presence of a catalytic amount of $Pd(OAc)_2$, indoles, indolines, oxindoles, quinolones, isoquinolines and isoquinolones were formed in high yields in short reaction times under mild temperatures (Scheme 2.106).

In 1991, Yamanaka and co-workers reported the synthesis of indoles from 2-bromoanilines and (*Z*)-1-ethoxy-2-tributylstannylethene in a two-step manner.[187] Indoles were formed in low yields by treating the Stille coupling product with TsOH in benzene. 1*H*-Isochromen-1-ones can be produced by adding 6 N HCl to the Stille coupling product with 2-bromobenzoates as substrates.

A palladium-catalyzed reaction of 2-chloro-*N*-allylanilines and 2-chloro-*O*-allylphenols with imidazolium salt (IPr) as the ligand was developed by Caddick and Kofie in 2002.[188] In this procedure, the addition of *tetra-n*-butylammonium salts (TBAB or TBAN) was found to be important in

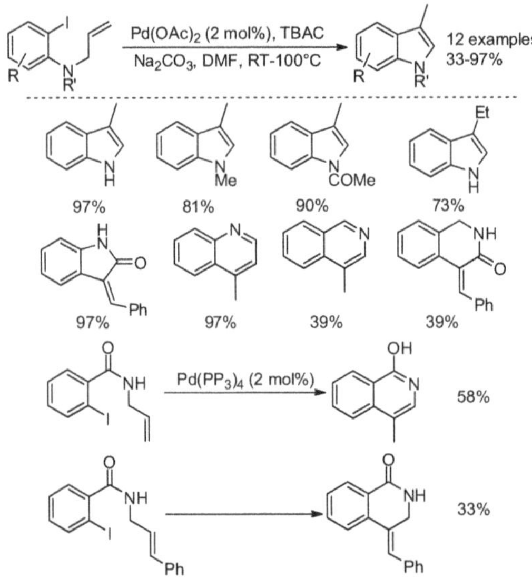

Scheme 2.106 Pd-catalyzed cyclization of 2-iodo-*N*-allylanilines.

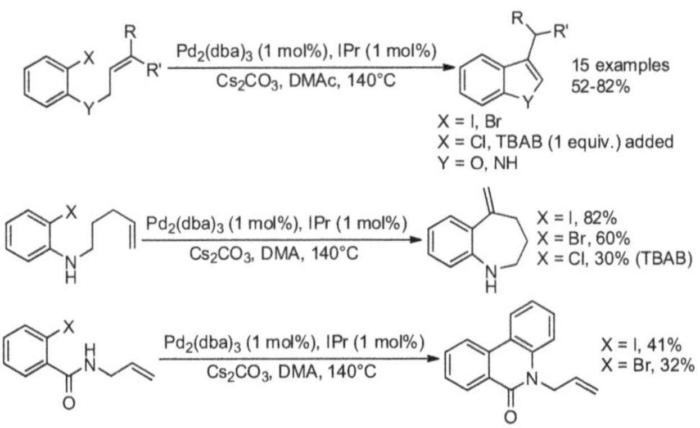

Scheme 2.107 Pd/NHC-catalyzed cyclization of 2-halo-*N*-allylanilines.

facilitating the reactivity of aromatic chlorides. For the corresponding iodide and bromide derivatives, no additive was necessary. All the desired products were formed in good yields (Scheme 2.107). Later on in 2009, Beck's group reported a one-pot *N*-alkylation–Heck cascade of 2-chloroanilines to access substituted indoles.[189] The reaction is general and mild, tolerating several functional groups. Several steric and electronic substituents are tolerated. Low to moderate yields of the products were produced.

In 2011, a palladium-catalyzed Heck-type reaction between allylic esters and aryl halides was developed by Deng, Li and their co-workers.[190] When 2-iodoanilines and 2-iodophenol were applied as substrates with allyl acetate, indole and benzofuran derivatives were selectively produced

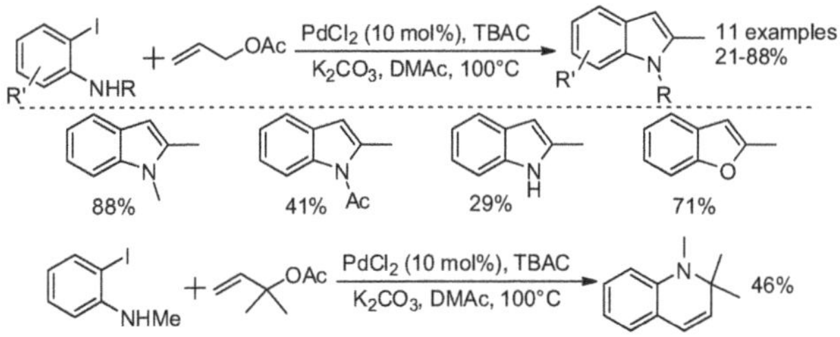

Scheme 2.108 Pd-catalyzed cyclization of 2-iodoanilines with allylic esters.

Reaction Procedure (Scheme 2.108): To a Schlenk tube were added allylic esters (0.2 mmol), aryl halides (0.3 mmol), $PdCl_2$ (10 mol%), *n*-Bu_4NCl (1.5 equiv.), K_2CO_3 (2.0 equiv.) and anhydrous DMAc (2 mL). Then the tube was stirred at 100 °C (oil bath temperature) in air for the indicated time until complete consumption of the starting material as monitored by TLC and GC-MS analysis. After the reaction was finished, the reaction mixture was cooled to room temperature, diluted in diethyl ether, and washed with brine. The aqueous phase was re-extracted with diethyl ether. The combined organic extracts were dried over Na_2SO_4 and concentrated under vacuum, and the resulting residue was purified by silica gel column chromatography (hexane–ethyl acetate) to afford the indoles or benzofurans.

(Scheme 2.108). 2-Iodo-*N*-methylaniline provides a much higher yield than 2-iodoaniline (88% *vs.* 29%).

The intramolecular cyclization of *N*-(2-haloaryl)methyl-substituted enamines can lead to indole derivatives as well. In 1982, a copper-mediated cyclization of *N*-(2-haloaryl)methyl-substituted enaminones was developed.[191] By using stoichiometric amounts of CuI, indole derivatives were produced in excellent yields from the corresponding iodide and bromide substrates (Scheme 2.109). But the corresponding chloride substrates only provide traces of the desired products.

Since then, numbers of catalytic systems were developed. Regarding the intramolecular cyclization, in 2004, Nishida and co-workers reported a new type of palladium-catalyzed cyclization, which proceeds *via* selective isomerization of a double bond in the enamine structure followed by 5-*endo* cyclization. Various fused indoles were produced in good yields (Scheme 2.110).[192] Interestingly, this procedure was also applied for the synthesis of the azocinoindole skeleton.

In 2013, Kurth's group developed a rapid and effective microwave mediated route to 3-nitroindoles from *N*-aryl β-nitroenamines by a palladium-catalyzed intramolecular arene alkene coupling reaction (Scheme 2.111).[193] This method utilizes catalytic amounts of relatively inexpensive $Pd(PPh_3)_4$ as the catalyst. The enamines required for this transformation are readily

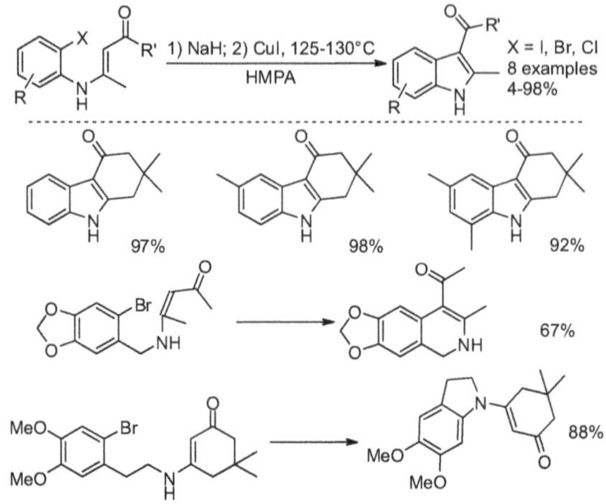

Scheme 2.109 Cu-mediated cyclization of *N*-(2-haloaryl)methyl-substituted enaminones.

> **Reaction Procedure** (Scheme 2.109): To a stirred suspension of sodium hydride (80 mg of a 60% suspension in mineral oil, 2 mmol; triturated with dry hexane) under a nitrogen atmosphere in 5 ml of HMPA at 25 °C was added 3-(2-iodo-4-methylanilino)-5,5-dimethylcyclohex-2-en-1-one (1 mmol). When hydrogen evolution ceased, copper(I) iodide (380 mg, 2 mmol) was added. The mixture was heated at 125–130 °C with stirring for 2 hours. After cooling, the mixture was diluted with water and the product was extracted with ether. The extract was dried over anhydrous Na_2SO_4 and evaporated; the pure product was isolated by chromatography on silica gel.

available in one synthetic step from commercially available *o*-bromoanilines. This technology accesses indoles with functionalization at the 2-position, accommodating alkyl, aryl, or heterocyclic substituents. Substrate studies have established that the chemistry reported here affords good functional group tolerance and can also provide access to 3-carboalkoxy- and 3-cyanoindoles.

This type of cyclization reaction was applied in the synthesis of β-carbolinones[194] and total synthesis of dipyrrolobenzoquinone (±)-terreusinone as well.[195] In 2005, Barluenga and co-workers reported a novel approach for the synthesis of indoles from *o*-bromoanilines and alkenyl halides *via* a palladium-catalyzed cascade process (Scheme 2.112).[196] The reaction involves an alkenyl amination followed by an intramolecular Heck reaction. As the authors demonstrated, this process represents the first example of the participation of alkenyl amination reactions in Pd-catalyzed cascade reactions. Initially, the relative reactivity of aryl and alkenyl bromides and chlorides towards Pd-catalyzed amination was investigated. Competition

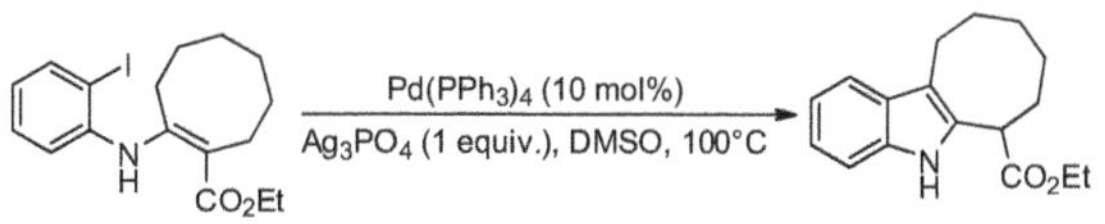

Scheme 2.110 Pd-catalyzed cyclization of enamines.

Reaction Procedure (Scheme 2.110): A mixture of enamine (0.28 mmol), Ag_3PO_4 (0.28 mmol), and $Pd(PPh_3)_4$ (32 mg, 28 μmol) was heated (18 h) with stirring in DMSO (1.0 mL) at 100 °C under Ar. The mixture was diluted with Et_2O at rt, and filtered through a Celite pad. The filtrate was concentrated and the residue was purified by silica gel column chromatography (EtOAc–*n*-hexane, 1 : 5) to give the pure product.

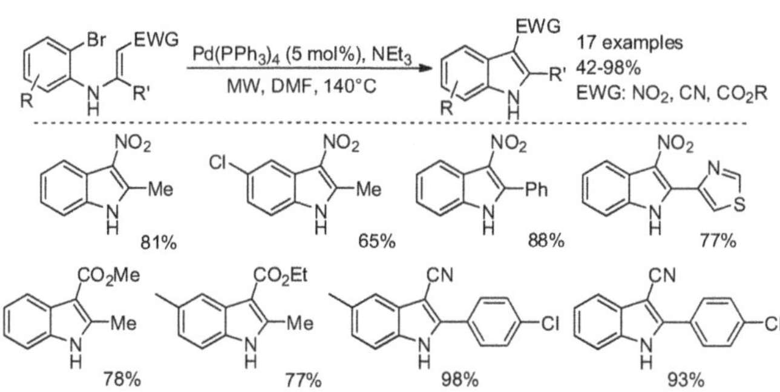

Scheme 2.111 Pd-catalyzed synthesis of 3-nitroindole derivatives.

Reaction Procedure (Scheme 2.111): Enamine (0.60 mmol, 1 equiv.) and tetrakis(triphenylphosphine)palladium(0) $Pd(PPh_3)_4$ (0.030 mmol, 5 mol%) were introduced into an oven-dried Biotage® 5–10 mL vial containing a Teflon-coated stir bar. The vial was then sealed with a Teflon-lined septum, flame-dried, and purged with nitrogen. DMF (6.0 mL, 0.1 M) and Et_3N (3.00 mmol, 5 equiv.) were successively added *via* syringes. The resulting mixture was irradiated by microwave at 140 °C for 90 min (absorption level: normal; fixed hold time, TLC monitoring). The reaction mixture was then cooled to room temperature and concentrated under reduced pressure. The crude material was purified by flash chromatography using a mixture of hexane and ethyl acetate as eluent to afford the desired indoles.

experiments were carried out in the presence of primary and secondary amines, and these revealed the reactivity order alkenyl bromides > aryl bromides > alkenyl chlorides > aryl chlorides, as well as very high chemoselectivity; the more reactive halide was always favored. Thereafter, optimized reaction conditions for the sequential alkenyl amination–Heck cyclization to

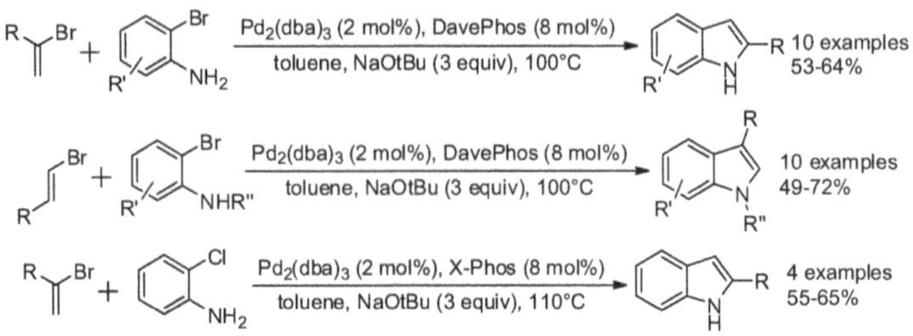

Scheme 2.112 Pd-catalyzed synthesis of indoles from *o*-bromoanilines and alkenyl halides.

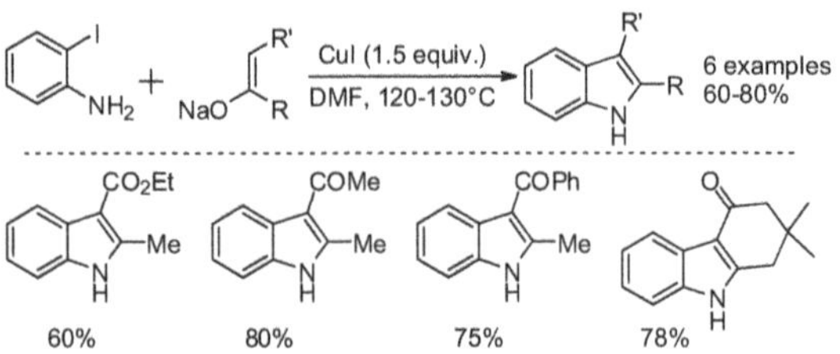

Scheme 2.113 Cu-mediated synthesis of indoles from *o*-iodoanilines and enolates.

give indoles were investigated with the model reaction of *o*-bromoaniline with α-bromostyrene. An extensive screening of ligands, bases, and reaction conditions revealed that the [$Pd_2(dba)_3$]/DavePhos, NaO*t*Bu, toluene combination at 100 °C were the optimized reaction conditions to carry out the cascade process. The reaction proceeds with aryl, alkyl, and functionalized substituents in both starting reactants. The cyclization was also studied with *N*-substituted *o*-bromoanilines (which would give rise to *N*-substituted indoles); however, in this case, indole formation occurred only with 1-substituted-2-bromoalkenes. Finally, the application of this methodology to *o*-chloroanilines required further optimization. Although the catalyst based on DavePhos failed to promote the cascade process, a catalytic combination based on [$Pd_2(dba)_3$]/X-Phos promoted the formation of the indole ring also from the less reactive chloroanilines.

In the procedures mentioned for indole synthesis, enamines are the corresponding intermediates which could be prepared by reacting 2-haloanilines with ketones as well. As early as 1984, a copper-mediated indole synthesis from 2-iodoaniline with acetoacetates was developed.[197] In this procedure, acetoacetates were treated with NaH to give enolates before the addition of CuI and 2-iodoaniline. The desired 2,3-disubstituted indoles were isolated in moderate to good yields (Scheme 2.113). In 1999, $FeBr_2$ or *hν* was reported to promote this reaction as well.[198] Using DMSO as a solvent and KO*t*Bu as a base, the desired indoles were formed in good yields.

Later on, the pre-formation of enolates was found to be unnecessary. In 1997, Chen and co-workers reported a palladium-catalyzed indole synthesis

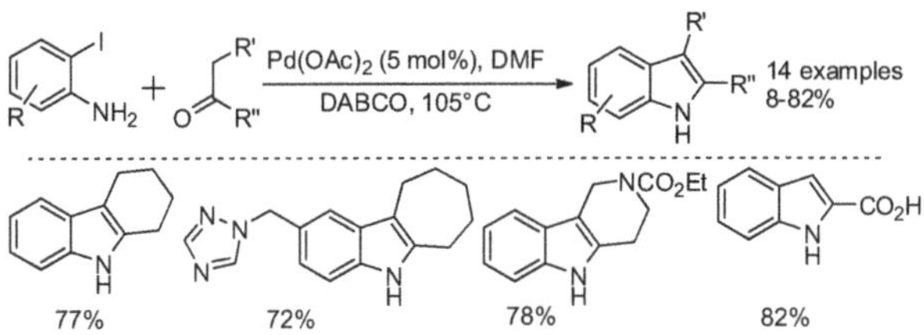

Scheme 2.114 Pd-catalyzed synthesis of indoles from *o*-iodoanilines and ketones.

Reaction Procedure (Scheme 2.114): A mixture of iodoaniline (10 mmol), cyclohexanone (30 mmol, 3 equiv.), DABCO (30 mmol, 3 equiv.), and $Pd(OAc)_2$ (112 mg, 0.5 mmol, 5 mol%) in dry DMF (30 mL) was degassed *via* vacuum–nitrogen purges and heated to 105 °C. The mixture was heated at 105 °C for 3 h or until the reaction was complete (usually <12 h). The reaction mixture was cooled to room temperature and partitioned between isopropyl acetate (150 mL) and water (50 mL). The organic layer was separated, washed with brine (50 mL), and concentrated under vacuum to dryness. The residue was chromatographed and crystallized from isopropyl acetate–heptane to give the pure product.

using 2-iodoaniline and ketone as substrates (Scheme 2.114).[199] The reaction tolerates a variety of functional groups, especially the acid-sensitive acetal, carbamate, or benzyl triazole. These intermediates, which would be unstable under the conditions of the traditional Fischer indole reaction, were conveniently synthesized using this method. The coupling of the indole nucleus onto a steroid was also achieved with 5α-cholestanone (1.0 equiv.), affording the corresponding indole derivative exclusively in 79% yield.

In 2006, Zhu and Jia developed an efficient synthesis of highly functionalized indoles by a palladium-catalyzed annulation reaction between *ortho*-haloanilines and aldehydes.[200] The reaction is very general, and a variety of 2-haloanilines with different electronic properties and aldehydes, including chiral ones, can be used to afford indoles in good to excellent yields (Scheme 2.115). Coupling of *ortho*-iodoaniline with aldehyde was realized under mild ligand-less conditions [$Pd(OAc)_2$, DABCO, DMF, 85 °C], whereas X-Phos was found to be the ligand of choice for coupling reactions involving *ortho*-chloroanilines/*ortho*-bromoanilines and aldehydes [$Pd(dba)_2$/X-Phos, KOAc, DMAc at 120 °C]. Coupling of (*S*)-2-*N*,*N*-di-*tert*-butoxycarbonyl-5-oxopentanoate, derived from L-glutamic acid, with *ortho*-haloanilines provided the corresponding ring-A-substituted tryptophans in good to excellent yields. This represents one of the shortest routes for the synthesis of tryptophan derivatives.

Several other types of more specialized carbonyl compounds were applied in the cyclization with 2-haloanilines as well. Cyclohexane-1,3-dione, cyclopentane-1,3-dione,[201] indole-3-butanal,[202] cyclopenta[*b*]indol-1-ones

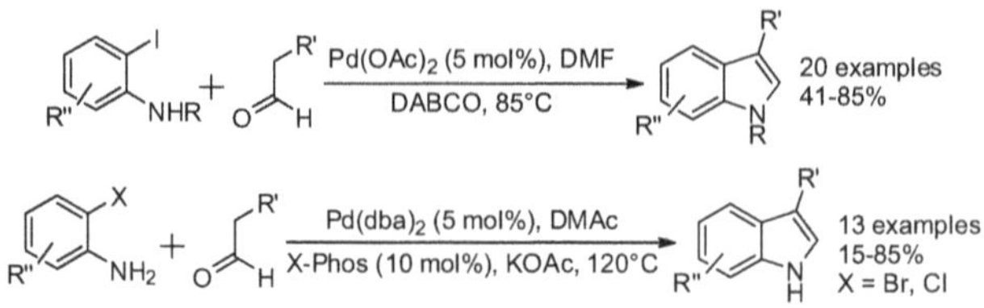

Scheme 2.115 Pd-catalyzed synthesis of indoles from *o*-haloanilines and aldehydes.

Reaction Procedure (Scheme 2.115): A mixture of *o*-iodoaniline (0.33 mmol), aldehyde (0.30 mmol), and DABCO (101.0 mg, 0.90 mmol) in dry DMF (1.5 mL) was degassed for 20 min. $Pd(OAc)_2$ (3.4 mg, 0.015 mmol) was added to the reaction, and the resulting reaction mixture was heated at 85 °C until the reaction was complete. The reaction mixture was cooled to room temperature and was diluted with water. The aqueous phase was extracted with EtOAc, and the combined organic phases were washed with brine, dried (Na_2SO_4), and evaporated to dryness under reduced pressure. Purification of crude product by flash column chromatography provided the desired product.

A mixture of *o*-chloroaniline/*o*-bromoaniline (0.30 mmol), aldehyde (0.30 mmol), and KOAc (88 mg, 0.90 mmol) in dry DMAc (1.5 mL) was degassed for 20 min. $Pd(dba)_2$ (8.0 mg, 0.015 mmol) and X-Phos (14.3 mg, 0.03 mmol) were added to the reaction, and the resulting reaction mixture was heated at 120 °C until the reaction was complete. The reaction mixture was cooled to room temperature and was diluted with water. The aqueous phase was extracted with EtOAc, and the combined organic phases were washed with brine, dried (Na_2SO_4), and evaporated to dryness under reduced pressure. Purification of crude product by flash column chromatography provided the desired product.

and 1,2-bis(3-indolyl)ethanes can be selectively produced. On the other hand, new catalytic systems were explored as well. For example, the combination of a palladium catalyst with microwave irradiation,[203] or a well-defined NHC–Pd complex [(IPr)Pd(acac)Cl] as a catalyst.[204] In the latter case, both *o*-bromoanilines and *o*-chloroanilines gave rise to efficient coupling under the reaction conditions. Related to acyclic ones, cyclic ketones coupled more effectively with *o*-haloanilines. With [$Pd(IPr)_2$] as a catalyst, base-sensitive groups, including OH and CO_2H groups, could be tolerated.

Regarding the 2-chloroanilines, a general palladium-catalyzed procedure was reported in 2004 by Nazaré and co-workers.[205] The procedure represents a very efficient method for the synthesis of polyfunctionalized indoles by direct reaction of substituted 2-chloroanilines with cyclic or acyclic ketones (Scheme 2.116). The procedure is simple to carry out and broadly applicable; all the desired products were isolated in good yields.

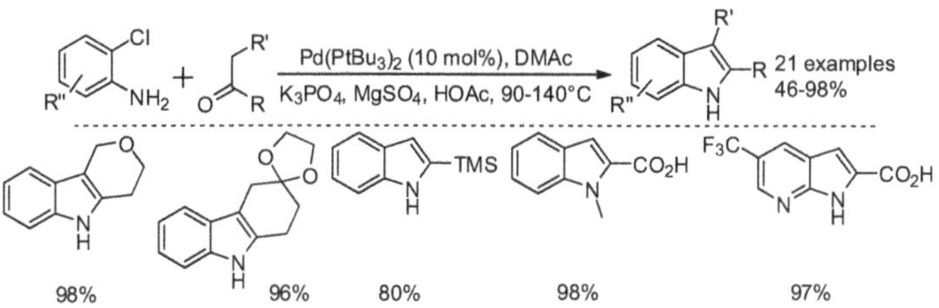

Scheme 2.116 Pd-catalyzed synthesis of indoles from *o*-chloroanilines and ketones.

Reaction Procedure (Scheme 2.116): The appropriate 2-chloroaniline (4 mmol), ketone (12 mmol), acetic acid (0.3 mL, 6 mmol) and $MgSO_4$ (240 mg, 2 mmol) were suspended in dimethylacetamide (12 mL) in a test tube. The tube was sealed with a septum, and argon was bubbled through the solution for 10 min. K_3PO_4 (1104 mg, 5.2 mmol) and $[Pd(tBu_3P)_2]$ (202 mg, 0.4 mmol) were added to the solution, and argon was bubbled through the mixture for an additional 5 min. The reaction mixture was heated to the temperature reported for the time reported in a preheated parallel synthesis block (HPLC-MS reaction control). After cooling to room temperature, the reaction mixture was filtered, water (30 mL) was added to the filtrate, and the mixture was extracted with ethyl acetate (3×50 mL). The combined organic layers were dried over $MgSO_4$ and filtered. After removal of the solvents under reduced pressure, the residue was purified either by flash chromatography on silica or by recrystallization.

In 2007, a copper-catalyzed synthesis of 2,3-disubstituted indoles was developed by Tanimori and co-workers.[206] They use readily available starting materials, 2-iodoaniline and various β-keto esters as substrates; the corresponding products were formed in moderate to good yields (Scheme 2.117a). The advantage of this method is the use of cheap catalysts and simple experimental procedures under mild reaction conditions. Later on, a ligand-free system was developed by using Cu_2O as the catalyst.[207] Polysubstituted indoles were prepared by the domino reaction of 2-haloanilines and 1,3-dicarbonyl compounds, 1,3-diketones, β-keto esters and β-keto amides (Scheme 2.117b).

Interestingly, Ma and co-workers developed a simple and convenient method to convert 2-halotrifluoroacetanilides into 2-(trifluoromethyl)indoles in 2008.[208] By varying the substituents of 2-halotrifluoroacetanilides and β-keto esters, a wide range of substituted 2-(trifluoromethyl)indoles were prepared, including two indoles that are useful precursors for known biologically important compounds (Scheme 2.118a). This reaction occurs *via* a new mechanism, and many functional groups, such as nitro, ketone, ester, hydroxy, iodo, chloro, and olefin, are tolerated under these conditions.

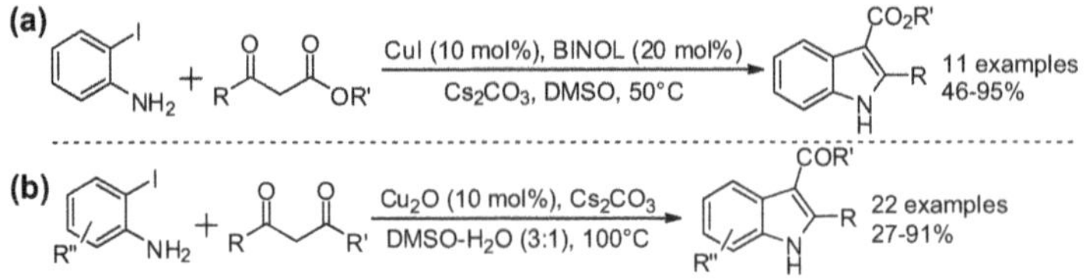

Scheme 2.117 Cu-catalyzed synthesis of indoles from *o*-iodoanilines and ketones.

Reaction Procedure (Scheme 2.117a): A mixture of 2-iodoaniline (0.46 mmol), ethyl acetoacetate (0.51 mmol), CuI (0.046 mmol), BINOL (17.5 mmol, 0.092 mmol) and Cs_2CO_3 (0.46 mmol) in DMSO (1 mL) was stirred at room temp. for 4 h under an atmosphere of nitrogen. The mixture was partitioned between ethyl acetate and saturated NH_4Cl, the organic layer was washed with brine, dried with $MgSO_4$ and concentrated *in vacuo*. The residue was purified by preparative TLC (hexane–EtOAc, 3 : 1) to provide the indole.

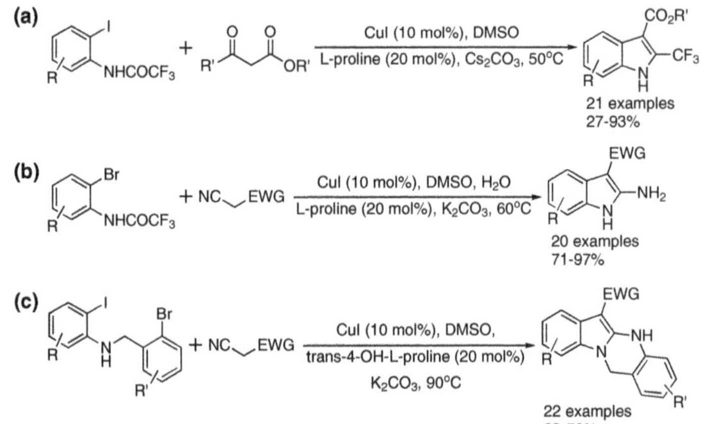

Scheme 2.118 Cu-catalyzed synthesis of indole analogues.

Reaction Procedure (Scheme 2.118a): A Schlenk tube was charged with aryl iodide (0.25 mmol), CuI (5 mg, 0.025 mmol), L-proline (6 mg, 0.05 mmol), and Cs_2CO_3 (326 mg, 1 mmol). Then β-keto ester (0.5 mmol) and DMSO (0.5 mL) were successively injected. After the reaction mixture was stirred at the corresponding temperature for 12–18 h, it was partitioned between ethyl acetate and saturated NH_4Cl. The organic layer was washed with brine, dried over Na_2SO_4 and concentrated *in vacuo*. The residue was purified by column chromatography on silica gel (eluting with 20 : 1 to 1 : 1 petroleum ether–ethyl acetate) to provide the desired product.

(Scheme 2.118b): A 10 mL round-bottomed flask was charged with a magnetic stirrer and DMSO (0.5 mL), water (0.5 mL) [water was not added until the Ullmann-type coupling was completed (TLC determination) when methyl 2-cyanoacetate was used as the substrate], substituted *N*-(2-halophenyl)-2,2,2-trifluoroacetamide (0.5 mmol), alkyl 2-cyanoacetates or malononitrile (0.6 mmol), L-proline (0.1 mmol, 12 mg) and K_2CO_3 (1 mmol, 138 mg). After stirring the mixture for 15 min under a nitrogen atmosphere, CuI (0.05 mmol, 10 mg) was added to the flask. The mixture was stirred at 60 °C for 9 h or 12 h under a nitrogen atmosphere. The resulting mixture was cooled to room temperature and filtered. The solid was washed with methanol twice (2×3 mL). The combined filtrate was concentrated on the rotary evaporator, and the residue was purified by column chromatography on silica gel using petroleum ether–ethyl acetate (1 : 1 to 1 : 2) as the eluent to give the desired product.

(Scheme 2.118c): Substrate (1 mmol), CuI (0.1 mmol), *trans*-4-OH-L-proline (0.2 mmol) and K_2CO_3 (3 mmol) were added into an oven-dried 10 ml microwave vial containing a magnetic stir bar. The tube was evacuated and backfilled with argon three times, then the tube was sealed with a rubber-stopper under a stream of argon. The corresponding functionalized acetonitrile (1.2 mmol) in DMSO (2 ml) was injected *via* a syringe; the resulting mixture was heated to 90 °C for 16 h. After cooling to room temperature, the reaction mixture was partitioned between ethyl acetate and water. The organic layer was separated, and the aqueous layer was extracted with ethyl acetate three times, the combined organic solution was washed with water, 6 N HCl, saturated aq. $NaHCO_3$, brine, dried over anhydrous Na_2SO_4, and concentrated under vacuum to give crude product. Purification by recrystallization or flash chromatography afforded the desired product.

In their mechanistic study, they demonstrated clearly that the β-keto group in the coupling product plays a crucial role for the transformation into indoles. In 2010, a simple and efficient method for the synthesis of 2-amino-1*H*-indole-3-carboxylate derivatives was developed.[209] The cascade reactions of substituted *N*-(2-halophenyl)-2,2,2-trifluoroacetamide with alkyl 2-cyanoacetate or malononitrile were performed well under mild conditions, the corresponding indole derivatives containing amino and carboxylate groups were obtained in good yields, and they are biologically active molecules and important intermediates for the construction of medicinal molecules (Scheme 2.118b). Recently, a domino synthesis of 5,12-dihydroindolo[2,1-*b*]quinazoline derivatives *via* copper-catalyzed Ullmann-type intermolecular C–C and intramolecular C–N couplings was reported by Zhao, Dong and their co-workers.[210] Good yields of various 5,12-dihydroindolo[2,1-*b*]quinazoline derivatives were obtained (Scheme 2.118c). In these three procedures, the reactions started with Ullmann-type C–C bond formation.

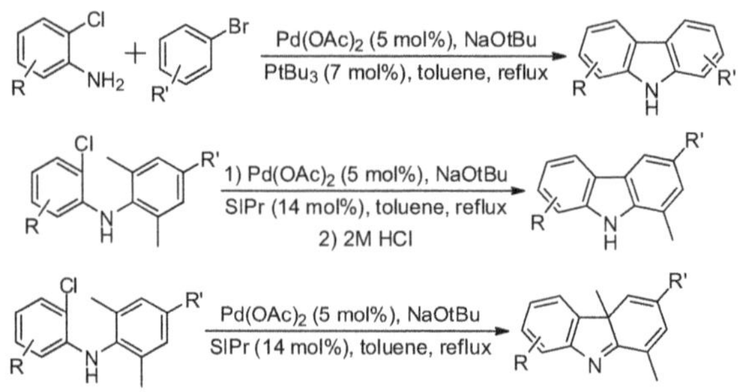

Scheme 2.119 Pd-catalyzed synthesis of carbazoles.

Besides indole derivatives, carbazoles could be produced when arenes were applied as the coupling partners. Bedford's group performed a systematic study on the reactions between 2-chloroanilines with aryl bromides to give carbazoles.[211–214] In the presence of palladium and ligand, the desired carbazoles were isolated in good yields (Scheme 2.119). As well as nitrogen-substituted 2-chloroanilines, nitrogen-free anilines can also be applied as substrates. In the case of sterically hindered substrates, 4a-alkyl-4a*H*-carbazoles can be generated by a catalytic dearomatisation process and their reactivity was demonstrated by facile dealkylation and highly unusual cyclodimerisation (2 + 2) processes. Here, the acidic work-up can affect the outcome of the product structure.

Maes and co-workers performed the intramolecular direct arylations at high temperature.[215] The reaction involves oxidative addition of aromatic C–Br or activated azaheteroaromatic C–Cl bonds and C(sp^2)–H activation at high reaction temperatures (180–200 °C). This allowed the fast (10–30 min) synthesis of a variety of azaheteroaromatic ring systems (dibenzo[*f*,*h*]phthalazine, dibenzo[*f*,*h*]cinnoline, benzofuro[2,3-*d*]pyridazine, 5*H*-pyridazino[4,5-*b*]-indole, 7*H*-indolo[2,3-*c*]quinoline and 5*H*-δ-carboline) in moderate to good yields. $PdCl_2(PPh_3)_2$ (1 mol%) was used as the pre-catalyst with NaOAc · $3H_2O$ as the base in DMA at 180 °C (μW). The short reaction times required, the high conversions and good yields make the procedure especially interesting for fast library design in medicinal and agrochemical research programmes.

In 2003, the group of Maes published a new method for the synthesis of 11*H*-indolo[3,2-*c*]quinoline using commercially available 4-chloroquinoline and 2-chloroaniline as substrates.[216] An excellent yield of the desired product was achieved. In 2008, they succeeded in performing this reaction in a one-pot one-step manner. Moderate to good yields of the corresponding products were given (Scheme 2.120a). This procedure was later applied in the synthesis of indoloquinoline alkaloids.[217] Their antiparasitic and cytotoxic activities were tested as well. Interestingly, Catellani's group offers another reaction pathway.[218] The new procedure developed is based on the combination of palladium and norbornene. Sequential synthesis of 1-substituted carbazoles by C–C and C–N cross-coupling starts from simple and readily available *o*-substituted iodoarenes and *N*-sulfonylated or *N*-acetylated

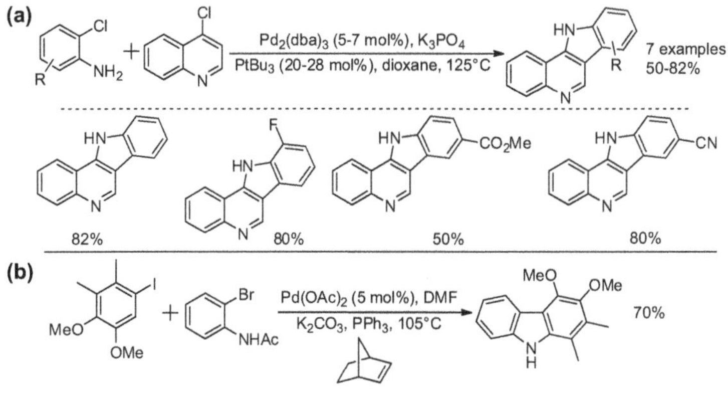

Scheme 2.120 Pd-catalyzed synthesis of carbazoles from aryl chlorides and alternative procedure.

Reaction Procedure (Scheme 2.120a): A 50 mL round-bottomed flask was charged with $Pd_2(dba)_3$ (0.05 mmol), followed by dry freshly distilled 1,4-dioxane (20 mL). To this solution, $P(tBu)_3$ (0.2 mmol, 0.2 mL of 1 M solution in toluene) was added *via* a syringe. The obtained mixture was flushed with argon for 15 min under magnetic stirring. Meanwhile an 80 mL pressure tube was charged with 4-chloroquinoline (2.0 mmol), aniline (2.4 mmol) and finely ground K_3PO_4 (20.0 mmol). To this mixture, the preformed Pd catalyst was added under an argon flow. The 50 mL flask was subsequently rinsed with freshly distilled 1,4-dioxane (2×10 mL). Then the resulting mixture was flushed with argon for 5 min, closed and heated at 125 °C under vigorous magnetic stirring for 24 h. After cooling down to room temperature the crude reaction mixture was filtered through a pad of Celite which was rinsed with CH_2Cl_2–MeOH or CH_2Cl_2. The filtrate was evaporated under reduced pressure and purified by flash column chromatography on silica gel.

(Scheme 2.120b): A Schlenk-type flask equipped with a magnetic stir bar was charged with $Pd(OAc)_2$ (5 mg, 0.022 mmol), norbornene (10 mg, 0.11 mmol), the desired aryl iodide and bromide (0.48 and 0.44 mmol, respectively) in DMF (10 mL) and K_2CO_3 (138 mg, 1.0 mmol). The reaction mixture was stirred at 105–120 °C for the time needed for Pd black precipitation. When needed, TPP (11.5 mg, 0.044) was added. After cooling to room temperature, the organic layer was diluted with EtOAc (30 mL) and extracted three times with a solution of NaCl (25 mL). The organic layer was dried over Na_2SO_4, the solvent was removed under reduced pressure and the resulting residue was purified by flash chromatography on silica gel using mixtures of hexane–EtOAc as eluent (97 : 3 for sulfonylated carbazoles, 95 : 5 for unprotected carbazoles).

o-bromoanilines. Besides the mechanistic interest, the present procedure offers a valid complementary alternative to the previously reported methods. A well-known member of this class, carbazomycin A, was successfully prepared as well (Scheme 2.120b).

Sakamoto and co-workers reported an alternative procedure for the preparation of carbazoles.[219] By using *ortho,ortho'*-dibromo-substituted anilinoarenes as substrates, in the presence of a palladium catalyst and with $(SnBu_3)_2$ as an additive, the desired carbazoles were formed in good yields.

In 2004, Larock and Liu reported an efficient route to synthesize a variety of carbazoles and dibenzofurans from *o*-iodoanilines or *o*-iodophenols and silylaryl triflates.[220] The reaction involves two steps: forming the *N*- or *O*-arylated products in the presence of CsF, and a subsequent cyclization using a palladium catalyst to carbazoles and dibenzofurans. Good to excellent yields of the desired products were produced (Scheme 2.121a). This chemistry tolerates a variety of functional groups. Several new and multi-substituted carbazoles and dibenzofurans were synthesized. Later on, a one-pot one-step procedure was developed by the same group by using 2-haloacetanilides as substrates.[221] *N*-Acylcarbazoles were synthesized in moderate to good yields by the annulation of *in situ* generated arynes with 2-haloacetanilides in the presence of a palladium catalyst and CsF (Scheme 2.121b). Both C–C and C–N bonds were formed simultaneously, and a variety of functional groups are tolerated in this reaction. Interestingly, a different reaction mechanism was proposed for this procedure.

In 2009, Buchwald and co-workers reported the first asymmetric palladium-catalyzed intramolecular dearomatization reaction.[222] Benzocarbazole derivatives were obtained in high yields and enantioselectivities from naphthalene derivatives (Scheme 2.122). Here, it is important to note that dearomatization of arenes is recognized as a fundamental chemical transformation for organic chemists which allows efficient access to alicyclic frameworks present in many biologically active compounds.

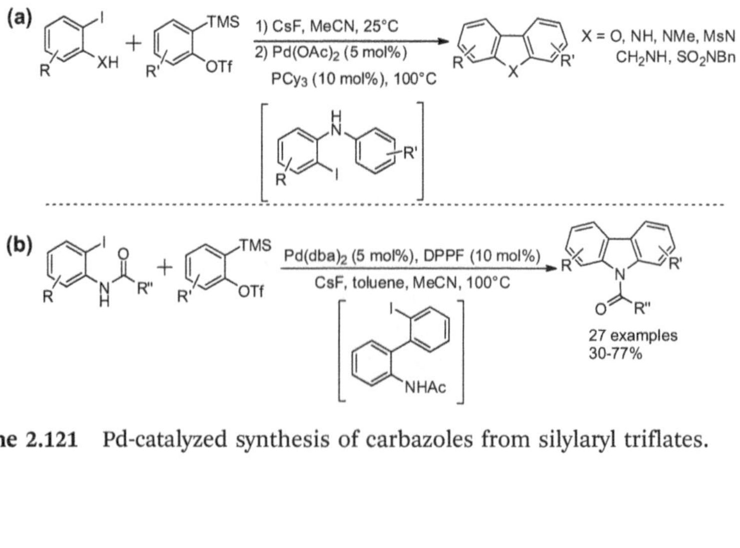

Scheme 2.121 Pd-catalyzed synthesis of carbazoles from silylaryl triflates.

Reaction Procedure (Scheme 2.121a): In a 4-dram vial, the silylaryl triflate (0.275 mmol) and CsF (0.75 mmol) were added to a solution of the *o*-iodoaniline (0.25 mmol) in acetonitrile (4 mL). The solution was allowed to stir at room temperature for 10 h in air. The vial was then flushed with argon and $Pd(OAc)_2$ (5 mol%) and PCy_3 (10 mol%) were added to the reaction, which was heated to 100 °C for 1 d. The resulting solution was washed with brine (20 mL) and extracted with diethyl ether (20 mL). The combined ether fractions were dried over Na_2SO_4 and concentrated under reduced pressure. The residue was purified by flash chromatography on silica gel to afford the desired product.

(Scheme 2.121b): The 2-iodoacetanilide (0.25 mmol), the 2-(trimethylsilyl)aryl triflate (2.0 equiv.), CsF (5.0 equiv.), $Pd(dba)_2$ (5 mol%), DPPF (10 mol%), 8 mL of toluene, and 2 mL of MeCN were placed in a 4-dram vial, and the vial was sealed. The reaction mixture was stirred, first at room temperature for 1 min and then heated to 110 °C for 24 h. The mixture was allowed to cool to room temperature, diluted with ethyl acetate, washed with brine, dried over anhydrous $MgSO_4$, and concentrated under reduced pressure. The product was isolated by flash chromatography on silica gel using hexanes–EtOAc as the eluent.

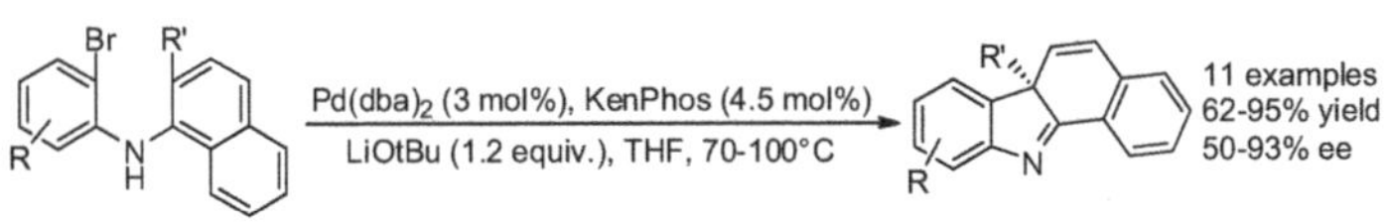

Scheme 2.122 Pd-catalyzed dearomatization of arenes.

Reaction Procedure (Scheme 2.122): An oven-dried resealable test tube containing a stir bar was charged with starting material (0.5 mmol, 1.0 equiv.), $Pd(dba)_2$ (8.6 mg, 15 μmol) and KenPhos (11.1 mg, 22.5 μmol). The test tube fitted with a screw cap with a pierceable teflon septum was then evacuated and backfilled with dry argon (this sequence was repeated three times) and lithium *t*-butoxide in THF (0.15M, 5.0 mL, 1.5 equiv.) was then added by syringe. The tube was then placed in an oil bath preheated at the desired temperature for 15 h. After cooling to room temperature the contents were diluted with EtOAc and the mixture filtered through a plug of Celite. The solution was then concentrated under reduced pressure and the residue purified by flash chromatography on silica gel.

Kan and co-workers developed a novel synthetic strategy for carbazoles based on a Pd-catalyzed tandem reaction of a Suzuki–Miyaura coupling and amination.[223] The phosphine ligand they developed themselves proved to be the most suitable ligand. 2-Bromophenylboronic acid and 2-iodoanilines were applied as substrates; the corresponding products were formed in

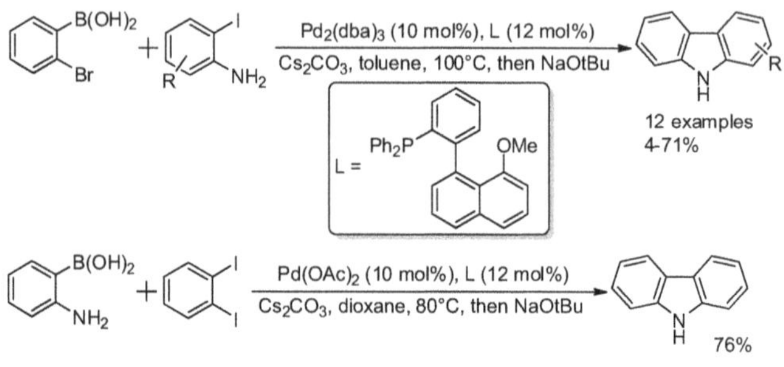

Scheme 2.123 Pd-catalyzed synthesis of carbazoles from 2-bromophenylboronic acid.

Reaction Procedure (Scheme 2.123): To a solution of 2-bromophenylboronic acid (0.15 mmol) and 2-iodoaniline (0.14 mmol) in toluene (0.5 mL) were added $Pd_2(dba)_3$ (13 mg, 0.01 mmol), ligand (6.9 mg, 0.02 mmol), and Cs_2CO_3 (134 mg, 0.41 mmol) at rt under an argon atmosphere, then the mixture was stirred at 100 °C for 2 h. The mixture was cooled to rt, and then NaO*t*-Bu (16 mg, 0.17 mmol) was added and stirred at 100 °C for 19 h. After stirring, H_2O was added, and extracted with EtOAc. The organic layer was washed with H_2O, brine, dried over $MgSO_4$ and evaporated. The residue was purified by column chromatography on silica gel (benzene–EtOAc, 99 : 1) to afford the 9*H*-carbazole.

To a solution of 2-aminophenylboronic acid (0.11 mmol) and 1,2-diiodobenzene (0.10 mmol) in 1,4-dioxane (0.33 mL) was added $Pd(OAc)_2$ (2.2 mg, 0.01 mmol), ligand (5.0 mg, 0.01 mmol), and Cs_2CO_3 (0.30 mmol) at rt under an argon atmosphere, then the mixture was stirred at 80 °C for 24 h. The mixture was cool to rt, and then NaO*t*Bu (12 mg, 0.12 mmol) was added and stirred at 80 °C for 24 h. After stirring, H_2O was added, and extracted with EtOAc. The organic layer was washed with H_2O, brine, dried over $MgSO_4$, and evaporated. The residue was purified by column chromatography on silica gel (benzene–EtOAc, 99 : 1) to afford the 9*H*-carbazole.

moderate yields (Scheme 2.123). Alternatively, the coupling reaction of 2-aminophenylboronic acid and 1,2-diiodobenzene could give carbazole as well. The advantage of this method is that the target carbazole framework is constructed in one pot, and new C–C and C–N bonds are formed in a regiospecific manner.

1,3-Dihydrobenzimidazol-2-ones are an important class of compounds with biological activities including selective vasopressin 1α receptor antagonists, HIV-1 RT non-nucleoside inhibitors, CGRP receptor antagonists, p38 MAP kinase inhibitors, respiratory syncytial virus fusion inhibitors, and progesterone receptor antagonists. Regarding their importance, Ma and co-workers reported a copper-catalyzed 1,3-dihydrobenzimidazol-2-one

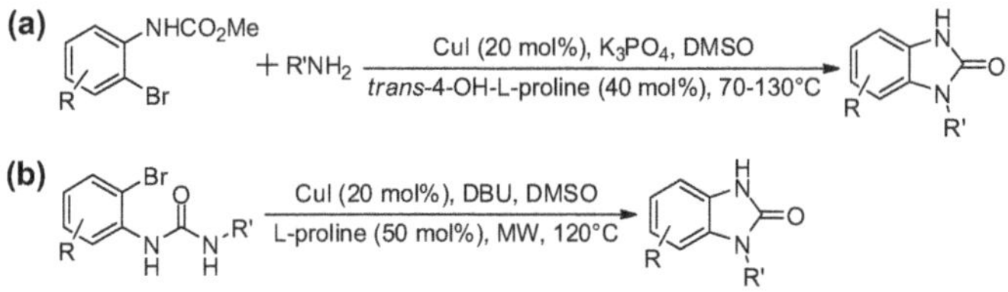

Scheme 2.124 Cu-catalyzed synthesis of 1,3-dihydrobenzimidazol-2-ones.

Reaction Procedure (Scheme 2.124a): A Schlenk tube was charged with aryl bromide (0.5 mmol), CuI (0.1 mmol), *trans*-4-hydro-L-proline (0.2 mmol), and K_3PO_4 (1.0 mmol), evacuated and backfilled with argon. Amine (0.5 mmol) and DMSO (1 mL) were successively added. The reaction mixture was stirred at 70 °C until the coupling was completed, detected by TLC. The solution was then heated at 130 °C until the coupling product was consumed, monitored by TLC. The cold mixture was partitioned between ethyl acetate and saturated NH_4Cl. The organic layer was washed with brine, dried over Na_2SO_4 and concentrated *in vacuo*. The residue was purified by column chromatography on silica gel to provide the desired product.

(Scheme 2.124b): To a solution of substrate (0.5 mmol) in DMSO, CuI (0.1 mmol), and DBU (1 mmol) were added. The vial was sealed and the mixture was then irradiated for 20 min at 120 °C. The cold mixture was partitioned between ethyl acetate and 1 M HCl. The organic layer was washed with brine, dried with Na_2SO_4 and concentrated *in vacuo*. The residue was purified by column chromatography on combiflash to provide the desired product.

synthesis in 2007.[224] *N*-Substituted 1,3-dihydrobenzimidazol-2-ones were achieved starting from methyl *o*-haloarylcarbamates *via* a CuI–amino acid catalyzed coupling with amines and subsequent condensative cyclization. A number of functional groups are tolerated by these reaction conditions, including vinyl, nitro, carboxylate, amide, ester, ketone, and silyl ether groups. Good yields of the desired products were isolated (Scheme 2.124a). Later on in 2008, Liu and co-workers demonstrated another efficient method to generate *N*-substituted 1,3-dihydrobenzimidazol-2-ones.[225] Using *N*′-substituted-*N*-(2-halophenyl)ureas as substrates, which can be easily prepared from commercial *o*-haloanilines on reaction with triphosgene and different kinds of amines or isocyanates, the corresponding products were isolated in good yields (Scheme 2.124b). A variety of functional groups can be tolerated.

2-Oxoindolines are pharmacologically important compounds that display great potential utilizations in many major therapeutic areas, such as oncology, inflammation, CNS, immunology, and endocrinology. For example, SU11248 was commercialized by Pfizer, Inc. in 2006 to treat renal cell

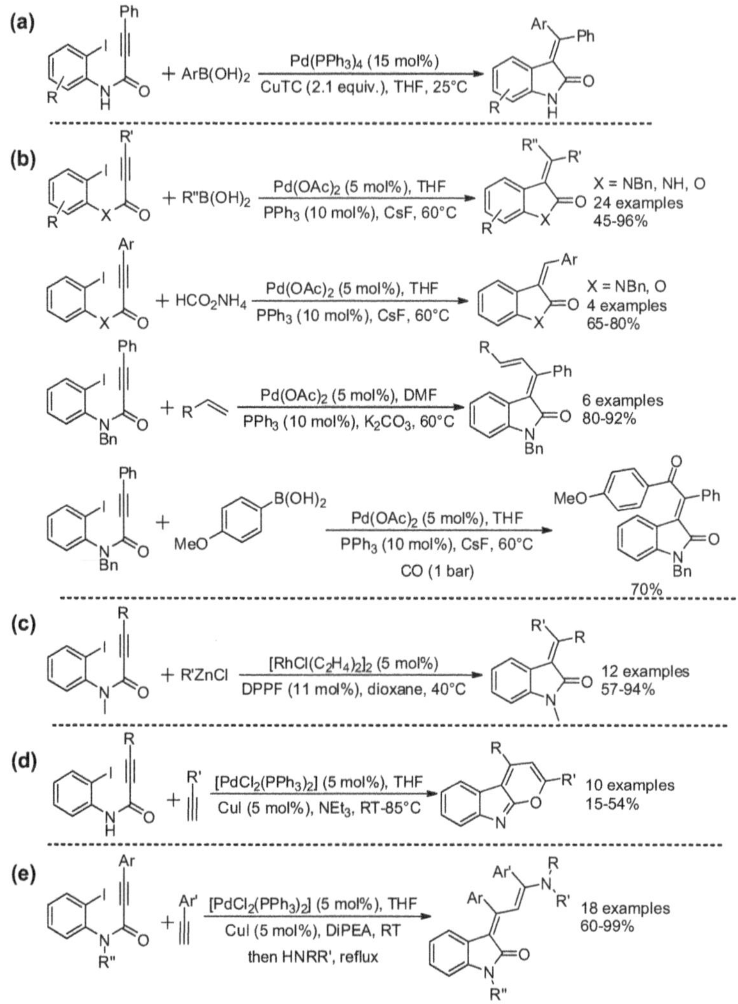

Scheme 2.125 Synthesis of 2-oxoindolines from 2-haloanilides.

carcinoma and gastrointestinal stromal tumors. Among the known procedures, the most direct pathways are methodologies based using 2-haloanilines as starting materials. In 2005, Player and co-workers developed and optimized a tandem Heck–Suzuki coupling reaction for the syntheses of asymmetric diarylmethylidenyl indolinones.[226] At ambient temperatures, the process is high yielding, highly regioselective, and can be accelerated by judicious choice of solvent mixtures. The desired (*E*)-3,3-(diarylmethylene)-indolinones were prepared in good yields *via* this palladium-catalyzed tandem Heck-carbocyclization–Suzuki-coupling sequence (Scheme 2.125a). In the same year, Takemoto and co-workers developed efficient catalytic methods for the stereoselective and diversity-generating synthesis of various

(*E*)-, (*Z*)-, and disubstituted 3-alkylideneoxindoles and 3-alkylidenebenzofuran-2-ones *via* palladium-catalyzed Heck/Suzuki–Miyaura, Heck/Heck, and Heck/carbonylation/Suzuki–Miyaura domino reactions (Scheme 2.125b).[227] This type of reaction was further studied by Seo and co-workers in 2013.[228] By adding Sonogashira-coupling for the preparation of the substrates followed by the reaction with arylboronic acids, a C–H activation version of this reaction was developed as well. The reactions were performed in acetic acid with $PhI(OAc)_2$ as an oxidant at 80 °C; the corresponding (2-oxoindolin-3-ylidene)methyl acetates were formed in good yields.[229] A rhodium catalyst was applied in this type of multicomponent-coupling reaction as well.[230] By applying an organozinc reagent as the coupling partner, the desired 2-oxoindolines were isolated in good yields (Scheme 2.125c). Through several experimental mechanistic investigations, the reaction has been demonstrated to most likely proceed *via* a carborhodation–oxidative addition–reductive elimination pathway, which clearly contrasts to the corresponding palladium-catalyzed processes. In 2010, Müller and co-workers reported a domino synthesis of 2,4-diarylpyrano[2,3-*b*]indoles, a class of compounds with unique photophysical properties, from alkynes and 2-iodoanilides (Scheme 2.125d).[231] Displaying no fluorescence as a free base, halochromic green fluorescence can be switched on upon protonation, methylation, or complexation with various metal cations, also displaying considerable fluorescence sensitivity. Interestingly, different types of indolones were produced by using *N*-substituted 2-iodoanilides as substrates (Scheme 2.125e).[232] Based on this consecutive three-component insertion–coupling–addition sequence, 4-aminoprop-3-enylidene indolones with a flexible substitution pattern in yields as high as 99% were obtained.

Alternatively, 2-oxoindolines can be prepared from 2-haloanilides *via* intramolecular cyclization as well. In 2006, a palladium-catalyzed intramolecular cyclization of 2-iodoanilides was reported.[233] In the presence of $Pd_2(dba)_3$ and (*R*)-BINAP, with Ag_3PO_4 as a base in DMAc, excellent conversion and low to moderate enantioselectivity were observed. The nickel-catalyzed electrochemical cyclization of 2-iodoanilides was described too.[234] With the assistance of $NiBr_2$ and electrochemistry, the desired 1,3-dimethylindolin-2-one was produced in 40% yield from *N*-(2-iodophenyl)-*N*-methylacrylamide. A sequential process based on carbonylation was reported in 1996 by Grigg's group.[235] They used 2-haloanilines and 2-halophenols with vinyl halides or vinyl triflates as substrates, in the presence of palladium and CO (1 bar); the desired 3-spiro-2(3*H*)-benzofuranones and 2-oxoindolines were isolated in moderate yields. The reaction started with palladium-catalyzed aminocarbonylation of vinyl halides with 2-haloaniline to provide 2-haloanilindes as the intermediates, followed by intramolecular cyclization to give the terminal products. In 2001, Hartwig and Lee performed a systematic study on the cyclization of 2-haloanilides (Scheme 2.126).[236] In their catalytic system, the catalysts comprising $Pd(OAc)_2$ and either PCy_3 or sterically hindered N-heterocyclic carbene ligands provide fast rates for the palladium-catalyzed synthesis of oxindoles by

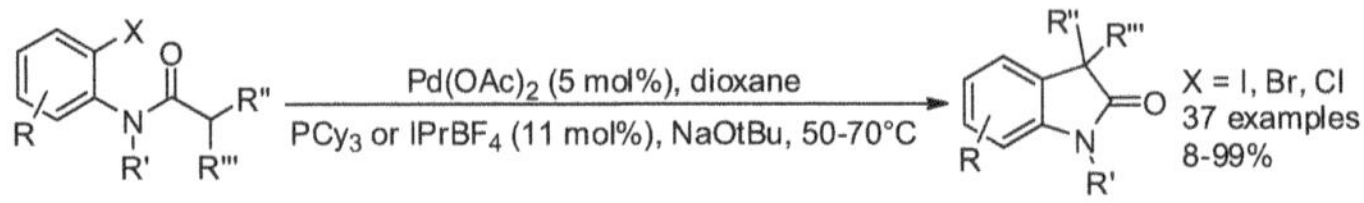

Scheme 2.126 Palladium-catalyzed synthesis of 2-oxoindolines from 2-haloanilides.

Reaction Procedure (Scheme 2.126): In a drybox, $Pd(OAc)_2$ (0.05 mmol), PCy_3 (0.0500 mmol), or carbene ligand $IPr \cdot BF_4$ (0.05 mmol) was combined with Na*Ot*Bu (1.50 mmol) in a small round-bottomed flask. 1,4-Dioxane (10 mL) was added, and the flask was sealed with a septum. The resulting mixture was allowed to stir for 1 min, at which time the 2-bromo or 2-chloroanilide substrate was added. After being removed from the drybox, the flask was placed in an oil bath at the appropriate temperature (50 or 70 °C) until the starting material was consumed, as determined by GC and TLC. The reaction was poured into 20 mL of saturated aqueous ammonium chloride and extracted (3×20 mL) with Et_2O. The combined ether extracts were washed with brine (60 mL), dried over $MgSO_4$, and filtered. The solvent was removed under vacuum, and the resulting crude product was purified by flash chromatography on silica gel. The product was eluted with 15% ethyl acetate in hexanes.

amide α-arylation. This catalyst system allowed for room-temperature reactions in some cases and reactions of aryl chlorides at 70 °C. Most importantly, reactions occurred in high yields under mild conditions to form the quaternary carbon in α,α-disubstituted oxindoles. The combined inter- and intramolecular reaction afforded an efficient synthetic method for the formation of α-aryloxindole derivatives. Surprisingly, catalysts containing *tert*-butylphosphine ligands, which have been most reactive for ketone arylations, were less active than those containing PCy_3. Use of new, optically active heterocyclic carbene ligands gave substantial enantioselectivity in the formation of α,α-disubstituted oxindoles. In contrast, a variety of optically active phosphine ligands that were tested gave poor enantioselectivity. Mechanistic studies showed that the reaction involves rate-limiting oxidative addition of aryl halide. Base-induced formation of and reductive elimination from an arylpalladium enolate intermediate were both faster than oxidative addition. Deprotonation of the tethered amide appeared to be faster than reductive elimination of the resulting palladium enolate to form the oxindole product.

As an interesting multicomponent reaction, an efficient palladium-catalyzed protocol for the synthesis of 3-arylidene-2-oxindoles was developed in 2010 by Balalaie and co-workers (Scheme 2.127).[237–239] In this approach, a sequential one-pot six-component reaction *via* Ugi–Heck carbocyclization–Sonogashira–nucleophilic addition was used for the synthesis of the desired skeleton.

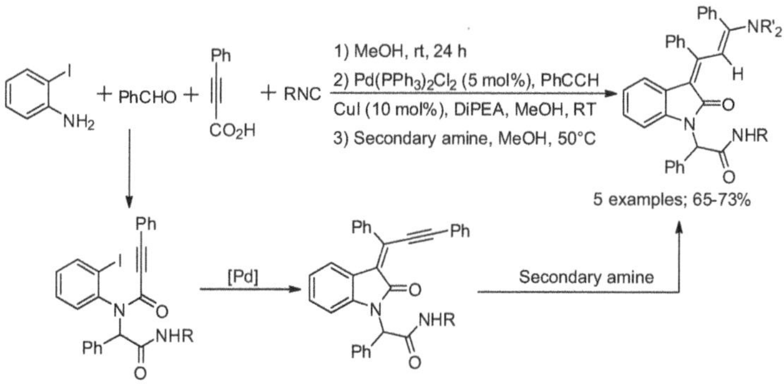

Scheme 2.127 Palladium-catalyzed multicomponent synthesis of 2-oxoindolines.

Reaction Procedure (Scheme 2.127): 2-Iodoaniline (1 mmol) and benzaldehyde (1 mmol) were mixed together in MeOH (5 mL) and stirred for 30 min. Then, phenyl propiolic acid (1 mmol) and after 15 min isocyanide (1 mmol) were added, and the mixture was stirred for 24 h. CuI (0.1 equiv.), $PdCl_2(PPh_3)_2$ (0.05 equiv.), phenylacetylene (1.5 mmol), and DiPEA (5 mmol) were simultaneously added to the solution of Ugi adduct in MeOH. The mixture was stirred at room temperature for 1 h. After this time, the secondary amine (1.5 mmol) was added as the sixth starting material in one portion, and the resulting solution was heated at 50 °C for 8 h. After cooling to room temperature, the reaction mixture was diluted with brine (30 mL) and extracted with ethyl acetate (3×20 mL). The combined organic layers were dried with sodium sulfate, concentrated to dryness *in vacuo*, and purified by column chromatography on silica gel.

As early as 1990, a model study on the carbonylative cyclization of 2-iodoaniline with phenylacetylene was reported.[240] Benzylidene derivatives can be selectively formed in good yields. In 2011, a heterogeneous catalyst system was developed for the selective one-pot synthesis of 2-benzylidene-indoxyl and 2-phenyl-4-quinolone.[241] For this purpose, heterogeneous palladium-, amine-, and phosphine-catalysts were prepared by post-synthetic grafting onto SBA silica. The state of the hybrid materials was characterized using a wide variety of molecular and solid-state techniques. These materials exhibit high activities as 2-benzylidene-indoxyl was obtained in 81% yield through {[Pd]@SBA-15 + PPh_3} catalysis while 2-phenyl-4-quinolone was prepared by a fully heterogeneous {[Pd]@SBA-15 + [AMINE]@SBA-3} protocol in 61–75% isolated yield. For the latter, the authors demonstrated that their catalyst mixture could be reused up to three runs without strong deactivation. Grigg and co-workers found that bicyclopropylidene can react with *N*-(2-iodophenyl)-4-methylbenzenesulfonamide and carbon monoxide to give the dispiro compound under mild conditions in the presence of a novel palladium catalyst (Scheme 2.128).[242] The structure of the product was

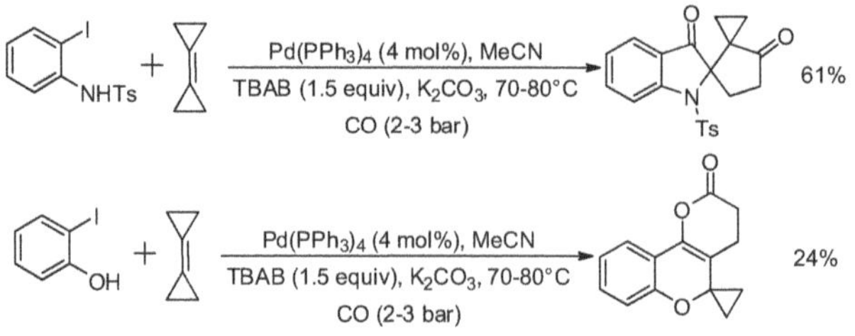

Scheme 2.128 Palladium-catalyzed carbonylative transformation of bicyclopropylidene.

determined by X-ray analysis. With *ortho*-iodophenol as the substrate, a mixture of the desired product (20%) and 3,4-dihydro-2*H*,5*H*-pyrano[3,2-*c*]chromen-2-one (11%) was obtained. Mechanistically this cascade is interpreted in terms of the acylation of iodo compounds followed by a cyclopropylcarbinyl–homoallyl rearrangement, carbon monoxide insertion and intramolecular trapping of a nucleophile. Various thiophenols reacted with bicyclopropylidene and CO (2–3 bar) to give the thiocarbonylated products as well. In 1986, Yamamoto and co-workers reported the double carbonylation of *N*-(2-halophenyl)acetamide with diethylamine.[243] The corresponding α-keto amides were isolated in moderate to good yields under high CO pressure (52 bar). By treating with HCl (3 N), isatins were produced in excellent yields.

Substituted benzotriazoles, one of the most privileged heterocyclic subunits, display many interesting properties including anticancer, antifungal, anti-inflammatory, anti-depressant activities and exhibit an utility as synthetic auxiliaries. In 2010, an efficient and highly versatile method for the synthesis of diverse regiospecific 1-arylbenzotriazole, by the copper-catalyzed intramolecular *N*-arylation of diazoaminobenzenes of 2-haloaryldiazonium salts in PEG-water, was developed.[244] This procedure represents a very simple and efficient reaction protocol; large numbers of substrates were produced in excellent yields (Scheme 2.129).

Benzimidazoles are a class of privileged core structures that are found in a broad spectrum of biologically active compounds such as nonpeptide luteinizing hormone-releasing hormone (LHRH) antagonist, lymphocyte specific kinase (Lck) inhibitor, *N*-methyl-D-aspartate (NMDA) antagonist, neuropeptide Y Y1 receptor antagonist, nonpeptide thrombin inhibitor, 5-lipoxygenase inhibitor, factor Xa (FXa) inhibitor, and poly(ADP-ribose)polymerase (PRAP) inhibitor. In 2002, Brain's group reported an intramolecular palladium-catalyzed amination reaction to form benzimidazoles.[245] By using (*o*-bromoaryl)amidines as precursors, the corresponding products were isolated in moderate to good yields (Scheme 2.130a). *N*-(*o*-Haloaryl)imidoyl chlorides and the corresponding imidates were applied as substrates in palladium-catalyzed benzimidazole synthesis as well. Under the same reaction conditions, quinazolinones can be produced in the presence of CO. The desired products were formed in good yields in general (Scheme 2.130b).[246] In 2011, a versatile and efficient method for the

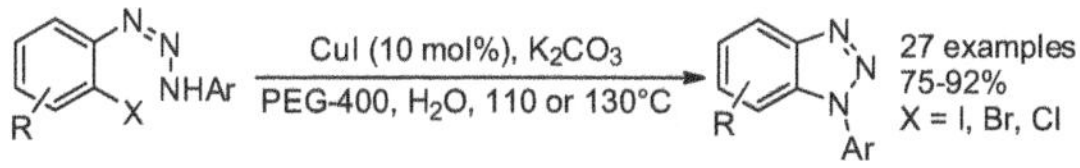

Scheme 2.129 Copper-catalyzed synthesis of 1-aryl benzotriazoles.

Reaction Procedure (Scheme 2.129): Diazoaminobenzene (1 mmol), CuI (0.1 mmol), K_2CO_3 (2 mmol), PEG-400 (0.2 mL) and H_2O (2 mL) were taken up in an Erlenmeyer flask and the resulting mixture was heated at 110 °C (130 °C for diazoaminobenzenes of 2-chloroanilines) until the disappearance of reacting materials (monitored by TLC). The reaction mixture was then cooled and poured into a beaker containing crushed ice. After stirring the mixture for 10 min with a glass rod, it was filtered and extracted through Celite by EtOAc. The solvent was evaporated under reduced pressure and was further purified by column chromatography using silica gel (60–120 mesh) to obtain the pure products.

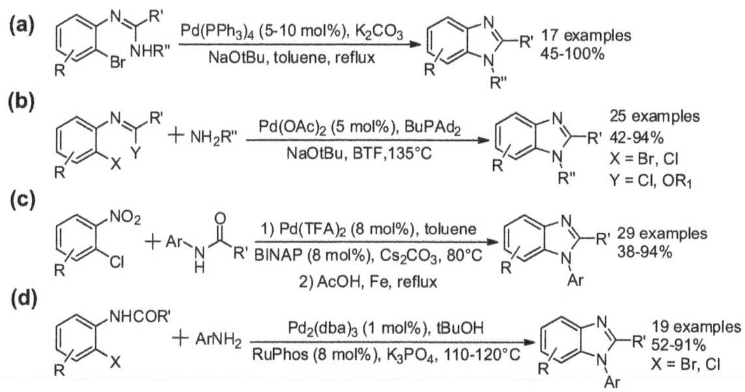

Scheme 2.130 Palladium-catalyzed synthesis of benzimidazoles.

Reaction Procedure (Scheme 2.130a): To a solution of substrate (0.2 mmol) in toluene (5 ml) under argon were added K_2CO_3 (1.6 equiv.), NaO*t*Bu (1.6 equiv.) and $Pd(PPh_3)_4$ (5–10 mol%) sequentially and the resulting mixture was heated under reflux for 18 h. The reaction mixture was filtered through Celite and the solvent was removed *in vacuo*. The crude residue was purified either by column chromatography or preparative HPLC to give the pure product.

(Scheme 2.130b): To a 5 mL microwave vial were added imidoyl chloride (1.0 equiv.), palladium(II) acetate (5 mol%), $BuPAd_2$ (7 mol%), and sodium *tert*butoxide (2.2 equiv.). This was capped and purged with N_2 three times, and anhydrous benzotrifluoride (0.4 M) and *N*-nucleophile (1.5 equiv.) were then added. This was heated in a microwave for 2 h at 135 °C (unless otherwise stated). After being cooled to room temperature, the reaction mixture was diluted with diethyl ether and filtered through a Celite pad, washing with diethyl ether. This was then concentrated *in vacuo*. Purification was conducted by column chromatography.

(Scheme 2.130c): The 2-halonitroarene (0.5 mmol), the amide (0.6 mmol), palladium trifluoroacetate (0.04 mmol), (*R*)-BINAP (0.04 mmol), and caesium carbonate (0.7 mmol) were placed in a reaction tube, which was then purged with dry argon. Dry toluene (3 mL) was added, and the mixture was heated at 80 °C for 18 h. The reaction mixture was hydrolyzed with water (3 mL), and filtered through a pad of Celite and rinsed with ethyl acetate. The crude product was dissolved in glacial acetic acid (10 mL) and was heated at reflux for 30 min in the presence of iron powder (279 mg). The acid was removed under vacuum, and the residue was suspended in saturated sodium hydrogen carbonate solution and extracted with ethyl acetate. The obtained crude product was purified by preparative HPLC (C18 reverse-phase column, elution with a water–MeCN gradient with 0.1% TFA). The fractions containing the product were evaporated and lyophilized to yield a white solid.

(Scheme 2.130d): An oven-dried Schlenk tube containing a Teflon-coated stir bar was charged with [$Pd_2(dba)_3$] (0.005 mmol, 2.0 mol% Pd), ligand XPhos or RuPhos (0.04 mmol, 8 mol%), *ortho*-haloanilides (0.5 mmol), aromatic amines (0.75 mmol) and K_3PO_4 (1.25 mmol). The Schlenk tube was capped with a Teflon screw cap and then evacuated and backfilled with argon (3 cycles). *t*BuOH (1.0 mL) was added to the Schlenk tube under a positive flow of argon (liquid aromatic amines were added to the Schlenk tube at this point). The Schlenk tube was sealed and put into a pre-heated oil bath at 110 °C. After stirring for 18 h, the reaction mixture was allowed to cool to room temperature and diluted with dichloromethane (*ca.* 4 mL). The diluted mixture was filtered through Celite with the aid of dichloromethane. The filtrate was concentrated under vacuum to give a residual that was purified by flash chromatography on silica gel.

regioselective synthesis of polyfunctional benzimidazoles from commercially available 2-halonitrobenzene derivatives and secondary amides was developed.[247] The readily applicable process allows the preparation of this important class of heterocycles in good to excellent yields (Scheme 2.130c). In addition, starting from *ortho*-nitro-pyridines, the method is also well suited for the synthesis of imidazopyridines. In 2007, Buchwald and co-workers reported a palladium-catalyzed procedure based on using 2-haloanilides and anilines as starting materials.[248] By using RuPhos as the ligand, moderate to excellent yields of the *N*-aryl benzimidazoles were isolated (Scheme 2.130d).

Pyrido[1,2-*a*]benzimidazoles as a kind of fused benzimidazole were prepared as well. In 2006, Ila and co-workers developed a new high-yielding route for benzimidazo[1,2-*a*]quinolines *via* palladium-catalyzed intramolecular heterocyclization of readily accessible 2-(2′-bromoanilino)quinolines.[249]

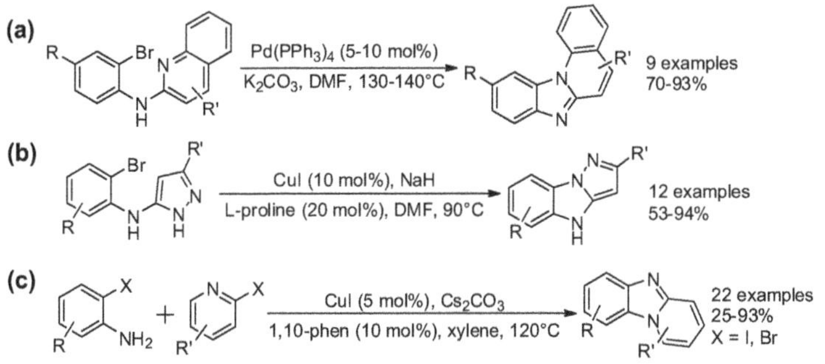

Scheme 2.131 Pyrido[1,2-*a*]benzimidazoles synthesis.

Reaction Procedure (Scheme 2.131a): A mixture of 2-(2′-bromoanilino)-quinolines (2.0 mmol), freshly prepared $Pd(PPh_3)_4$ (10 mol%), and anhydrous K_2CO_3 (4.4 mmol) in 5 mL of DMF was heated at 130–140 °C in a pressure tube for 10–12 h with constant stirring. The reaction mixture was cooled and filtered, and the filtrate was diluted with $CHCl_3$ (25 mL) and washed with water (2×25 mL) followed by brine (20 mL). The organic layer was dried over anhydrous Na_2SO_4 and concentrated in a vacuum to give crude benzimidazo[1,2-*a*]quinolines which were purified by silica gel column chromatography using hexanes–EtOAc (3 : 1) as eluent.

(Scheme 2.131c): An oven-dried Schlenk tube was charged with 2-iodoaniline (0.25 mmol), Cs_2CO_3 (0.75 mmol), CuI (0.05 mmol), and 1,10-phenanthroline (0.1 mmol). The tube was evacuated and backfilled with N_2, and then 2-iodopyridine (0.3 mmol) and xylene (0.5 mL) were added. The reaction mixture was stirred at 120 °C for 12 h and then allowed to cool to room temperature. The mixture was diluted with water and extracted with ethyl acetate. The extracts were combined and then dried with anhydrous Na_2SO_4. The solvent was removed under vacuum, and the residue was purified by silica gel column chromatography to afford the corresponding product.

The overall process involves Buchwald–Hartwig intramolecular aryl amination in which a heteroarene ring nitrogen participates in N–C bond formation. Good to excellent yields of the desired products were formed (Scheme 2.131a). Later on, they developed a copper-catalyzed cyclization of 1-unsubstituted 5-(2-bromoanilino)pyrazole to form pyrazolo[1,5-*a*]benzimidazoles (Scheme 2.131b).[250] In 2011, an intermolecular version of this reaction was reported.[251] By using 2-haloanilines and 2-halopyridines as starting materials, the corresponding pyrido[1,2-*a*]benzimidazoles were selectively produced with a catalytic amount of copper catalyst (Scheme 2.131c). A wide range of functional groups were well tolerated under the reaction conditions, and a series of pyrido[1,2-*a*]benzimidazoles with substituents at different positions were generated in moderate to excellent yields.

Copper catalysts have the advantage of being cheap and abundant, and the application of copper catalysts in heterocycle synthesis is interesting and attractive. In 2007, Ma and co-workers reported a CuI/L-proline-catalyzed cyclization of 2-iodoanilides with primary amines.[252] This procedure represents a novel and highly practical method for elaborating benzimidazoles; all the desired products were isolated in good to excellent yields (Scheme 2.132a). Variation at the 1- and 2-positions of the benzimidazole is possible when different primary amines are employed and with variation in the amido groups of the 2-haloacetanilides. This reaction was studied further with other copper systems, such as CuI/2,2′-biimidazole[253] and sulfonato/Cu(salen) complex in water.[254] Instead of primary amines,

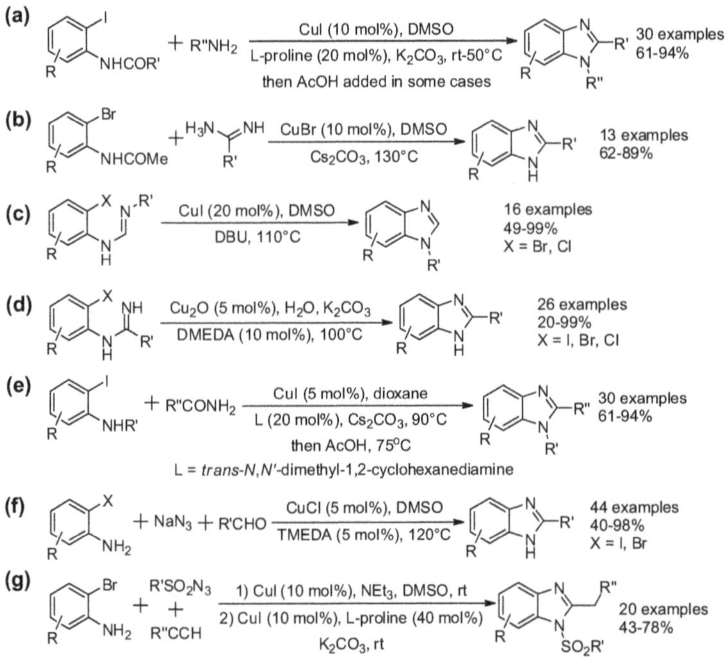

Scheme 2.132 Copper-catalyzed synthesis of benzimidazoles.

Reaction Procedure (Scheme 2.132a): A Schlenk tube was charged with aryl iodide (0.5 mmol), CuI (0.05 mmol), L-proline (0.1 mmol), and K_2CO_3 (1.0 mmol), evacuated and backfilled with argon. Amine (0.75 mmol) and DMSO (1 mL) were successively added. Then the reaction mixture was stirred at room temperature until the aryl iodide disappeared, monitored by TLC. The mixture was partitioned between ethyl acetate and saturated NH_4Cl, the organic layer was washed with saturated $NaHCO_3$ and brine, dried over Na_2SO_4, and concentrated *in vacuo*. The residue was purified by column chromatography on silica gel to provide the desired product.

(Scheme 2.132b): A flask was charged with CuBr (0.1 mmol), Cs_2CO_3 (3 mmol) in 2 mL of DMSO, and *o*-haloacetanilide (1 mmol) and amidine hydrochloride (1.2 mmol) were added to the flask at room temperature under a nitrogen atmosphere. The mixture was stirred at 60–90 °C under a nitrogen atmosphere. After the coupling reaction, the resulting solution was raised to 120 °C, and the reaction was maintained for 48 h. The reaction system was cooled to room temperature, and then the workup was performed for synthesis of 2-substituted benzimidazoles. The cooled mixture was partitioned between water and ethyl acetate. The organic layer was separated, and the aqueous layer was extracted with ethyl acetate. The combined organic layers were dried over Na_2SO_4, and concentrated *in vacuo*. The residue was purified by column chromatography on silica gel with petroleum ether–ethyl acetate. For synthesis of disubstituted benzimidazoles, aryl iodide (2 mmol) and 3,4,7,8-tetramethyl-1,10-phenanthroline (0.2 mmol) were added to the resulting solution, and the reaction was carried out at 120 or 130 °C for 36 h under a nitrogen atmosphere.

(Scheme 2.132c): Formamidine (0.50 mmol, 1 equiv.) was dissolved in 2 mL of DMSO. CuI (0.1 mmol, 20 mol%) and DBU (1.0 mmol, 2 equiv.) were added and the reaction was stirred for 1 h and 20 min at 110 °C. H_2O (20 mL) and EtOAc (20 mL) were added and the layers were separated. The aqueous layer was extracted with EtOAc (2×20 mL), and the combined organic layers were treated with brine and dried over Na_2SO_4. Purification by column chromatography (*n*-pentane–EtOAc, 3 : 1) yielded the benzimidazoles.

(Scheme 2.132d): A 10 mL Schlenk tube equipped with a magnetic stir bar was charged with (*o*-haloaryl)benzamidine substrates (1.0 mmol, 1.0 equiv.), Cu_2O (0.05 mmol), DMEDA (0.1 mmol), and K_2CO_3 (2.0 mmol, 2.0 equiv.), and then 1.5 mL of H_2O was added *via* syringe at room temperature. The tube was sealed and put into a preheated oil bath at 100 °C for 30 h. The reaction mixture was cooled to room temperature, quenched with water (3 mL), and diluted with ethyl acetate (5 mL). The layers were separated, and the aqueous layer was extracted with (2×5 mL) ethyl acetate. The combined organic extracts were dried over anhydrous sodium sulfate, filtered, and concentrated *in vacuo*. The crude product was then purified by flash chromatography on silica gel, eluting with 5–10% ethyl acetate–petroleum ether.

(Scheme 2.132f): CuCl (0.1 mmol), 2-iodoaniline (2.0 mmol), NaN_3 (4.0 mmol), TMEDA (0.1 mmol), and aldehydes (2.4 mmol) were reacted in 6.0 mL of DMSO. The reaction mixture was heated to 120 °C for 12 h. After cooling, the mixture was poured into the EtOAc (50.0 mL), washed with brine (25.0 mL) and water (2×25.0 mL), dried over $MgSO_4$, and passed through a Celite. Evaporation of the solvent under reduced pressure provided the crude product, which was purified by column chromatography (hexane–EtOAc, 1 : 1) to afford the final product.

amidine hydrochlorides were found to react with 2-haloacetanilides to give benzimidazoles as well.[255] The protocol uses 10 mol% CuBr as the catalyst, Cs_2CO_3 as the base, and DMSO as the solvent; no ligand was required. The procedure proceeds *via* the sequential coupling of *o*-haloacetoanilide derivatives with amidines, hydrolysis of the intermediates (amides), and intramolecular cyclization with the loss of NH_3 to give 2-substituted 1*H*-benzimidazoles (Scheme 2.132b). Zhou and co-workers found that the reaction between 2-haloaniline and amidine hydrochlorides can provide the same products as well.[256] In this case, Cu_2O was used as a catalyst and DMEDA (*N*,*N*′-dimethylethane) as a ligand, with Cs_2CO_3 as a base in DMF and the reaction was heated to 140 °C. In 2009, Glorius and co-workers reported a copper-catalyzed cyclization of 2-bromoarylamidines to substituted benzimidazoles (Scheme 2.132c).[257] This procedure was studied further with CuO nanoparticles as a catalyst under ligand-free conditions.[258] Additionally, this catalyst was explored in the cyclization of *o*-bromoarylguanidine to 2-aminobenzimidazoles; *o*-bromoarylthiourea to 2-aminobenzothiazoles and *o*-bromoanilide to 2-arylbenzoxazoles. An intermolecular version based on 2-bromoaniline and imidoyl chlorides was reported as well.[259] In the presence of CuI/1,10-Phen, with Cs_2CO_3 as a base, NMP as a solvent, at 100 °C, the transformation proceeded smoothly. In 2011, a copper-catalyzed cyclization of *N*-(2-haloaryl)benzamidines was described (Scheme 2.132d).[260] With Cu_2O (5 mol%) as the catalyst, DMEDA (10 mol%) as the ligand, and K_2CO_3 as the base, benzimidazoles were formed in high yields. Remarkably, the reaction was exclusively carried out in water as the solvent. In 2007, Buchwald and Zheng reported a copper-catalyzed procedure for the preparation of *N*-alkylbenzimidazoles in regioisomerically pure form starting from *o*-haloanilines.[261] The method utilizing CuI and *trans*-*N*,*N*′-dimethyl-1,2-cyclohexanediamine allows the preparation of *N*-alkylbenzimidazoles in good to excellent yields (Scheme 2.132e). Recently, the usage of 2-iodoanilines and benzamides to the corresponding 2-substituted benzimidazoles was developed.[262] Based on the CuI/DMEDA-catalyzed coupling reaction and post-cyclization with glacial acetic acid, a variety of benzimidazoles were isolated in moderate to good yields in short reaction times. In 2011, Lee's group found that benzimidazoles could be synthesized by the copper-catalyzed, one-pot, three-component reaction of 2-haloanilines, aldehydes, and NaN_3.[263] The reaction was optimized when 2-iodo- or 2-bromoanilines (1.0 equiv.), aldehydes (1.2 equiv.), NaN_3 (2.0 equiv.), 5 mol% of CuCl, and 5 mol% of TMEDA were reacted in DMSO at 120 °C for 12 h. Good yields resulted, and the reaction showed tolerance toward functional groups such as ester, nitro, and chloro. Aliphatic and heteroaromatic aldehydes also afforded the desired products in moderate to good yields (Scheme 2.132f). The reaction was proposed to start with condensation of 2-haloaniline and aldehyde affording 2-halobenzimine. Then copper-catalyzed coupling of 2-halobenzimine with azide gave the final product. The reaction between sulfonyl azides, alkynes, and 2-bromoanilines with CuI as a catalyst was reported to give benzimidazole derivatives as well.[264] Moderate to good yields of the desired products were produced (Scheme 2.132g). The reaction was proposed to start with CuI/NEt_3-catalyzed reaction of sulfonyl azide with terminal alkyne to form the ketenimine species, which was quickly attacked by 2-bromoaniline to generate *N*-sulfonylamidine and then sulfonamide as the intermediate.

Methodologies based on other substrates were developed as well. In 2009, a novel, efficient, and versatile strategy for the synthesis of various 2-hetero-benzimidazoles was developed by Bao and Lv.[265] A number of *N*-substituted 2-aminobenzimidazoles (or 2-aminopyridoimidazoles), 2-imidazylbenzimid-azoles, and 2-phenoxylbenzimidazoles, were smoothly synthesized from *o*-haloarylcarbodiimides and *N*- or *O*-nucleophiles *via* the Cu-catalyzed one-pot addition/C–N coupling process (Scheme 2.133a). Xi and co-workers found that 2-aminobenzimidazoles could be synthesized *via* copper-catalyzed cascade addition and cyclization of *o*-haloanilines and carbodiimides.[266] A variety of 2-animobenzimidazole derivatives were prepared in good yields (Scheme 2.133b). Here, *o*-haloanilines could be *o*-iodoaniline, *o*-bromoaniline, and *o*-chloroaniline derivatives. Carbodiimides could be symmetrical and unsymmetrical substrates with aryl or alkyl substituents. The reaction exhibited a good regioselectivity when unsymmetrical carbodiimides were employed. When bis-(*o*-haloaryl)carbodiimides were applied as starting materials, more complicated products could be produced depending on the nucleophiles used (Scheme 2.133c).[267,268] More recently, transition metal-free systems were developed as well. Based on radical chemistry, the corresponding benzimidazoles were produced in good yields.[269]

In 2010, Bao and Shen reported a novel and efficient procedure for the synthesis of benzoxazole and benzimidazole derivatives.[270] The one-pot cascade process is based on CuI under ligand-free conditions. A variety of carbodiimides coupled with *o*-halophenols or *o*-haloanilines to give the products in moderate to excellent yields under the mild conditions (Scheme 2.134).

Batey's group reported a general method for the formation of benzoxazoles *via* a copper-catalyzed cyclization of *ortho*-haloanilides in 2006 (Scheme 2.135).[271–273] This approach complements the more commonly

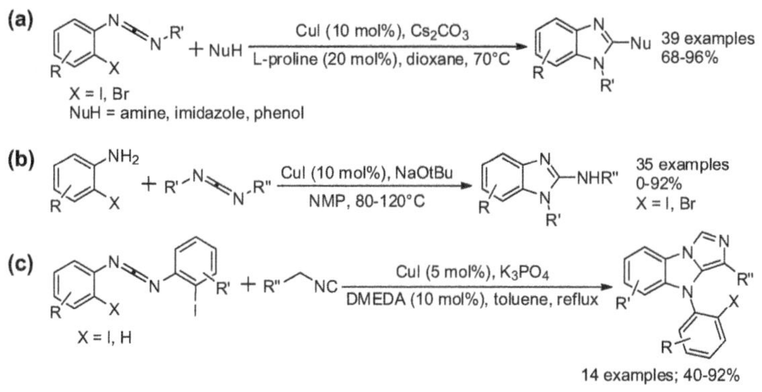

Scheme 2.133 Copper-catalyzed cyclization of carbodiimides.

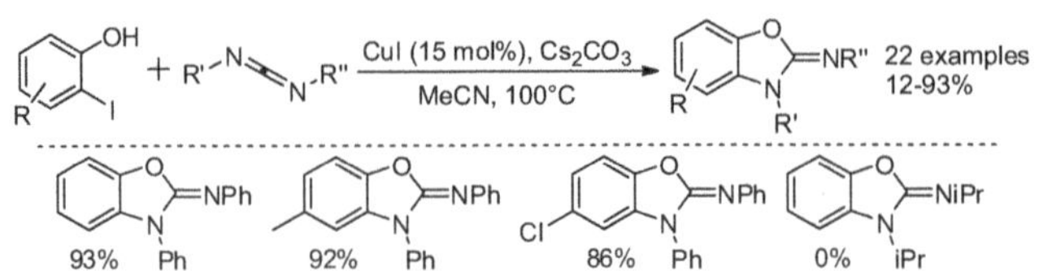

Scheme 2.134 Copper-catalyzed synthesis of benzoxazoles from carbodiimides.

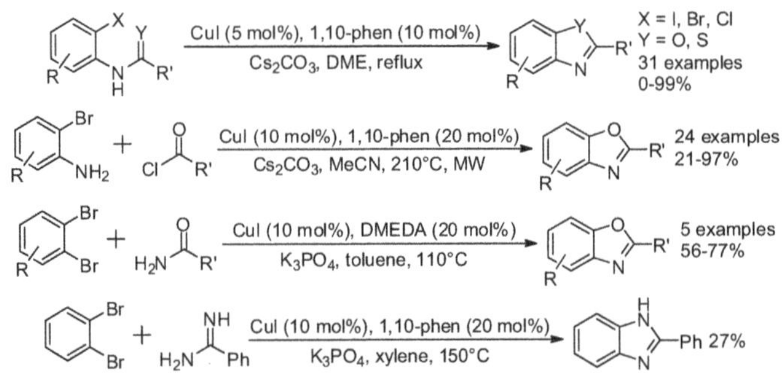

Scheme 2.135 Copper-catalyzed synthesis of benzoxazoles from *ortho*-haloanilides.

used strategies for benzoxazole formation which require 2-aminophenols as substrates. The reaction involves an intramolecular C–O cross-coupling of the *ortho*-haloanilides and was believed to proceed *via* an oxidative insertion–reductive elimination pathway through a Cu(I)/Cu(III) manifold. The reaction is also applicable to the formation of benzothiazoles. A variety of ligands including 1,10-phenanthroline and *N*,*N*′-dimethylethylenediamine were shown to provide ligand acceleration/stabilization in the reaction. Optimal conditions for cyclization consisted of a catalyst combination of CuI and 1,10-phenanthroline (10 mol%). The method was amenable to a parallel-synthesis approach, as demonstrated by the synthesis of a library of benzoxazoles and benzothiazoles substituted at various positions in the ring. Most examples utilized the cyclization of *ortho*-bromoanilides, but *ortho*-iodoanilides and *ortho*-chloroanilides also undergo a reaction under these conditions. The rate of reaction of the *ortho*-haloanilides follows the order I > Br > Cl, consistent with oxidative addition being the rate-determining step. Later on, the same group reported two alternative domino annulation approaches for benzoxazole synthesis. In the first approach, copper-catalyzed intermolecular cross-coupling of 1,2-dihaloarenes with primary amides initially forms the Ar–N bond of the benzoxazole ring, followed by copper-catalyzed intramolecular cyclization to form the Ar–O bond. Benzoxazoles were formed in good yields for the reaction of 1,2-dibromobenzene, but the reaction was not regioselective for the reaction of 3,4-dibromotoluene. Furthermore, the method is limited by the availability of 1,2-dihaloarenes. As a result of these limitations, another alternative more versatile one-pot domino annulation strategy was developed involving the reaction of 2-bromoanilines with acyl chlorides in the presence of Cs_2CO_3, catalytic CuI, and the non-acylatable ligand 1,10-phenanthroline. Under these conditions, initial acylation of the aniline was followed by copper-catalyzed intramolecular cyclization of the resultant 2-haloanilide to form the Ar–O bond of the benzoxazole ring. Optimized conditions using microwave irradiation achieved much shorter reaction times than conventional heating (*i.e.*, 210 °C for 15 min *versus* 95 °C for 24 h) and were applied to the synthesis of a small library of benzoxazoles. These copper-catalyzed approaches complement existing strategies for benzoxazole synthesis, which typically utilize 2-aminopheonls as precursors. The copper-catalyzed cyclization of *ortho*-haloanilides can be performed in water as well.[274] With CuCl as a catalyst and TMEDA as a ligand and base, the desired benzoxazoles were produced in moderate yields in water. Additionally, heterogeneous copper fluorapatite catalyst was reported for benzoxazole synthesis as well.[275] A variety of *ortho*-haloanilides (iodo-, bromo- and chloroanilides) were cyclized to the corresponding benzoxazoles, demonstrating the versatility of the reaction. Various other heterogeneous copper-supported catalysts such as Cu-HAP (copper hydroxyapatite), Cu/NaY, Cu/SiO_2, Cu/Al_2O_3, and Cu/TiO_2 were screened and their efficiency determined to be Cu-FAP > Cu-HAP > Cu/Al_2O_3 > Cu/TiO_2 > Cu/NaY > Cu/SiO_2. However, the recyclability was not demonstrated. More recently, CuO nanoparticles were explored in the cyclization of *o*-haloanilides in water.[276] *o*-Halo (I, Br) phenylbenzamides in the presence of CuO nanoparticles and Cs_2CO_3 in water at 100 °C provided *o*-hydroxyphenyl benzamides as the major product. However, a complete change in selectivity was observed in the presence of an organic base/ligand (TMEDA), giving 2-arylbenzoxazole as the exclusive product. These selectivities were not clearly distinct when the corresponding alkylamides were treated either in the presence or absence of the ligand. A number of *o*-halophenyl alkylamides provided either exclusively *o*-dehalogenated products or a mixture of *o*-dehalogenated and *o*-hydroxylated products, but none gave 2-alkylbenzoxazoles. The recyclability of the catalyst was studied; an 89% yield could be observed at the fifth run.

Substituted benzothiazoles are an important class of heterocyclic compounds that exhibit a wide range of biological properties such as inhibitors of stearoyl-coenzyme A δ-9 desaturase, antitumor, antimicrobial, LTD_4 receptor antagonist, *etc.* Among the transition metal catalysts, palladium and copper catalysts have been extensively investigated. In 1987, a palladium-catalyzed reaction between 2-iodoanilines and thioamides to 2-alkylbenzothiazoles was developed.[277] The reaction started with a palladium-catalyzed arylation of thioamides at the sulfur atom followed by intramolecular transamination to give the final product. Excellent yields of the desired products were produced (Scheme 2.136).

Due to the importance of 2-aminobenzothiazoles, Li and co-workers found a procedure by using 2-haloaniline derivatives and tetraalkylthiuram disulfides (TMTD) as substrates.[278] In the presence of CuBr and Cs_2CO_3, a variety of 2-haloanilines underwent the reaction with tetramethylthiuram disulfide efficiently to afford the corresponding 2-aminobenzothiazoles in

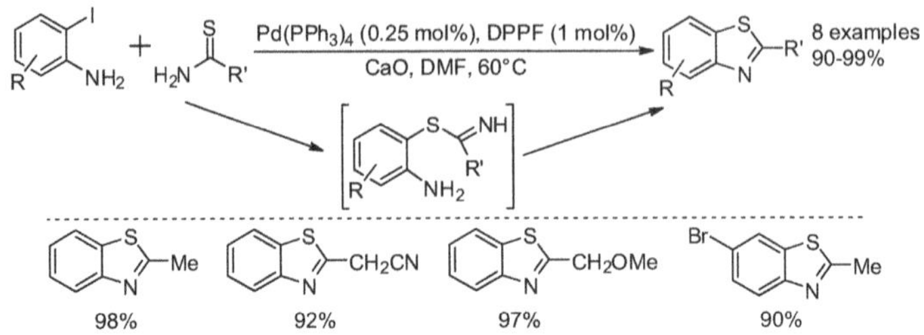

Scheme 2.136 Pd-catalyzed synthesis of benzothiazoles from thioamides.

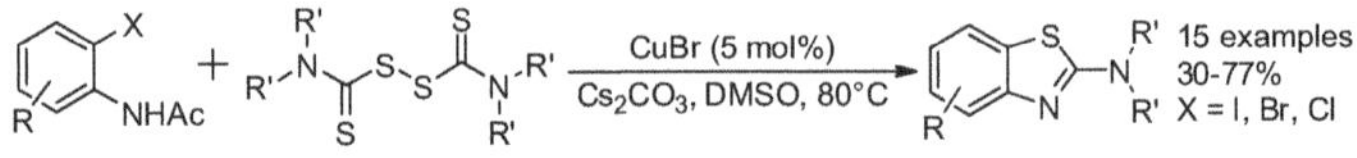

Scheme 2.137 Cu-catalyzed synthesis of 2-aminobenzothiazoles from TMTD.

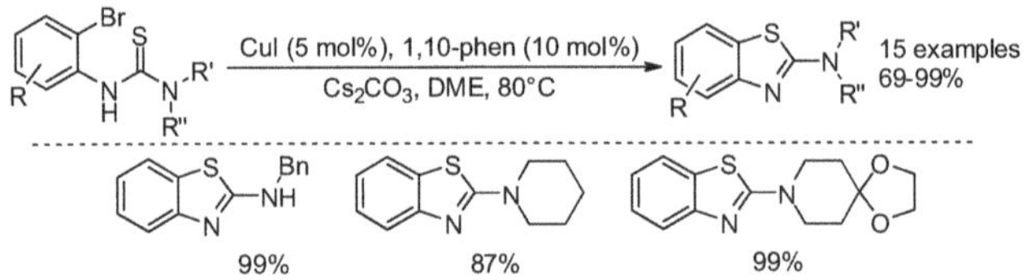

Scheme 2.138 Cu-catalyzed synthesis of 2-aminobenzothiazoles from thioureas.

moderate to excellent yields (Scheme 2.137). The present process allows the construction of 2-aminobenzothiazoles from a wide range of 2-haloaniline derivatives, including 2-iodoanilines, 2-bromoanilines, and 2-chloroanilines.

The most direct reaction pathway for benzothiazole synthesis is using the corresponding *o*-haloarylthiamide analogues as starting materials. In 2003, a palladium-catalyzed cyclization of *o*-bromophenylthioureas and *o*-bromophenylthiamides to 2-amino-, and 2-alkyl-benzothiazoles was developed.[279] By applying $Pd_2(dba)_3$/JohnPhos as the catalyst system, moderate to good yields of the products were formed in 1,4-dioxane with Cs_2CO_3 as a base. In 2004, a copper-catalyzed transformation was reported.[280] Excellent yields were achieved (Scheme 2.138). A palladium-catalyzed system using *N*-arylthioureas *via* C–H activation was developed by the same group. Here, MnO_2 was used as an additive and O_2 as an oxidant, with $Pd(PPh_3)_4$ as the precursor in MeCN.

More recently, a procedure for the synthesis of 2-substituted benzothiazoles *via* cyclization of *o*-iodothiobenzanilides was developed.[281] The reaction uses inexpensive, readily available Pd/C as the catalyst without the addition of a ligand or additive, and works at room temperature in DMF. A variety of 2-aryl-, 2-alkyl-, and 2-aminobenzothiazoles were prepared in good yields. However, the recycling of this Pd/C catalyst was not tested by the authors. In 2007, Itoh and Mase demonstrated the synthesis of benzothiazoles from 2-bromoanilides and thiol by using $Pd_2(dba)_3$/Xantphos as the

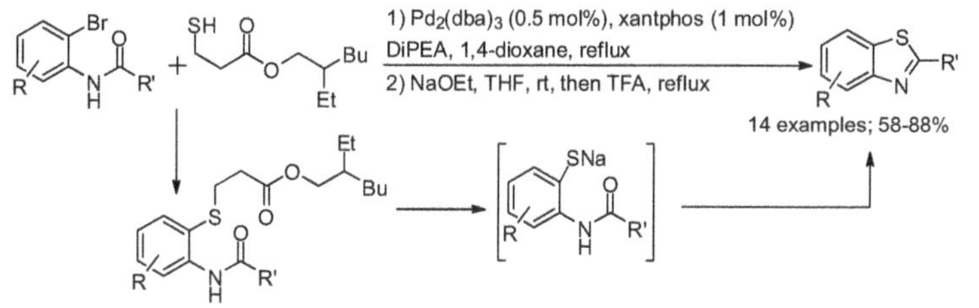

Scheme 2.139 Pd-catalyzed synthesis of benzothiazoles from thiol.

Reaction Procedure (Scheme 2.139): To a round-bottomed flask were added *ortho*-bromobenzanilide (1 mmol), *i*-Pr_2NEt (2 mmol) and 1,4-dioxane (5 mL). The mixture was evacuated and backfilled with nitrogen (3 cycles). Catalyst $Pd_2(dba)_3$ (0.005 mmol), Xantphos (0.01 mmol) and 2-ethylhexyl 3-mercaptopropionate (1 mmol) were added and then the batch was degassed twice more. The reaction was gently refluxed for 12 h and then allowed to cool to room temperature after checking consumption of starting material by TLC analysis. The mixture was filtered through a glass paper filter and washed with $CHCl_3$. The filtrate was concentrated to dryness and to the residue was added THF (5 mL) and 20% NaOEt in EtOH (2 mmol) then the mixture was aged for 1 h at 50 °C. After checking the consumption of sulfide starting material by TLC analysis, TFA (5 mmol) was carefully added at 0 °C to the mixture, which was then aged for 4 h at room temperature (or at reflux). After checking the consumption of thiol starting material by TLC analysis, to the slurry were added EtOAc and 10 wt% NaCl in water then the organic layer was separated and was concentrated to dryness *in vacuo*. The residue was purified by column chromatography to afford the benzothiazole.

catalyst system.[282] As the authors stated, this was the first report of the use of thiol surrogates in the synthetic approach for the benzothiazole. The various 2-bromoanilides were reacted with an alkyl thiolate in high yields using a palladium catalyst. The resulting sulfides were easily converted to the corresponding benzothiazoles *via* the simultaneous generation of thiols and condensation under basic or acidic conditions (Scheme 2.139).

Iron salts, which have the advantage of being cheap, environmentally benign, less toxic and so on, were applied in 2-aminobenzothiazole synthesis without exception.

2-Iodoanilines and isothiocyanates were applied as substrates, iron(III) halides (FeF_3 or $FeCl_3$) as the catalyst, and the desired products were formed in excellent yields.[283,284] Later on, the same reaction was performed under metal-free conditions.[285] In the presence of base (DABCO or Cs_2CO_3), the reactions proceeded smoothly. We believe the reaction either undergoes a base-mediated SET (single-electron transfer) mechanism or the impurities in the base or iron salt catalyze the reaction.

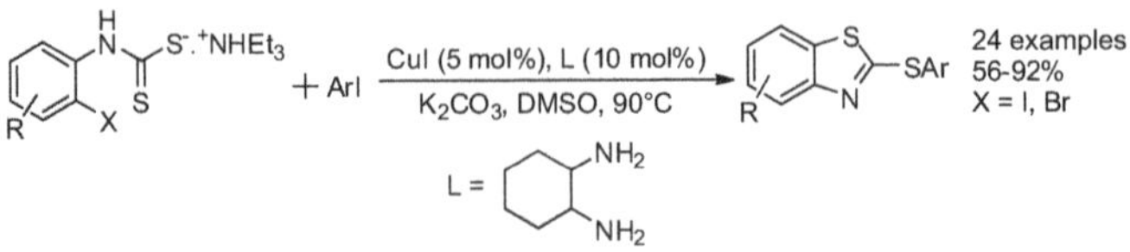

Scheme 2.140 Cu-catalyzed synthesis of arylthiobenzothiazoles.

Reaction Procedure (Scheme 2.140): An oven-dried flask was charged with CuI (5 mol%), cyclohexane-1,2-diamine (10 mol%), dithiocarbamate (1 mmol), iodobenzene (1.1 equiv.), K_2CO_3 (3 equiv.), and solvent DMSO (1 mL). The flask was kept in a pre-heated oil bath at 90 °C. Heating was continued for 4 h after which the reaction mixture was cooled and admixed with water (5 mL). The product was extracted with ethyl acetate (2×10 mL), the organic layer was dried over Na_2SO_4, concentrated under reduced pressure, and purified over a column of silica gel to give the product.

In 2009, Patel and co-workers developed two ligand-assisted Cu(I)-catalyzed sequential intra- and intermolecular *S*-arylations for the synthesis of arylthiobenzothiazoles in a one-pot manner without an inert atmosphere (Scheme 2.140).[286] Low catalyst loading, inexpensive metal catalyst and ligand, lower reaction temperature, and shorter reaction times make this method superior to all reported methods for the synthesis of arylthiobenzothiazole. In detail, the reaction started with an intramolecular *S*-arylation of a dithiocarbamate salt and yielded benzothiazol-2-thiol or MBT (2-mercaptobenzothiazole), which was then followed by an intermolecular C–S coupling giving 2-arylthiobenzothiazoles directly. In the communication, the authors found the following order of reactivity in aryl iodides *p*-NO_2 > *o*-NO_2 > *o*-COOMe > *m*-Cl > H > *p*-Me > *p*-OMe.

Sekar and co-workers found that by combining $CuCl_2$ with 1,1′-binaphthyl-2,2′-diamine in MeCN with Cs_2CO_3 as a base, *N*-(2-chlorophenyl)-benzothioamides can be cyclized at 82–110 °C.[287] Moderate to excellent yields of benzothiazoles were isolated. Using a copper catalyst, *N*-benzothiazol-2-yl-amides can be produced by cyclization of 1-acyl-3-(2-bromophenyl)-thioureas. Dihydroimidazo-, dihydropyrimido-, tetrahydrodiazepino–thiazoles, and -benzothiazoles can be produced as well.[288]

A concise method to synthesize *N*-substituted-2-aminobenzothiazoles under ligand-free copper(I)-catalyzed conditions was developed by Bao and co-workers in 2009.[289] The advantages of this procedure are that it is a "one-pot" operation, there are readily available starting materials, it is ligand-free, and is carried out under relatively mild conditions. Various *N*-substituted-2-aminobenzothiazoles were conveniently synthesized in moderate to excellent yields (Scheme 2.141a). In the same period, Patel and co-workers developed another efficient cascade process for the synthesis of a library of

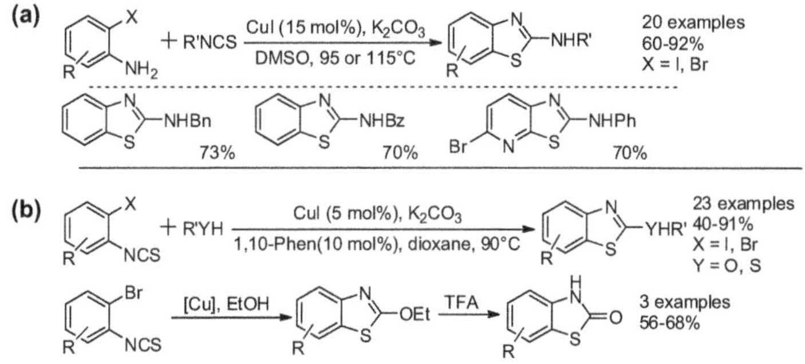

Scheme 2.141 Cu-catalyzed synthesis of benzothiazoles from isothiocyanates.

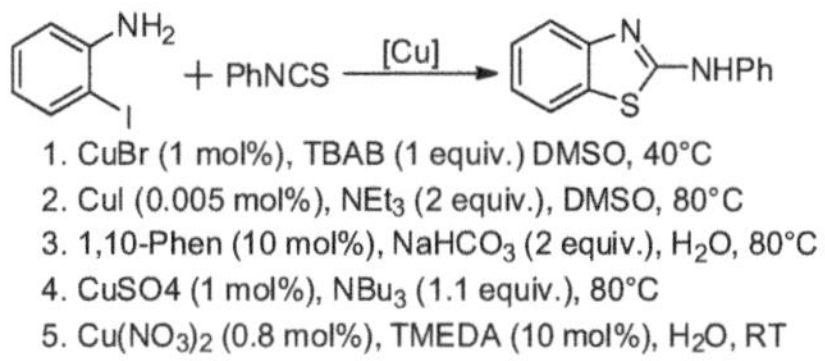

Figure 2.2 Procedures for Cu-catalyzed benzothiazole synthesis.

2-substituted 1,3-benzothiazoles.[290] The thiocarbamate or dithiocarbamate generated *in situ* by the reaction of 2-haloaryl isothiocyanates with *O* or *S* nucleophiles undergoes CuI/1,10-Phen-catalyzed intramolecular C–S bond formation giving substituted benzothiazoles (Scheme 2.141b). Both phenols and thiophenols reacted with equal ease. On the other hand, alcohols and thiols were less reactive. The rate of the reaction was faster and gave better yields when electron-withdrawing substituents were present in either of the coupling partners. 1,3-Benzothiazolones were prepared in one pot using ethanol as the solvent and nucleophile. This transformation was realized later on by using CuO nanoparticles as the catalyst in water.[291] The biggest advantage of this procedure is the recyclability of the catalyst system. No obvious decrease in activity was observed after five runs.

Recently, several alternative catalytic systems based on copper catalysts were developed (Figure 2.2), such as the CuBr/TBAB-base-free system,[292] PPM loading of CuI system,[293] 1,10-Phen-catalyzed system,[294] $CuSO_4$-solvent-free system,[295] $Cu(NO_3)_2 \cdot 3H_2O$/TMEDA in water system,[296] and supported copper systems.[297–299] From our understanding, because of the high nucleophilicity of sulphur, the role of the copper catalyst is not so obvious, especially in cases using aryl iodides as substrates.

By replacing isothiocyanates with isoselenocyanates, 2-aminobenzoselenazoles could be selectively produced. In 2011, a convenient and successful one-pot preparation of 2-aminobenzoselenazoles *via* phenylselenoureas was developed by Sashida and co-workers.[300] By using $Cu(OTf)_2$ as the catalyst under ligand-free conditions, 2-iodoanilines and isoselenocyanates were reacted to provide the desired products in good yields (Scheme 2.142).

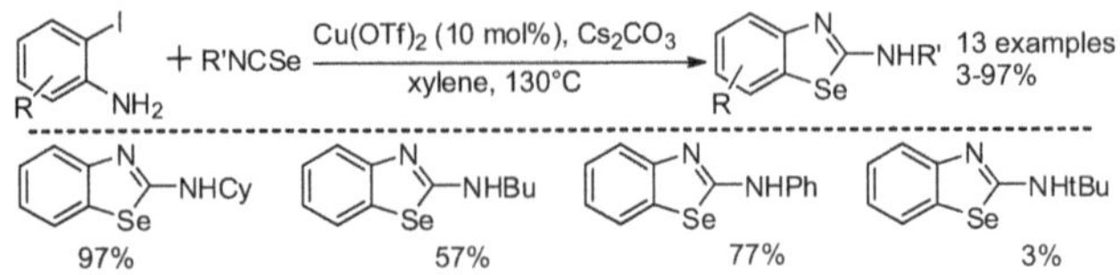

Scheme 2.142 Cu-catalyzed synthesis of benzoselenazoles.

Notably, the intermediate selenoureas were isolated and quantitatively transformed into the selenazoles with a copper catalyst.

Regarding the importance of benzothiazoles, alternative procedures were developed as well. In 2009, Ma, Jiang and co-workers developed a novel method for the synthesis of substituted benzothiazoles, which relies on an unprecedented CuI-catalyzed coupling reaction of aryl halides with metal sulfides.[301] This approach has definite advantages in terms of the scope and the conditions of the reaction. By using 2-haloanilides as substrates, various benzothiazoles were produced in moderate to good yields (Scheme 2.143a). Under the same reaction conditions, benzothiazolones could be produced by using ethyl 2-iodophenylcarbamates and sodium sulfide as starting materials. More recently, a copper-catalyzed procedure based on using 1-iodo-2-nitrobenzenes and aldehydes was developed (Scheme 2.143b).[302] This new method is best performed in acetic acid at 100 °C with copper(I) iodide and 1,10-phenanthroline as the catalyst system. The reaction appears to tolerate a wide range of functional groups on both the 1-iodo-2-nitrobenzene and the aldehyde. The overall efficiency of the reaction might help it find numerous applications in organic synthesis. Regarding the reaction mechanism, the authors propose that the sequence begins with the reaction of the 1-iodo-2-nitrobenzene with hydrogen sulfide in the presence of the copper catalyst to form the corresponding 2,2′-disulfanediyldianiline. This then undergoes condensation with the benzaldehyde to form the corresponding imine, which subsequently undergoes cyclization and oxidation to form the desired benzothiazole. More recently, a new, highly efficient procedure for the synthesis of benzothiazoles from readily available *N*-benzyl-2-iodoaniline and potassium sulfide was developed.[303] The results showed copper-catalyzed double C–S bond formation *via* a traditional cross-coupling reaction and an oxidative cross-coupling reaction. Moderate to excellent yields were achieved (Scheme 2.143c).

Lee and co-workers developed a copper-based catalytic system for the one-pot syntheses of benzothiazoles from 2-iodoanilines, aldehydes, and $NaSH \cdot nH_2O$ (Scheme 2.144a).[304] This one-pot reaction system represents a significant advantage in that it does not require the isolation of the intermediate prior to execution of C–S bond formation and in most instances does not require the preparation of the starting materials. As the authors stated, this was the first example of the employment of $NaSH \cdot nH_2O$ as a sulfur source, not only in the synthesis of the benzothiazole, but also in transition-metal-catalyzed C–S bond formation. In addition, the

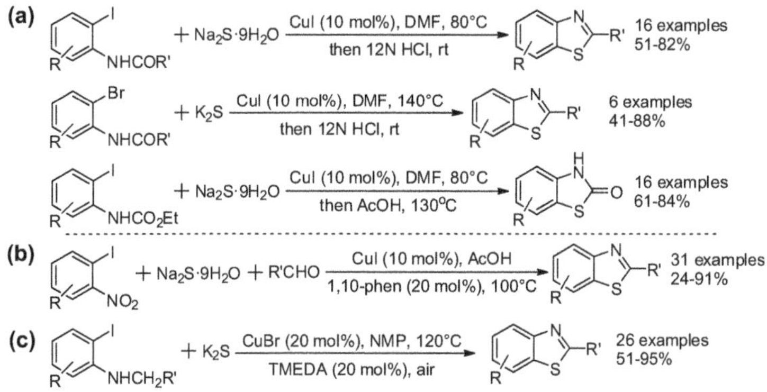

Scheme 2.143 Cu-catalyzed synthesis of benzothiazoles with $Na_2S \cdot 9H_2O$.

Reaction Procedure (Scheme 2.143a): A mixture of CuI (0.1 mmol), *o*-iodobenzamide (1 mmol), and $Na_2S \cdot 9H_2O$ (or K_2S) (3 mmol) in DMF (2 mL) was stirred at 80 °C for 12 h. The reaction mixture was cooled to rt and then 0.8 mL of conc. HCl was added and the reaction mixture stirred for 5–10 h. After adding 10 mL sat. aq. $NaHCO_3$, it was extracted with ethyl acetate and then purified by silica gel chromatography to furnish the desired product.

(Scheme 2.143b): A 25 mL tube was charged with a 1-iodo-2-nitrobenzene (1.0 mmol), an aldehyde (0.5 mmol), $Na_2S \cdot 9H_2O$ (2.5 mmol), CuI (0.05 mmol), and 1,10-phenanthroline (0.1 mmol). Glacial AcOH (2 mL) was added, the tube was capped in air, and the mixture was stirred at 100 °C for 12 h. The mixture was then cooled to rt and diluted with EtOAc (40 mL). The resulting mixture was washed sequentially with sat. aq. $NaHCO_3$ (3×10 mL) and H_2O (2×10 mL), and the organic layer was dried (Na_2SO_4) and concentrated under reduced pressure. The residue was purified by column chromatography (silica gel, PE–EtOAc, 12 : 1 to 4 : 1).

employment of $NaSH \cdot nH_2O$ offers economic and environmental advantages, due to its low cost and odor-free nature. Compared to the cases mentioned earlier, no acidification was needed. Later on, a procedure for the synthesis of substituted benzothiazoles from aromatic ketones and anilines was developed by Wu's group in 2013.[305] In this method, six reactions were assembled in one pot. Low to moderate yields of the corresponding products were produced (Scheme 2.144b).

CS_2 as a type of liquid sulfur source was applied in benzothiazole synthesis as well. In 2011, Ma and co-workers found that copper salt can catalyse the reaction between 2-haloanilines and dithiocarbamate, which was generated *in situ* from amines and carbon disulfide (Scheme 2.145a).[306] 2-*N*-Substituted benzothiazoles were produced in good yields. Later on, they found that thiols can be applied as well. Condensation of carbon disulfide with thiols in

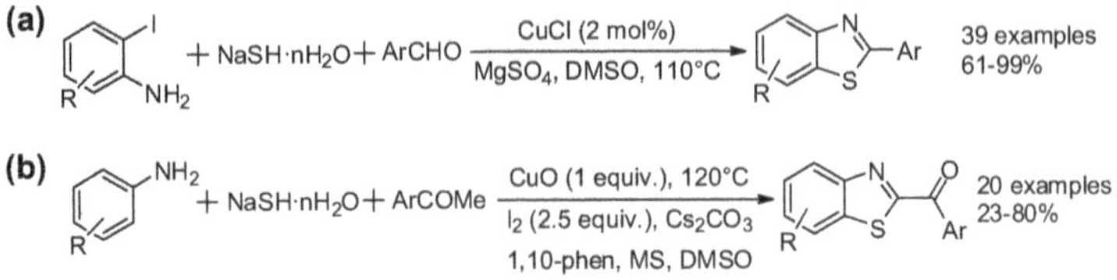

Scheme 2.144 Cu-catalyzed synthesis of benzothiazoles with NaSH · nH_2O.

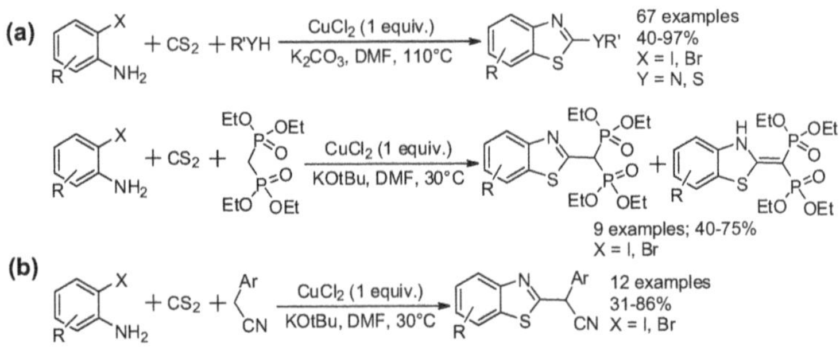

Scheme 2.145 Cu-mediated synthesis of benzothiazoles with CS_2.

the presence of K_2CO_3 generates carbonotrithioate salts *in situ*, which undergo coupling with 2-iodoanilines and subsequent intramolecular condensation and elimination with the assistance of CuBr to afford 2-thiosubstituted benzothiazoles. Both aliphatic and aromatic thiols are compatible with this process, delivering the corresponding heterocycles with good diversity. CuO nanoparticles were found to be efficient catalysts for this reaction too.[307] More recently, a straightforward route to various 2-*C*-substituted benzothiazoles using 2-haloanilines as starting materials and carbon disulfide as a thiol surrogate was developed (Scheme 2.145b).[308] Two series of medicinally useful 2-*gem*-bisphosphonate-substituted benzothiazoles and α-benzothiazol-substituted nitriles were synthesized *via* a copper-promoted domino condensation–*S*-arylation–heterocyclization process. These synthetic 2-*C*-substituted benzothiazoles bearing cyano and bisphosphonate groups allow further formation of diverse medicinal derivatives.

Xanthate was applied as a thiol surrogate in the one-pot synthesis of benzothiazoles and benzothiophenes by Sekar's group.[309] Using $Cu(OAc)_2$ (10 mol%) as the catalyst, in DMF at 80 °C, benzothiazoles can be produced by using 2-iodoanilides as substrates and benzothiophenes can be formed by applying 2-haloalkynylbenzenes as starting materials. Moreover, potassium thioacetate was studied and applied as well.[310]

All in all, the most direct and efficient pathway will use a sulfur atom as a thiol surrogate. In 2012, Zhou's group developed an efficient, simple, and environmentally friendly three-component reaction of *o*-iodoaniline, aldehydes, and sulfur powder to form benzothiazoles in a simple one-pot procedure (Scheme 2.146a).[311] In this catalytic system, neat water was used as

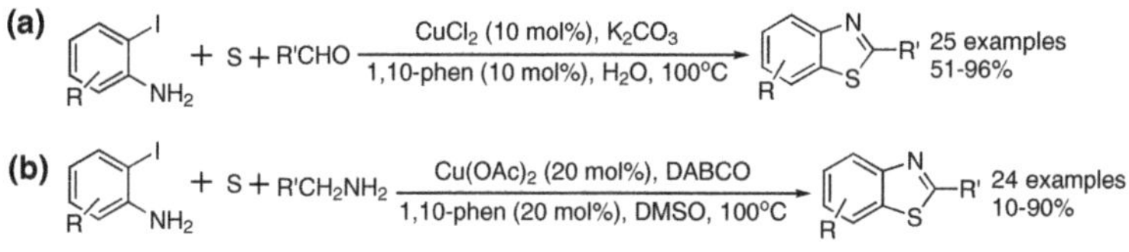

Scheme 2.146 Cu-catalyzed synthesis of benzothiazoles with S powder.

Reaction Procedure (Scheme 2.146a): 2-Iodoaniline (0.3 mmol), benzaldehyde (0.4 mmol), sulfur powder (0.9 mmol), $CuCl_2 \cdot 2H_2O$ (0.03 mmol), 1,10-Phen (0.03 mmol), and K_2CO_3 (0.6 mmol) in water (3 mL) were put into a Teflon septum screw-capped tube. The reaction mixture was stirred at 100 °C for 24 h without an inert gas atmosphere and then cooled to room temperature and extracted with ethyl acetate. The organic layer was dried over Na_2SO_4 and the solvent was removed under reduced pressure. The residue was purified by silica gel column chromatography to afford the corresponding product.

(Scheme 2.146b): A mixture of 2-iodoaniline (1 mmol), benzylamine (1.2 mmol), sulfur powder (6 mmol), DABCO (2 mmol), $Cu(OAc)_2 \cdot H_2O$ (0.02 mmol), and 1,10-phenanthroline (0.02 mmol) was stirred in DMSO (5 mL) at 100 °C. After completion of the reaction as indicated by TLC, the mixture was cooled to room temperature. Water (20 mL) was added, and the aqueous solution was extracted with ethyl acetate (3×15 mL). The organic layers were combined, dried over anhydrous $MgSO_4$, the filtrate was concentrated under vacuum and then the residue was purified by column chromatography (eluent: petroleum ether–ethyl acetate, 10 : 1 to 14 : 1) on silica gel to provide the desired product.

the solvent. The tolerance of diverse functional groups makes the present system attractive. This was the first example of the use of sulfur powder as the sulfur source in the synthesis of substituted benzothiazoles in water. Regarding the reaction mechanism, the copper-catalyzed coupling of 2-iodoaniline with sulfur powder provides a diaryl disulfide product at the first stage. Then, a nucleophilic addition–elimination reaction of diaryl disulfide with an aldehyde to provide the corresponding imines followed by intramolecular nucleophilic addition of the disulfides to the C=N bond give the target product after further oxidation. Afterwards, a procedure replacing aldehydes with amines was developed (Scheme 2.146b).[312]

2.4 1-Carbon-2-haloarenes

In 1985, Xu and Zhang reported a palladium-catalyzed synthesis of benzoxaphosphacycloalkanes *via* intramolecular cyclization.[313] By using the corresponding alkane or benzenephosphonous acid mono(σ-*o*-bromophenylalkyl)esters as substrates, moderate yields of the desired products were

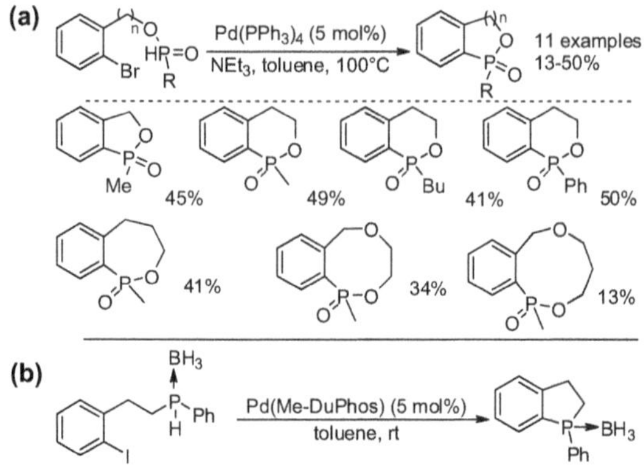

Scheme 2.147 Palladium-catalyzed C–P bond formation.

isolated (Scheme 2.147a). Twenty years later, a methodology on the enantioselective synthesis of P-stereogenic benzophospholanes *via* palladium-catalyzed intramolecular cyclization was developed.[314] Functionalized secondary phosphines or their borane adducts were prepared in an enantioselective or diastereoselective manner by using chiral Pd(diphosphine) complexes as catalysts. The desired *P*-stereogenic benzophospholanes were formed in up to 70% ee (Scheme 2.147b). These results provide a new method for the synthesis of chiral phospholanes, which are valuable ligands in asymmetric catalysis. Recently, a palladium-catalyzed synthesis of *ortho*,*ortho*′-bis(diphenylphosphino)biphenyls from *ortho*,*ortho*′-dihalobiphenyl precursors was reported. $HPPh_2$ was applied as the coupling partner and phosphafluorenes were formed as the by-product by removing one phenyl group from the phosphine atom.[315]

Benzisothiazol-3(2*H*)-ones are widely used pharmaceutical agents with fungistatic, antimicrobial, antithrombotic, and antipsychotic activities. In 2012, Xi, Chen and their co-workers developed a novel method to prepare benzisothiazol-3(2*H*)-ones based on the copper-catalyzed reaction of *o*-bromobenzamides with potassium thiocyanate (KSCN) in water.[316] The reaction proceeds *via* a tandem reaction with S–C bond and S–N bond formation. In detail, the first step involves oxidative addition of a CuI catalyst to the *o*-bromobenzamide and ligand exchange between Br^- and ^-SCN to form an aryl thiocyanate after reductive elimination, which undergoes an intramolecular nucleophilic substitution reaction to afford the target product. The ^-CN group was hydrolyzed by a base under the reaction conditions. Moderate to good yields of the desired products were produced (Scheme 2.148a). Meanwhile, a procedure using sulfur powder was developed.[317,318] In the presence of sulfur powder, copper iodide/1,10-phenanthroline or 2,2′-bipyridine and potassium carbonate as a base, the desired products were formed in excellent yields in general (Scheme 2.148b). In respect of the reaction pathway, the authors believe the reaction proceeded *via* an LCu–NR amide complex followed by insertion of sulfur into

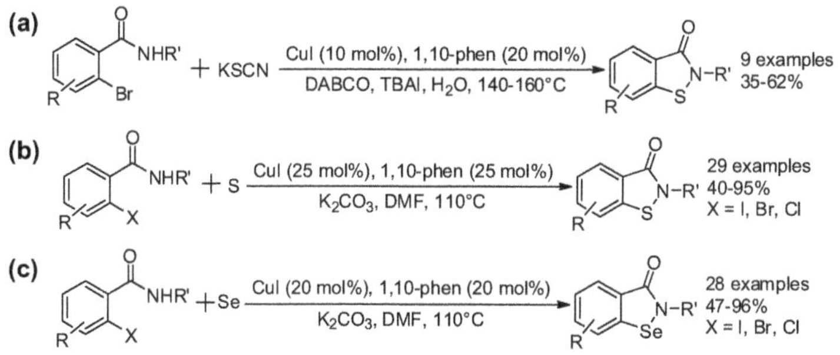

Scheme 2.148 Copper-catalyzed synthesis of benzisothiazol-3(2*H*)-ones and ebselen.

Reaction Procedure (Scheme 2.148a): A sealed tube was charged with a mixture of *o*-halobenzamide (0.5 mmol), potassium thiocyanate (1.0 mmol), CuI (0.05 mmol), 1,10-phenanthroline (0.1 mmol), DABCO (1.0 mmol), and Bu_4NI (0.1 mmol), which was then stirred in H_2O (1 mL) at room temperature under a nitrogen atmosphere. Half an hour later, the tube was sealed, and the mixture was allowed to stir at 140–160 °C for 48–72 h. After completion of the reaction, the mixture was cooled to room temperature, and then H_2O (5 mL) was added and the mixture extracted with EtOAc (5 mL×3) and dried by anhydrous Na_2SO_4. Evaporation of the solvent followed by purification on silica gel (petroleum ether–ethyl acetate, 2 : 1) provided the corresponding product.

(Scheme 2.148b): In a single neck flask (25 mL) containing DMF (5 mL), CuI (0.5 mmol) and 1,10-phenanthroline (0.5 mmol) were added and stirred for 15 min under N_2. After this, 2-iodo-*N*-benzylbenzamide (2.0 mmol), sulfur powder (2.4 mmol), and anhydrous K_2CO_3 powder (2.5 mmol) were added in the same sequence, stirred for 15 min at room temperature and then refluxed at 110 °C for 3 h under N_2. After this, the reaction mixture was poured into brine solution (50 mL), stirred for 3 h, and the reaction mixture together with brine was extracted with ethyl acetate (20 mL×3), dried over Na_2SO_4 (5.0 g), concentrated under vacuum to obtain a brown solid and then purified by column chromatography using hexane–ethyl acetate (8 : 2).

(Scheme 2.148c): Copper iodide (0.6 mmol) and 1,10-phenanthroline (0.6 mmol) were added into DMF (3 mL) in a single neck flask. The resulting brownish solution was stirred for 15 min and then 2-iodo-*N*-phenylbenzamide (3.0 mmol), selenium powder (3.6 mmol), and potassium carbonate powder (4.7 mmol) were added sequentially to the same reaction flask. The brown reaction mixture was refluxed at 110 °C using a refluxing condenser under a nitrogen atmosphere. The progress of the reaction was monitored by TLC. The reaction mixture was refluxed for 8 h. After this, the reaction mixture was poured over brine solution (60 mL) and stirred for 3 h. The product was precipitated as a white solid which was collected by filtration over a Buchner funnel. The product was washed with water (15 mL×2), dried in air, dissolved in ethyl acetate, concentrated in a rotary evaporator, resulting in a brown solid which was purified by column chromatography using hexane–ethyl acetate (8 : 2) over silica gel.

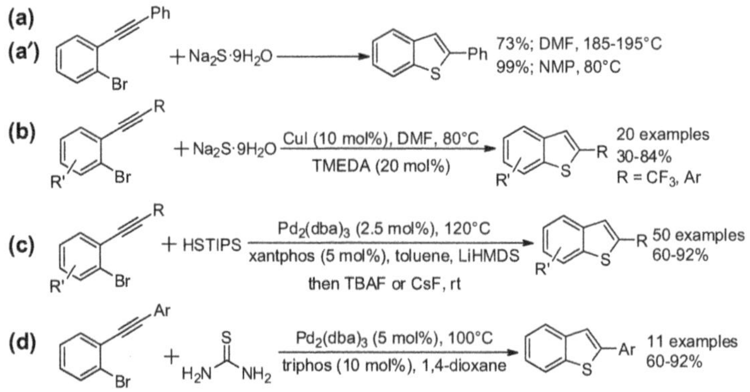

Scheme 2.149 Synthesis of benzothiophenes from *o*-bromo-ethynylbenzenes.

the LCu–NR bond. However, when 2,2′-bipyridine was used as a ligand, Ar–S bond formation was proposed to have priority. Additionally, sulfur can be replaced by selenium to produce related heterocycles. In 2010, Kumar and co-workers developed an efficient copper-catalyzed method for the synthesis of biologically important ebselen and related analogues.[319] This was the first report of a catalytic process of selenation and Se–N bond formation. Various functional groups such as amides, hydroxyls, ethers, nitro, fluorides, and chlorides are tolerated and provide the corresponding products in excellent yields (Scheme 2.148c). The best results were obtained by using a combination of potassium carbonate as a base, iodo- or bromo-arylamide substrates, selenium powder, and copper iodide as a catalyst.

Benzo[*b*]thiophenes represent an important class of fused-thiophene compounds in the field of bioactive materials as well as optoelectronic materials. In 2009, a convenient one-pot procedure for the synthesis of benzo[*b*]thiophenes and selenophenes from readily available *o*-halo-ethynylbenzene precursors was described (Scheme 2.149a).[320] Regardless of the substituent on the acetylene terminus or the number of cyclization moieties on the precursors, various benzo[*b*]thiophenes and selenophenes, including not only the parent, alkyl-, and phenyl-substituted derivatives but also benzo[1,2-*b*:4,5-*b*′]dithiophenes and diselenophenes and benzo[1,2-*b*:3,4-*b*′:5,6-*b*″]trithiophenes and triselenophenes can be prepared in good to high yields. Notably, no catalyst was needed here even though the reaction temperature was high (185–195 °C). Later on, in 2011, in a report on copper-catalyzed

C–S and C–Se bond formation between aryl iodides and $Na_2S\cdot 9H_2O$ to diaryl chalcogenides, the group mentioned one example of *o*-bromo-ethynylbenzene to benzothiophene.[321] 99% of 2-phenylbenzothiophene was isolated without a catalyst in NMP at 80 °C (Scheme 2.149a′). In the same period, a systematic study on *o*-bromo-ethynylbenzenes with $Na_2S\cdot 9H_2O$ with CuI/TMEDA as the catalyst was described.[322] Good yields of the desired products were produced (Scheme 2.149b). Palladium catalysts were applied in the annulation of *o*-halo-ethynylbenzenes as well. In 2011, Sanz and co-workers reported a palladium-catalyzed benzothiophene synthesis from *o*-bromo-ethynylbenzenes and HSTIPS (triisopropylsilanethiol) (Scheme 2.149c).[323] This novel one-pot procedure involves a Pd-catalyzed C–S bond formation using a hydrogen sulfide surrogate followed by a heterocyclization reaction. Moreover, *in situ* functionalization with selected electrophiles further expands the potential of this methodology to the preparation of the corresponding highly substituted sulfur heterocycles. Thiourea was found to be a suitable sulfur source and was applied in the cyclization of *o*-bromo-ethynylbenzenes to benzothiophenes as well.[324] By the proper combination of palladium and ligand, benzothiophenes were produced in moderate to excellent yields (Scheme 2.149d).

Xi and co-workers developed a series of copper-catalyzed double *S*-alkenylations of 1,4-diiodo-1,3-dienes.[325] Various thiophene and benzothiophene derivatives were selectively produced in good yields (Scheme 2.150). K_2S, $Na_2S\cdot 9H_2O$, and CS_2 were applied as sulfur sources for these transformations.

In 2009, Lautens and co-workers developed a general, efficient method for the catalytic synthesis of diversely functionalized benzothiophenes from *gem*-dihalovinyl thiophenols.[326] This reaction was both the first example of a

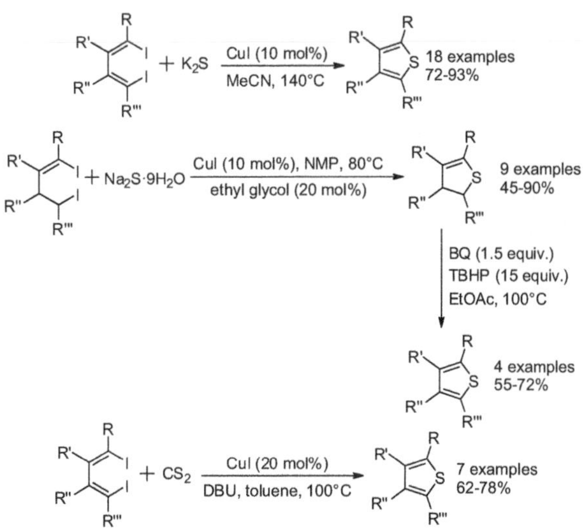

Scheme 2.150 Synthesis of thiophenes from 1,4-diiodo-1,3-dienes.

tandem catalytic process that incorporates C–S coupling, and the first example of the palladium-catalyzed vinylation of a thiol. Various nucleophiles can be applied as coupling partners and provide the corresponding benzothiophenes in good yields (Scheme 2.151a). Interestingly, this method

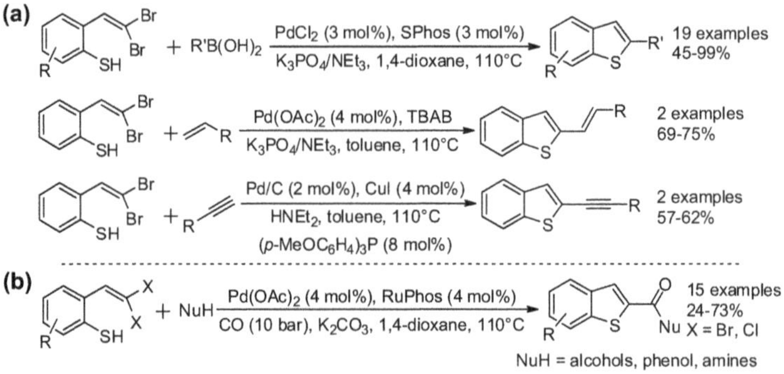

Scheme 2.151 Pd-catalyzed synthesis of benzothiophenes from *gem*-dihalovinyl thiophenols.

Reaction Procedure (Scheme 2.151a): Thiophenol (0.17 mmol), 4-fluorophenylboronic acid (0.29 mmol, 1.7 equiv.), K_3PO_4 (0.51 mmol, 3.0 equiv.), Et_3N (0.07 mL, 0.51 mmol, 3.0 equiv.), and 1,4-dioxane (2.5 mL) were placed in an oven-dried 2–5 mL Biotage microwave vial, and the vial was capped with a septum and degassed with argon for 5 min. The septum was removed, $PdCl_2$ (0.8 mg, 0.0051 mmol, 3 mol%) and SPhos (1.7 mg, 0.0051 mmol, 3 mol%) were added quickly, the septum was replaced, and the solution was degassed with argon for an additional 5 min. The septum was then replaced with a Teflon microwave cap, the vial was sealed, and the reaction mixture was stirred for 10 min and then heated in an oil bath for 16 h at 110 °C. The flask was then cooled to room temperature, and the solids were removed by vacuum filtration through a plug of silica gel and washed with EtOAc. The solution was concentrated, and the crude product was purified by column chromatography over silica gel (eluent: Et_2O–pentane, 0–2%).

(Scheme 2.151b): A mixture of substrates (0.50 mmol), base (1.50 mmol), $Pd(OAc)_2$ (0.020 mmol), Ruphos (0.020 mmol), and 6 mL of MeOH–THF (v/v = 1 : 1) were sequentially added to a 45 mL glass-lined autoclave. After sealing, the autoclave was purged three times with carbon monoxide, and then pressurized with 10 bar of CO. The resulting mixture was stirred at 110 °C for 15 h. The autoclave was cooled to room temperature, and the excess carbon monoxide was released. The reaction mixture was filtered. The filtrate was concentrated in a rotary evaporator. The residue was purified by flash chromatography on silica gel using a mixture of hexanes and ethyl acetate (7 : 1 to 4 : 1) as the eluent to afford the products.

could be extended from Suzuki–Miyaura coupling reactions to other palladium-catalyzed cross-coupling processes. The tandem C–S coupling–Heck-reaction of thiophenol with *N*-acryloylmorpholine or 1-decen-3-ol proceeded under Jeffrey-type conditions to give the corresponding product in 75 and 69% yield, respectively. The substrate was also amenable to a tandem C–S coupling–Sonogashira reaction. The treatment of thiophenol with a terminal alkyne under palladium-catalyzed conditions in the presence of CuI gave the product of the tandem reaction in moderate yield. Notably, Pd/C was applied as the catalyst here. Later on, a novel strategy for the synthesis of 2-carbonylbenzo[*b*]thiophene derivatives was developed by Alper and Zeng.[327] This procedure is based on a one-pot palladium-catalyzed intramolecular C–S coupling–intermolecular carbonylation reaction sequence. This procedure tolerates significant variation of both nucleophiles and the thiophenol backbone and provides a general, straightforward, and practical approach to benzo[*b*]thiophene-2-carboxylates and benzo[*b*]thiophene-2-carboxamides (Scheme 2.151b). Additionally, several metal-free procedures have been developed for benzothiophenes synthesis.[328] By using 2-halobenzaldehydes or 2-halobenzonitriles and thiols as substrates, the corresponding benzothiophenes were selectively formed. The reactions usually followed nucleophilic substitution and intramolecular condensation sequences.

In 1996, Buchwald and co-workers developed the first Pd-catalyzed synthesis of cyclic aryl ethers from alcohols and aryl halides.[329] Reasonably functional groups were tolerated and provided five-, six-, and seven-membered heterocycles in moderate to good yields (Scheme 2.152a). In this process, Tol-BINAP [(*S*)-(−)-2,2′-bis(di-*p*-tolylphosphino)-1,1′-binaphthyl] was found to be the best ligand with K_2CO_3 as the base. In the case using DPPF as the ligand, NaO*t*Bu was found to be the best base. In 1999, Hartwig's group reported a palladium-catalyzed C–O coupling of unactivated aryl halides with newly developed ferrocenyldi-*tert*-butylphosphine (FcP(*t*Bu)$_2$) as the ligand.[330] In this communication, one example of intramolecular cyclization to provide dihydrobenzofuran was mentioned as well (Scheme 2.152b). A detailed mechanistic study was performed too.

Cossy and co-workers developed a rapid convergent microwave-assisted procedure for the synthesis of benzoannulated spiroacetals based on a

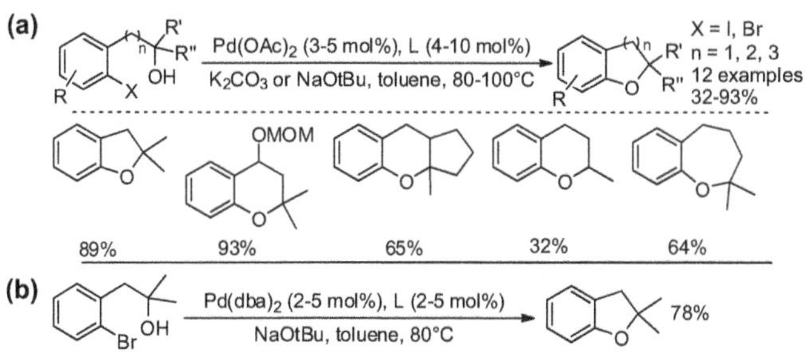

Scheme 2.152 Pd-catalyzed intramolecular C–O bond formation.

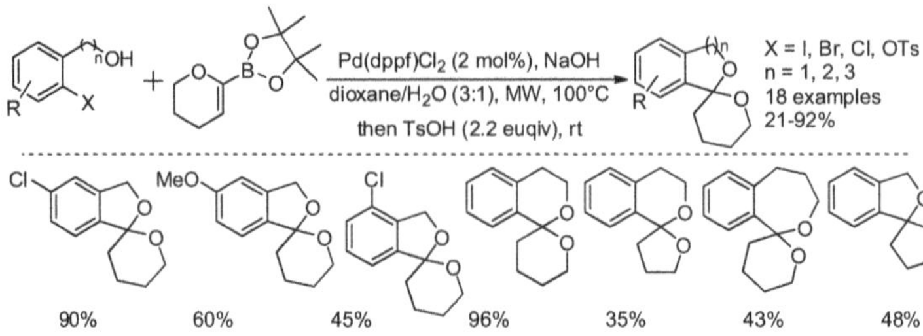

Scheme 2.153 Synthesis of benzoannulated spiroacetals.

Reaction Procedure (Scheme 2.153): To a suspension of an aryl bromide or aryl iodide (1 mmol, 1 equiv.), pinacol boronate (1.2 mmol, 1.2 equiv.) and Pd(DPPF)$Cl_2 \cdot CH_2Cl_2$ catalyst (0.02 mmol, 2 mol%) in 1,4-dioxane (1.5 mL) and 4 M aqueous NaOH (0.5 mL, 2 mmol, 2 equiv.) were added. The vial was flushed with argon and sealed. The reaction mixture was then heated under microwave irradiation at 100 °C for 10 min. After cooling to rt, the complete consumption of the starting aryl halide was checked by TLC. The vial was opened and solid $TsOH \cdot H_2O$ (2.2 mmol, 2.2 equiv.) was added in one portion. The resulting suspension was stirred at rt for 10 min (TLC control). The reaction mixture was then poured into saturated aqueous $NaHCO_3$ (50 mL), extracted with EtOAc (3×30 mL), and the combined organic layers were washed with brine and dried over Na_2SO_4. The product was isolated by flash chromatography on silica gel.

one-pot Suzuki–Miyaura coupling reaction and spiroacetalization.[331] By using (2-haloaryl)alkyl alcohols and dihydropyranyl or dihydrofuranyl pinacol boronates as substrates, substituted benzoannulated spiroacetals were prepared in moderate to excellent yields (Scheme 2.153).

A highly efficient Cu-catalyzed C–O bond-forming reaction of alcohol and aryl bromides was developed in 2009.[332] This transformation was realized through the use of copper(I) iodide as the catalyst, 8-hydroxyquinoline as the ligand, and K_3PO_4 as the base. A variety of functionalized substrates were found to react under these conditions to provide products in good to excellent yields. In the case of *ortho*-bromide aryl alcohols as starting materials, five- and six-membered cyclic ethers can be produced in good yields (Scheme 2.154a). Later on, a metal-free case was described by using CuI as the catalyst and LiO*t*Bu as the base in 1,4-dioxane (Scheme 2.154b).[333]

In 2007, Larhed and co-workers reported a palladium-catalyzed cyclization of *o*-halobenzyl cyclohexenyl ethers.[334] A number of new spiro[cyclohexane-1,1′-isobenzofuran]-based compounds were synthesized by palladium(0)-catalyzed 5-*exo* cyclization of a series of cyclohexenyl *o*-halobenzyl ethers. Controlled microwave heating was found to promote both product yield and reaction rate without compromising the selectivity. Heck cyclization of

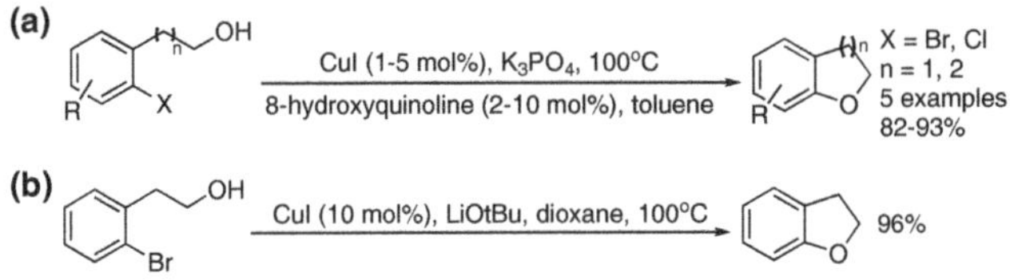

Scheme 2.154 Copper-catalyzed C–O bond formation.

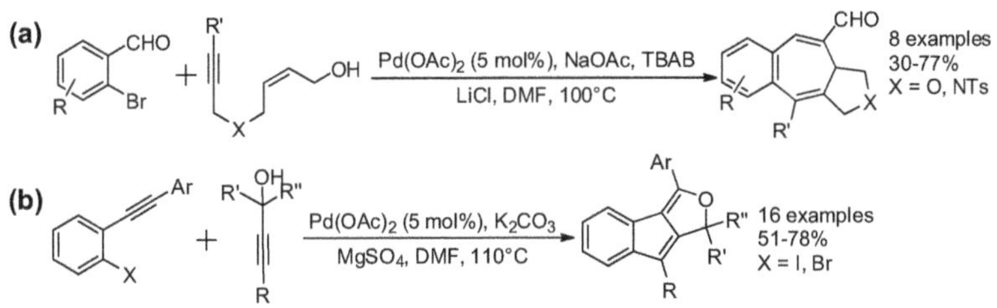

Scheme 2.155 Palladium-catalyzed synthesis of fused furans.

2-(2-iodobenzyloxy)cyclohex-2-enyl acetate proceeded selectively without involvement of the allylic acetate functionality. In 2010, Tong and Fang reported a domino sequence of cyclization of 1,6-enyne with 2-bromoarylaldehyde.[335] In the presence of 5 mol% of palladium catalyst, moderate yields of the desired products were formed (Scheme 2.155a). This sequence provides a convenient protocol to access polycyclic ring systems with readily available starting materials under simple Pd(0)-catalyzed conditions and the intramolecular aldol-condensation was proposed to be the key step for the catalytic cycle. Later in 2012, a general and efficient synthesis of indeno[1,2]furans *via* a Pd-catalyzed bicyclization reaction between 2-alkynyliodobenzenes and propargylic alcohols was described.[336] The procedure furnishes indeno[1,2]furans with moderate to excellent yields (51–78%) and a broad substrate scope (Scheme 2.155b). The cascade process combines the formation of one C–O bond and two C–C bonds in a single step.

In 1984, a study on the palladium-catalyzed cyclization of oximes was reported.[337] In this report, $(PPh_3)_2PdCl_2$/CuI/Et_3N/CH_2Cl_2 was found to be the best combination. In 2007, this experiment was repeated by Wailes, Maitra and their co-workers.[338] They found that the oxime remained unchanged in control experiments [either $(PPh_3)_2PdCl_2$ or $(PPh_3)_4Pd$)] in the absence of CuI. Interestingly, when only CuI (with no Pd) was used, the corresponding product was formed in 80% isolated yield. Then they reported a copper-catalyzed coupling of aryl halides with oximes. Here, potassium sodium tartrate was required to avoid deoximation. The intramolecular version was described as well (Scheme 2.156a). Three years later, Tois's group reported a systematic study of this transformation.[339] By adding DMEDA as a ligand and CuI as a catalyst, 1,2-benzisoxazoles were produced in good yields (Scheme 2.156b).

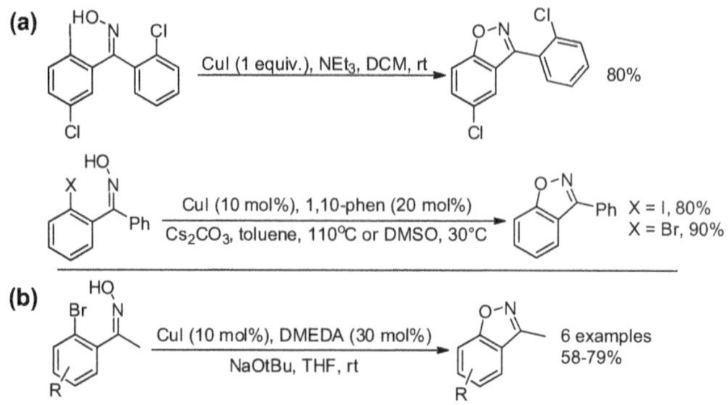

Scheme 2.156 Copper-catalyzed synthesis of benzisoxazoles.

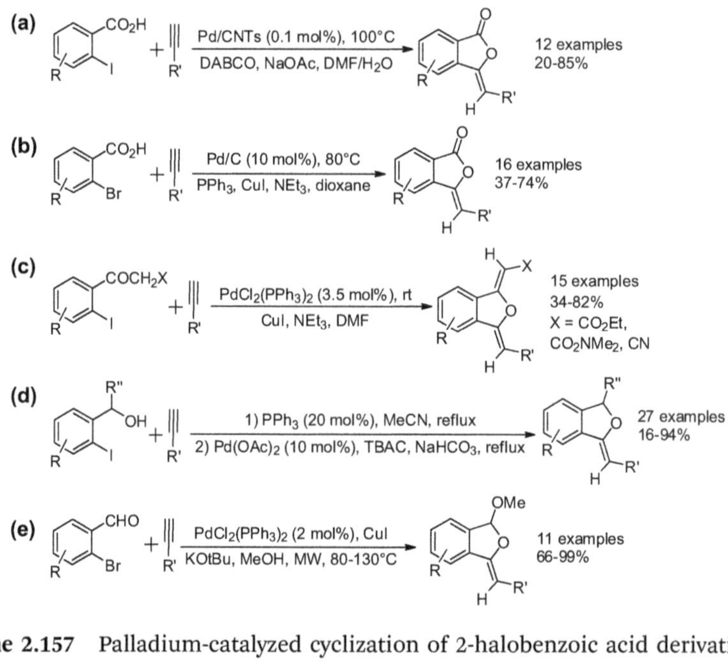

Scheme 2.157 Palladium-catalyzed cyclization of 2-halobenzoic acid derivatives.

In 1993, Kundu's group reported a palladium-catalyzed annulation of 2-iodo-benzoic acid with terminal alkynes.[340,341] Phthalides can be produced in moderate yields with $(PPh_3)_2PdCl_2$/CuI/Et_3N/DMF as the catalyst system. In 2007, Jiang and Zhou developed a phosphine- and copper-free condition for the synthesis of phthalides *via* a Pd/CNTs-catalyzed tandem coupling–cyclization process.[342] The palladium immobilized on CNTs (carbon nanotubes) showed a high catalytic effect on the reaction, and various phthalides were formed in moderate to good yields in the presence of 0.1% mmol Pd/CNTs (Scheme 2.157a). This protocol not only tolerates a range of functional groups but also does not require any additives such as phosphine ligands or CuI. Later on, Pal and co-workers applied Pd/C as a catalyst for the cyclization of 2-bromo benzoic acid with terminal alkynes.[343] They described a Pd/C–Cu-mediated, general, one-pot and direct synthesis of (*Z*)-3-alkylidenephthalides *via* the coupling–cyclization reaction of *o*-bromobenzoic acid with a range of terminal alkynes under mild conditions (Scheme 2.157b). The scope and limitations of the methodology were broad. However, the recyclability of the catalysts was not demonstrated with either reaction. Interestingly, when 2-halobenzoic acids were replaced with 3-(2-iodophenyl)-3-oxopropanoic acid derivatives, (*Z*)-2-[(*Z*)-3-alkylideneisobenzofuran-1(3*H*)-ylidene]acetic acid derivatives were formed (Scheme 2.157c).[344] By using dichlorobis(triphenylphosphine)palladium as the catalyst and CuI as an additive with NEt_3 as a base in DMF at room temperature, the corresponding products were produced in reasonable yields. 2-Halobenzyl alcohols can be applied as substrates and reacted with terminal alkynes as well. Benzofurans can be produced in good yields and selectivities. Depending on the reaction conditions, the reaction mechanisms are different. For example, the reaction underwent a Sonogashira reaction and a cyclic hydroalkoxylation sequence when the NHC–Pd system was applied;[345] while PPh_3-catalyzed nucleophilic addition followed by Pd-catalyzed Heck cyclization occurred when the phosphine/palladium system was applied (Scheme 2.157d).[346] Isonitriles can be applied as analogues of alkynes and cyclize with aryl bromides carrying pendant amine or alcohol groups on the *ortho* position.[347] The desired cyclic amidines or imidates were produced in good yields. Moreover, the reaction between *o*-bromoarylaldehydes, methanol and terminal alkynes could lead to other types of products.[348] A variety of substituted dihydroisobenzofurans were easily synthesized in high yields by this one-pot three-component approach (Scheme 2.157e). The reaction occurs through a cooperative palladium–base promoted coupling–addition–cyclization sequence.

Remarkably, a copper-catalyzed coupling reaction of 2-iodobenzoic acids and alkynes such as terminal acetylenes, alkynyl carboxylic acids, and trimethylsilylacetylene was reported by Lee and co-workers in 2013 (Scheme 2.158).[349] The reaction proceeded selectively to afford isocoumarins and phthalides in the presence of caesium carbonate (Cs_2CO_3) and dimethyl sulfoxide (DMSO). Among the regioselective products, only the 6-*endo-dig* product, isocoumarin, was formed at 100 °C, and the 5-*exo-dig* product, phthalide, was formed as a major product at 25 °C. A variety of alkynes produced the corresponding isocoumarins and phthalides in good yields. A mechanism was suggested in which the formation of 2-alkynylbenzoic acid as an intermediate *via* Sonogashira-type coupling was ruled out in the reaction pathway. During the work, the group found that: (i) the electronic properties of different substituents on the aryl rings did not affect the formation of isocoumarins at high temperature very much, but they did affect the formation of phthalide at low temperature; (ii) arylalkynyl carboxylic acid and trimethylsilylacetylene were also good substrates for the formation of isocoumarins; (iii) the survival of halide groups such as bromide and

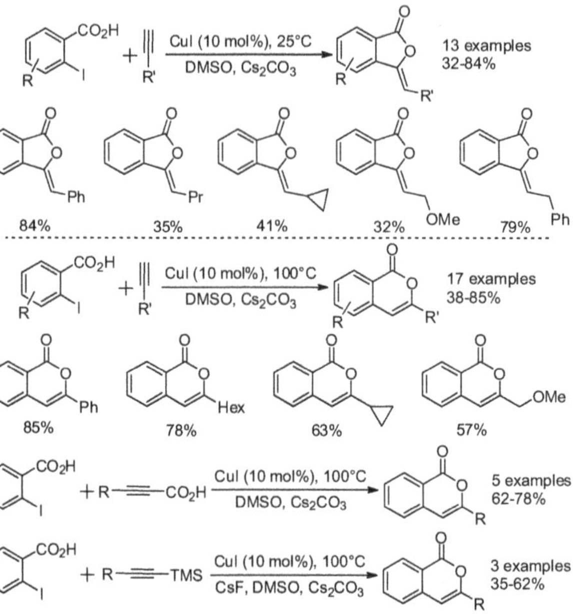

Scheme 2.158 Copper-catalyzed synthesis of isocoumarins and phthalides.

Reaction Procedure (Scheme 2.158): Synthesis of isocoumarin. 2-Iodobenzoic acid derivative (2.0 mmol) and Cs_2CO_3 (4.0 mmol) were added to a vial containing anhydrous DMSO (8 mL), followed by addition of the appropriate acetylene derivative (2.2 mmol) and CuI (0.2 mmol). The suspension was stirred for 12 h at 100 °C. After cooling, the mixture was poured into the EtOAc (50 mL) and washed with water (2×25 mL) and brine (2×25 mL), then dried over Na_2SO_4. Evaporation of the solvent under reduced pressure provided the crude product, which was purified by column chromatography (hexane–EtOAc, 20 : 1) or re-crystallized from hexane to afford the final product.

chloride offered an opportunity for further functionalization. Additionally, no ligand was needed by this copper-catalyzed reaction system.

Carbonylation procedures were explored and applied in phthalide synthesis as well. In 1986, Negishi and co-workers studied the palladium-catalyzed cyclization of *o*-iodoaryl alkenyl ketones. By using $Pd(dba)_2$ as the catalyst, bicyclic and polycyclic quinones could be produced in high yields.[350] Interestingly, phthalides were formed when using $Pd(OAc)_2/PPh_3$ as the catalyst system (Scheme 2.159). Afterwards, the group carried out systematic studies on this topic.[351–354] Lactones and lactams were produced selectively from various substrates.

In 1980, Stille and co-workers reported the synthesis of lactones by the palladium-catalyzed carbonylation reaction of halo alcohols.[355,356]

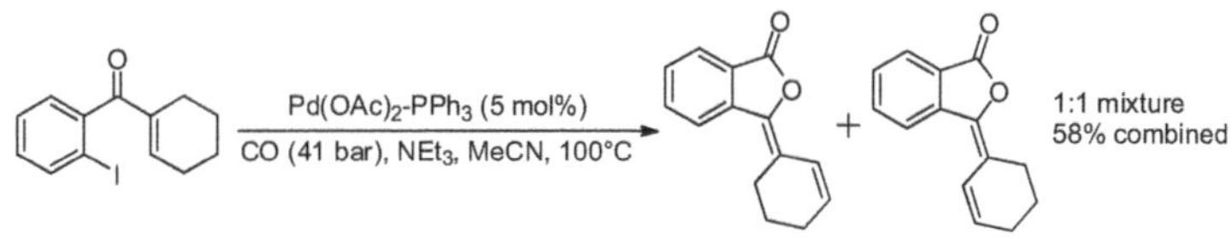

Scheme 2.159 Case study on Pd-catalyzed carbonylative synthesis of phthalides.

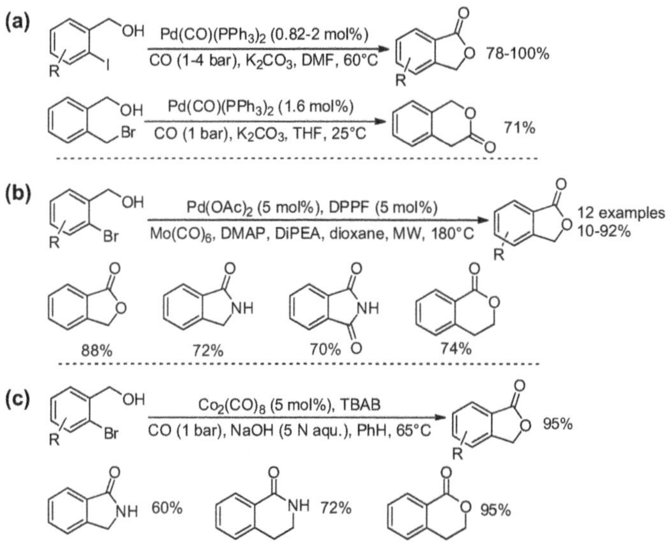

Scheme 2.160 Pd-catalyzed carbonylative synthesis of phthalides.

Under mild conditions (1–4 bar CO, 25–60 °C), high yields of the desired products were produced. Benzyl, allyl, aryl, and vinyl halides containing primary, secondary, or tertiary alcohol groups were readily converted to a variety of lactones, including phthalides and butenolides (Scheme 2.160a). Since then, 2-bromo benzyl alcohol has been tested as a substrate in various systems, such as carbonylation with a flow microreactor,[357] [^{11}C]-labelled compounds,[358] and microwave-promoted with $Mo(CO)_6$ as a carbon monoxide source (Scheme 2.160b).[359] Cobalt carbonyl as a cheaper catalyst was also applied in the intramolecular cyclization of aryl halides (Scheme 2.160c).[360,361]

During the end of the 20th century, Shim and co-workers developed a series of palladium-catalyzed carbonylative transformations of 2-bromo benzaldehydes with different nucleophiles. With the same catalyst system, 3-substituted phthalides were produced in good yields (Scheme 2.161).[362–364]

Larock's group studied the application of palladium catalysts in the carbonylative annulation of internal alkynes with 2-iodobenzyl alcohols.[365] Seven- and eight-membered ring lactones were synthesized and the scope and limitations of the process were examined. With a lower reaction temperature, non-alkyne-inserted phthalides were produced.

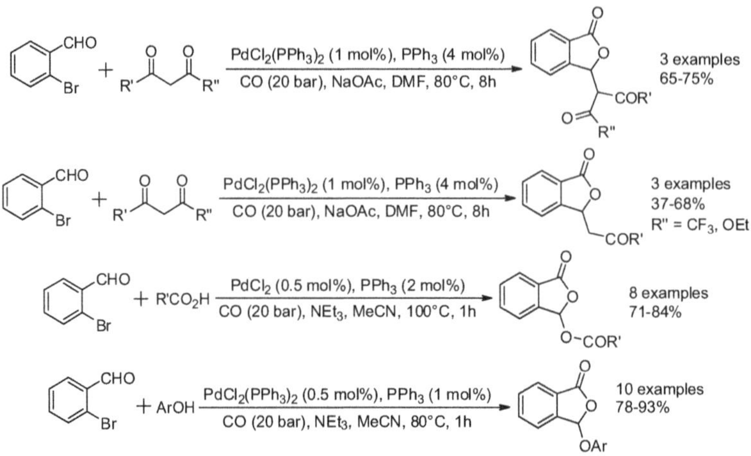

Scheme 2.161 Pd-catalyzed carbonylative synthesis of phthalides from 2-bromo benzaldehydes.

Reaction Procedure (Scheme 2.161): A mixture of *o*-bromobenzaldehyde (2 mmol), 1,3-dicarbonyl compound (2 mmol), $PdCl_2(PPh_3)_2$ (0.02 mmol), PPh_3 (0.08 mmol), and sodium acetate (2 mmol) in DMF was placed in a pressure vessel. After the system was flushed and then pressurized with carbon monoxide (20 atm), the mixture was stirred at 80 °C for 8 h. The reaction mixture was filtered through a short column (silica gel) and evaporated under reduced pressure. Preparative thin layer chromatography separation using ethyl acetate–hexane (10 : 3) as an eluent gives 3-alkylphthalides.

A mixture of *o*-bromobenzaldehye (2 mmol), phenylacetic acid (4 mmol), palladium(II) chloride (0.1 mmol), triphenylphosphine (0.4 mmol), and triethylamine (0.7 mL, 5 mmol) in dry acetonitrile (10 mL) was placed in a 50 mL stainless steel autoclave. After the system was flushed and then pressurized with carbon monoxide to 20 atm, the mixture was stirred at 100 °C for 1 h.

In 2004, Cheng and co-workers reported a nickel-catalyzed cyclization of 2-haloesters with aldehydes to phthalides.[366] Here, by using Ni(II) as the catalyst, phthalide derivatives were produced in excellent yields with high chemoselectivity under mild conditions (Scheme 2.162a). In addition to five-membered products, this methodology can be further applied to the synthesis of six-membered lactones. The reaction of methyl 2-(2-bromophenyl)-acetate with benzaldehyde under similar reaction conditions afforded a six-membered lactone in a 68% yield. A possible catalytic mechanism for this cyclization was also proposed. Reduction of nickel(II) to nickel(0) by using zinc powder is likely to initiate the catalytic reaction. Oxidative addition of aryl iodide to the nickel(0) species yields the nickel(II) intermediate.

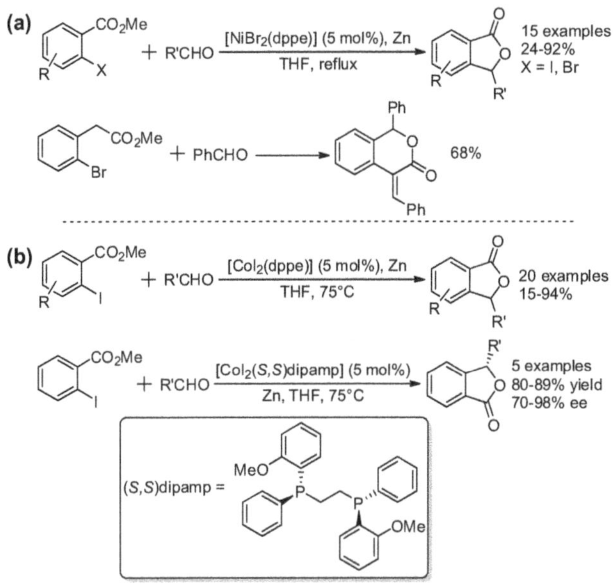

Scheme 2.162 Synthesis of phthalides from *o*-halobenzoates.

Reaction Procedure (Scheme 2.162a): A round-bottomed side-arm flask (25 mL) fitted with a reflux condenser containing [$NiBr_2$(dppe)] (0.050 mmol, 5.0 mol%) and zinc powder (2.75 mmol) was evacuated and purged with nitrogen gas three times. Freshly distilled THF (2.0 mL), *o*-bromobenzoate (1.50 mmol), and aldehyde (1.00 mmol) were sequentially added to the system, and the reaction mixture was stirred under reflux conditions for 24 h. The reaction mixture was cooled to rt, diluted with dichloromethane, and then stirred in air for 15 min. The mixture was filtered through a short Celite and silica-gel pad, and washed with dichloromethane several times. The filtrate was concentrated, and the residue purified on a silica-gel column by using hexanes–ethyl acetate as the eluent to afford the cyclization product.

(Scheme 2.162b): A sealed tube containing [CoI_2(dppe)] (0.050 mmol, 5.0 mol%) and zinc powder (2.75 mmol) was evacuated and purged with nitrogen gas three times. Freshly distilled THF (3.0 mL), *o*-iodobenzoate (1.20 mmol), and aldehyde (1.00 mmol) were sequentially added to the system and the reaction mixture was allowed to stir at 75 °C for 24 h. The mixture was filtered through a short Celite and silica-gel pad and washed with dichloromethane several times. The filtrate was concentrated and the residue was purified on a silica-gel column by using hexanes–ethyl acetate as the eluent to afford the cyclization product.

Coordination of an aldehyde molecule to the nickel center, followed by insertion into the nickel–carbon bond, affords a nickel–alkoxide intermediate. Attack of the coordinated alkoxy group at the ester moiety of the nickel–alkoxide intermediate gives the final phthalide derivative and nickel(II). The latter is reduced by zinc powder to regenerate the active nickel(0) species. Later on, the group applied a cobalt catalyst in this cyclization reaction as well.[367] In the presence of a cobalt catalyst, 2-iodobenzoates and aldehydes were reacted and afforded the substituted phthalide derivatives in one pot under mild reaction conditions with good to excellent yields (Scheme 2.162b). In addition, high enantioselectivity of the cyclization was obtained by employing cobalt complexes with a suitable bidentate chiral ligand. Compared with nickel, a Co(I) and Co(III) catalytic cycle was proposed. Reduction of cobalt(II) to (I) by Zn metal likely initiates the catalytic reaction. Oxidative addition of methyl 2-iodobenzoate with the cobalt(I) species yields an *o*-metalated methylbenzoate complex with both the *o*-carbon atom and the ester oxygen atom bonded to the cobalt(III) center. After the formation of the final product, the cobalt(III) species was eliminated, which will be reduced by zinc metal to regenerate the active cobalt(I) species.

In 2009, Liu, Fu and their co-workers reported a Pd-catalyzed decarboxylative cross-coupling of aryl iodides, bromides, and chlorides with potassium oxalate monoesters.[368] In their system, a cyclized product was produced when 2-bromoacetophenone was applied as a substrate (Scheme 2.163).

Alternatively, a ruthenium catalyst could be applied in phthalide preparation. In 2011, Ackermann's group developed a ruthenium-catalyzed cross-dehydrogenative C–H bond alkenylation reaction. The methodology used water as a solvent, benzoic acids and terminal alkenes as substrates; good yields of the desired phthalides were isolated (Scheme 2.164).[369] The reaction sequence consisted of cross-dehydrogenative alkenylation and a subsequent intramolecular oxa-Michael reaction. Mechanistic studies provided strong evidence that the oxidative alkenylation proceeds by an irreversible C–H bond metalation *via* acetate assistance.

An efficient copper-catalyzed cross-coupling of aryl iodides with terminal alkynes was reported in 2010 by Lee and co-workers.[370] In this procedure, a cyclization reaction occurred when 2-iodobenzyl alcohols were applied as substrates (Scheme 2.165a). Additionally, benzofuran was produced in a good

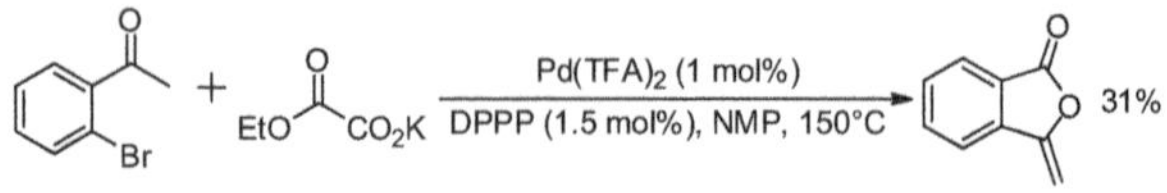

Scheme 2.163 Synthesis of phthalides from 2-bromoacetophenone.

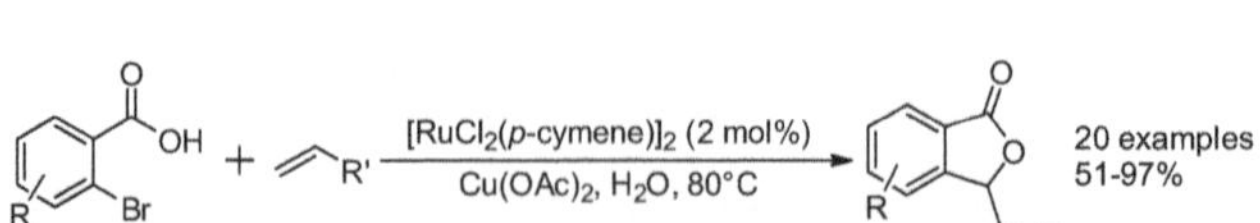

Scheme 2.164 Ru-catalyzed oxidative synthesis of phthalides.

Reaction Procedure (Scheme 2.164): Ethyl acrylate (0.22 mL, 2.02 mmol) was added *via* a syringe to a mixture of 2-methylbenzoic acid (136 mg, 1.00 mmol), $[RuCl_2(p\text{-cymene})]_2$ (12.2 mg, 2.0 mol%) and $Cu(OAc)_2 \cdot H_2O$ (400 mg, 2.00 mmol) in H_2O (5.0 mL), and the mixture was stirred at 80 °C for 16 h. At ambient temperature, the reaction mixture was diluted with a solution of saturated aq. NH_4Cl-NH_3 (1 : 1, 50 mL) and extracted with EtOAc (4×25 mL). The combined organic phase was washed with aq. NH_4Cl solution and conc. NH_3 (1 : 1, 2×20 mL) and dried over Na_2SO_4. After filtration and evaporation of the solvents *in vacuo*, the crude product was purified by column chromatography on silica gel (*n*-hexane–EtOAc, 7 : 1) to yield the pure product.

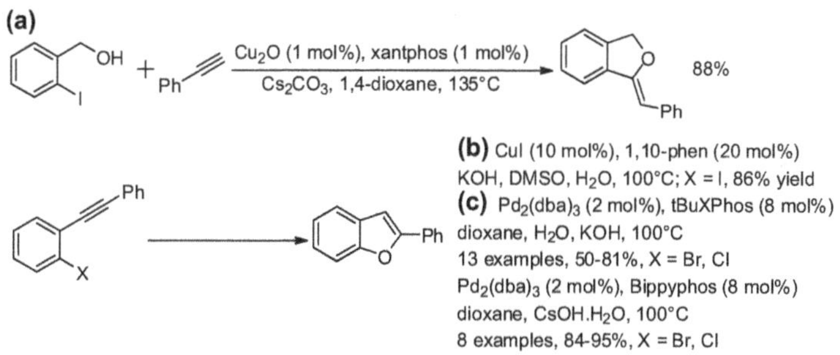

Scheme 2.165 Synthesis of benzofurans from alkynes.

yield if 2-iodophenol and phenylacetylene were used as starting materials. In a report from You's group on copper-catalyzed hydroxylation of aryl halides, 86% of benzofuran was formed when 2-iodoaryl alkyne was applied as a substrate (Scheme 2.165b).[371] As the hydroxylation of 2-haloaryl alkynes offers a straightforward pathway for benzofuran derivative preparation, palladium catalysts were explored in this topic as well (Scheme 2.165c).[372,373]

The cyclization of *o*-halobenzyl ketones provides another pathway for the synthesis of benzofurans. In 2004, Willis and co-workers reported a palladium-catalyzed synthesis of benzofurans through intramolecular *O*-arylation of enolates.[374] By the combination of $Pd_2(dba)_3$ and DPEphos, the corresponding benzofurans were isolated in good to excellent yields (Scheme 2.166a). Both cyclic and acyclic ketones were effectively reacted and gave the corresponding products. Additionally, thio ketones can be employed as well and provide the desired benzothiophenes in good yields. Later on, the combination of palladium–IPr · BF_4 was found to be as efficient as the previous conditions (Scheme 2.166b).[375] This allows the reactions to be performed under phosphine-free conditions. Copper catalysts were applied in this topic as well. In 2005, CuI (10 mol%) was reported to be an active catalyst for this reaction with K_3PO_4 as a base in DMF at 105 °C.[376,377] Good to excellent yields of the desired products were achieved. By adding TMEDA as a ligand and base, the reaction can be performed in water at 120 °C.[378]

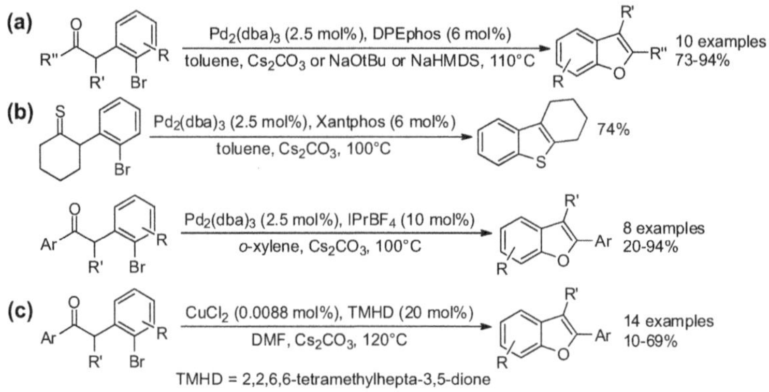

Scheme 2.166 Synthesis of benzofurans from *o*-halobenzyl ketones.

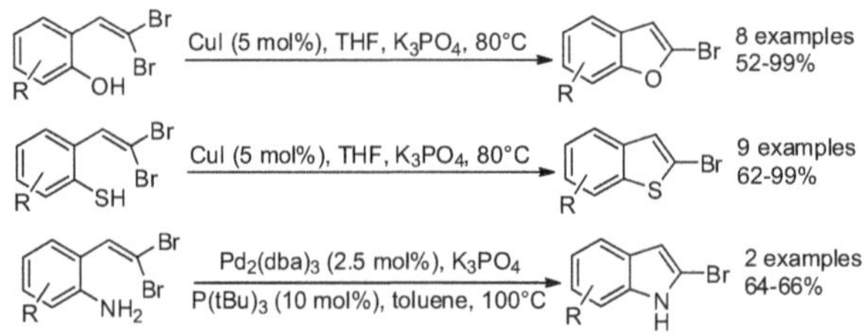

Scheme 2.167 Copper-catalyzed synthesis of 2-bromobenzofurans.

But still, 8.5 mol% of CuI was required. Recently, Bolm and co-workers found that using only 0.0088 mol% of $CuCl_2$ can provide the corresponding benzofurans in moderate to good yields (Scheme 2.166c).[379]

gem-Dihalovinyl compounds are versatile synthons in modern organic synthesis.[380] In 2004, the synthesis of 2-functionalized benzo[*b*]furans and indoles from readily available *o*-(2,2-dibromovinyl)-phenol, -aniline or -acetanilide was reported.[381] Diethyl phosphite, trimethyl(phenyl)tin, *p*-methoxyphenylboronic acid and phenylboronic anhydride were tested as coupling partners. Since then, procedures on this topic have been explored. In 2009, Lautens and co-workers reported a procedure on the synthesis of 2-bromobenzofurans and 2-bromobenzothiophenes.[382] By using CuI as the catalyst, good to excellent yields of the desired 2-bromo benzofused heterocycles were obtained (Scheme 2.167). Later on, this transformation was found to be promoted by TBAF[383] or the trace (PPM level) of Cu in Cs_2CO_3 (0.5 equiv.).[384] This methodology was applied in the synthesis of biological active molecules as well.[385]

In 2012, Wang and co-workers developed an interesting procedure for the preparation of 2-arylbenzofurans and 2-arylbenzothiophenes.[386] In the presence of tetra-(*n*-butyl)ammonium fluoride (TBAF) and $Pd(OAc)_2/PPh_3$, 2-(*gem*-dibromovinyl)phenols(thiophenols) reacted with phenyl(trialkoxy)-silanes smoothly to generate the corresponding products in good

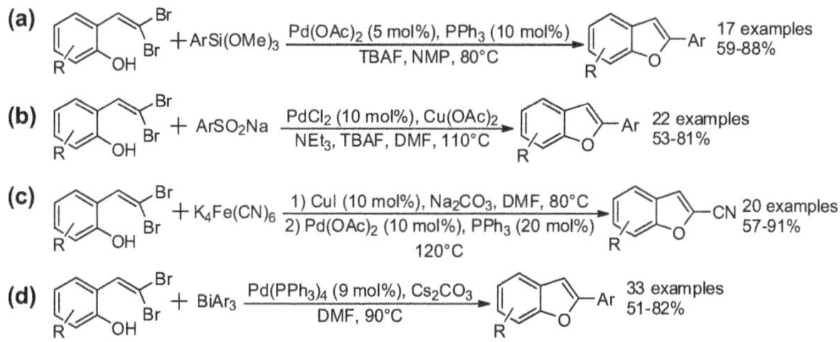

Scheme 2.168 Palladium-catalyzed synthesis of benzofurans from 2-(*gem*-dibromovinyl)phenols.

yields (Scheme 2.168a). The reaction goes through a tandem elimination–intramolecular addition–Hiyama cross-coupling reaction sequence. It is important to note that TBAF plays an important role in the tandem elimination, intramolecular addition and Hiyama coupling reactions. By slightly changing the catalytic system, sodium arylsulfinates could be applied as coupling partners.[387] In the presence of TBAF–$PdCl_2$–$Cu(OAc)_2$–NEt_3, the reactions of 2-(*gem*-dibromovinyl) phenols(thiophenols) with sodium arylsulfinates proceeded well to afford the desired products with good yields in one pot under ligand-free conditions *via* a tandem elimination–cyclization–desulfitative arylation reaction sequence (Scheme 2.168b). Meanwhile, the group succeeded in preparing 2-cyanobenzofurans and 2-cyanobenzothiophenes by a one-pot two-step manner as well (Scheme 2.168c).[388] A procedure based on using triarylbismuth reagents as coupling partners was reported in 2013 by Rao and co-workers.[389] Functionalized 2-arylbenzofurans were prepared in good yields *via* this one-pot, chemoselective methodology (Scheme 2.168d).

A copper(I)-catalyzed synthetic method for the preparation of 2-alkynylbenzofurans from *gem*-dihaloolefins was established recently.[390] By using DABCO and TBAF · $3H_2O$ as the catalytic system, the corresponding products were isolated in good yields (Scheme 2.169a). Coupling with heteroaryl azoles *via* C–H activation was also achieved with copper as the catalyst (Scheme 2.169b).[391,392] In the same period, a copper(I)-catalyzed tandem reaction of 2-(2,2-dibromovinyl)phenols with polyfluoroarene to produce 2-(polyfluoroaryl) benzofurans was developed as well (Scheme 2.169c).[393] During the reaction process, a copper-catalyzed intramolecular C–O bond formation and an alkenylation of the polyfluoroarene C–H bond were involved. More recently, an efficient synthesis of 2-selenyl(sulfenyl)benzofurans has been accomplished through a copper(I)-catalyzed tandem reaction of 2-(*gem*-dibromovinyl)phenols with diorganyl diselenides and disulfides in the presence of CuI–Mg–*t*-BuOLi in DMSO.[394] Using this protocol, a variety of 2-selenyl(sulfenyl)benzofuran derivatives were obtained in good yields (Scheme 2.169d). In all the processes mentioned, the pre-generation of 2-bromobenzofurans was proposed.

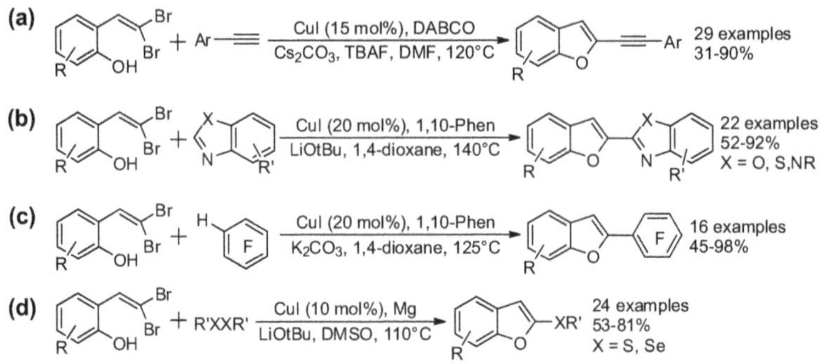

Scheme 2.169 Copper-catalyzed transformations of 2-(*gem*-dibromovinyl)phenols.

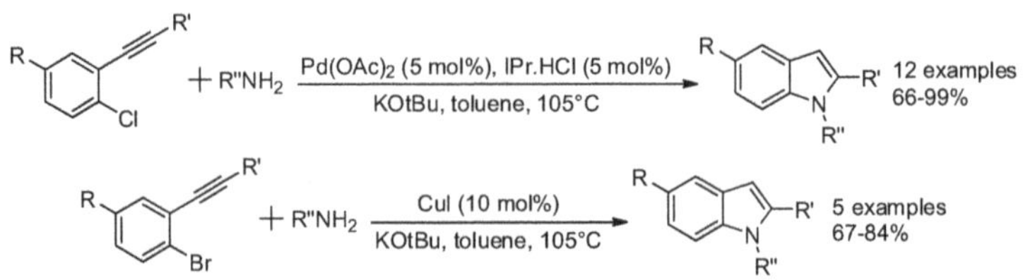

Scheme 2.170 Cyclization of *o*-alkynylhaloarenes with amines to indoles.

Reaction Procedure (Scheme 2.170): To a solution of $Pd(OAc)_2$ (5 mol%), IPrHCl (5 mol%), and KO*t*-Bu (3.0 mmol) in PhMe (3 mL) were added 1-chloro-2-phenylethynylbenzene (1.0 mmol) and 4-methylaniline (1.2 mmol) at room temperature. The resulting red mixture was stirred at 105 °C for 2 h, after which GC/MS analysis indicated complete conversion of the starting material. DCM (50 mL) and aq. HCl (2 N, 50 mL) were added to the cooled reaction mixture. The separated aqueous phase was extracted with DCM (2×50 mL). The combined organic layers were dried over $MgSO_4$ and concentrated *in vacuo*. The remaining residue was purified by column chromatography on silica gel (*n*-pentane–Et_2O, 200 : 1) to yield the pure product.

In 2005, Ackermann reported the cyclization of *o*-alkynylhaloarenes with amines to indole derivatives.[395,396] By using palladium as the catalyst, *o*-alkynylchloroarenes were transformed into the corresponding indoles with primary amines in good to excellent yields (Scheme 2.170). A copper catalyst was explored as well; a low yield was observed with chloro derivatives. However, good yields can be achieved with the corresponding bromide derivatives.

One year later, in 2006, Hu and Tang reported that $Pd(OAc)_2/PtBu_3$ is another active system for the transformation of *o*-alkynylhaloarenes.[397] By using KO*t*Bu as a base in toluene, the corresponding indoles were produced in excellent yields by coupling with various amines. With 1-chloro-2-iodobenzene

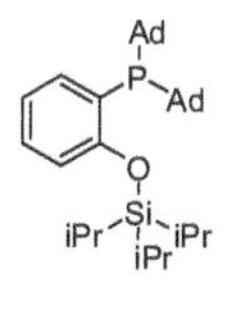

Figure 2.3 Structure of (silanyloxyphenyl)phosphine ligand.

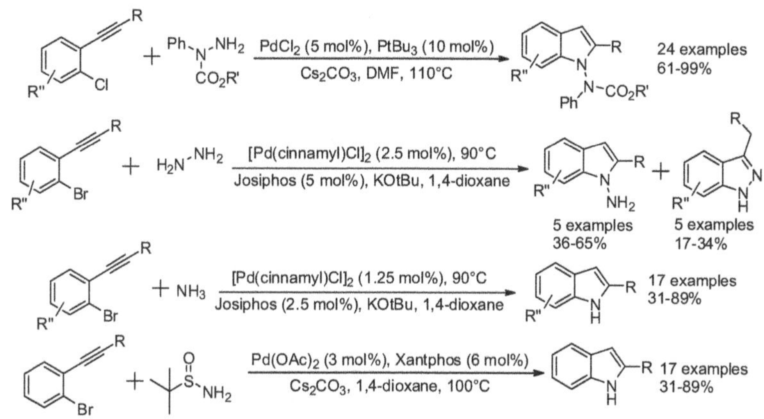

Scheme 2.171 Synthesis of N-H free indoles and 2-aminoindoles.

as the starting material, *via* a one-pot two-step reaction sequence, indoles can be produced in moderate yields as well. By using a newly developed (silanyloxyphenyl)phosphine (Figure 2.3) as the ligand, the reaction temperature can be decreased to 90 °C.[398]

In addition to amines, hydrazines, ammonia and *tert*-butyl sulfinamide were explored and applied as coupling partners as well.[399–401] In the presence of palladium catalysts, *N-H* free indoles and 2-aminoindoles were selectively produced (Scheme 2.171).

The cyclization of *o*-alkynylhaloarenes with ammonia to provide N-H free indole derivatives was reported with a copper catalyst as well. In 2011, a convenient, efficient and atom economic synthesis of free N-H 2-arylindoles from 2-arylhaloarenes *via* sequential amination and cyclization was reported.[402] By applying this protocol, a wide range of electron-rich and electron-deficient 2-phenylindoles were obtained in good yields (Scheme 2.172a). As nickel catalysts are low in price, they have been explored in the cyclization of *o*-alkynylhaloarenes with amines to indoles too.[403] Ackermann and co-workers found that the combination of $Ni(cod)_2$ and DPPF can promote the amination of a C–Br bond effectively followed by base-mediated hydroamination to provide the desired indoles in good to excellent yields (Scheme 2.172b). Notably, among the various phosphine and NHC ligands tested, DPPF gave a superior result compared to DPPE, DPPP, IPr · HCl and the others. Additionally, the cyclization of *o*-bromo-(propa-1,2-dien-1-yl)arenes to indoles was studied without exception.[404]

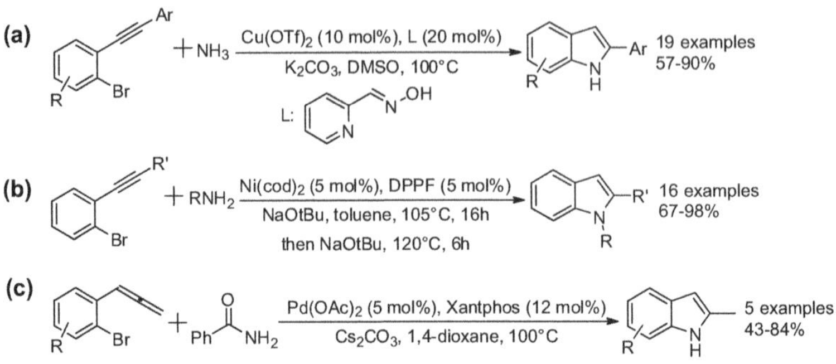

Scheme 2.172 Transition metal-catalyzed synthesis of indoles.

These interesting substrates exhibit novel and orthogonal reactivity under Pd catalysis in the presence of secondary amines to form enamines (concerted Pd insertion, intramolecular carbopalladation, and terminative Buchwald–Hartwig coupling) and of amides (benzamide or acetamide) to form indoles (addition, Buchwald–Hartwig cyclization, and loss of the acetyl group). The substrates for these reactions can be accessed in a reliable and highly selective two-step process from 2-bromo(bromo)methylbenzene (Scheme 2.172c).

Similar to the preparation of benzofurans from *gem*-dihalovinyl phenols mentioned earlier, the cyclization of *gem*-dihalovinyl anilines is a promising pathway for indole derivative synthesis. In 2010, Lautens and co-workers reported the synthesis of 2-bromoindoles from *gem*-dihalovinyl anilines with palladium as the catalyst.[405] In this procedure, the bulky phosphine $PtBu_3$ was required for the conversion of starting materials. The group proposed that the origin of this effect is the reversibility of oxidative addition into the carbon–bromine bond of the product. Various desired products were isolated in good yields; the 2-bromo benzofuran derivative was produced as well when the corresponding phenol was applied (Scheme 2.173). Interestingly, a novel synthesis of 2-halo-3-carboxyindoles from 2-(2,2-dihalovinyl)anilines was discovered in 2011.[406] The reaction takes only minutes when these substrates are heated in DMSO at 120 °C in the presence of caesium carbonate. However, the reaction is robust and takes place at a wide range of temperatures, is tolerant of aqueous reaction conditions, and can be performed in a variety of polar solvent–carbonate base combinations where the limiting factor is base solubility. A wide range of substituents is tolerated on the 2-(2,2-dihalovinyl)anilines, and yields are generally high, requiring only acidic aqueous work-up to obtain pure products. No catalyst is required for the transformation. The mechanism is believed to involve initial formation of an alkynyl bromide intermediate followed by ring closure and carbon dioxide trapping, leading to product formation.

Lautens and co-workers did a series of studies on the transformations of *gem*-dihalovinylanilines. In 2005, they developed a highly efficient method for indole synthesis using *gem*-dihalovinylaniline substrates and an

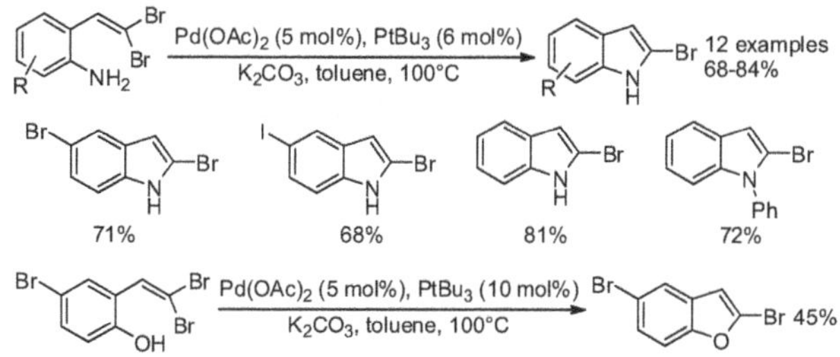

Scheme 2.173 Palladium-catalyzed synthesis of 2-bromoindoles.

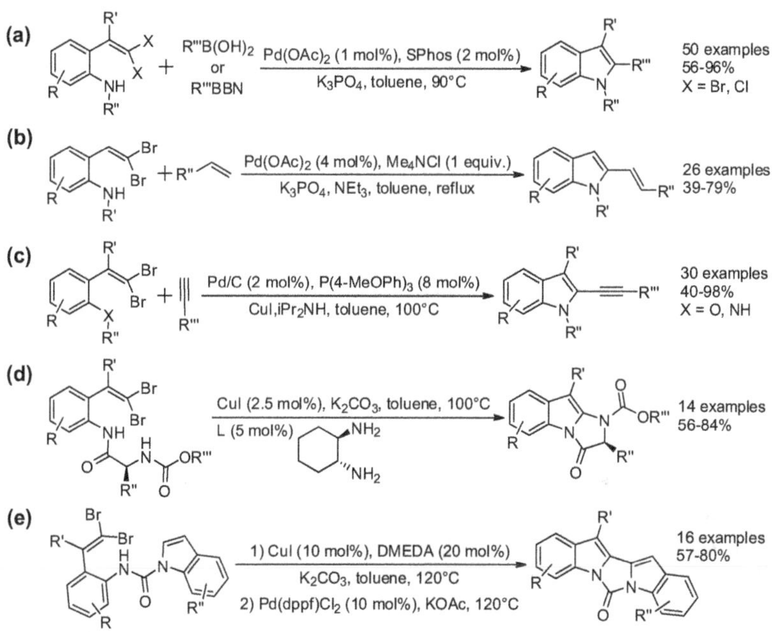

Scheme 2.174 Synthesis of indoles from *o-gem*-dibromovinyl anilines.

organoboron reagent *via* Pd-catalyzed tandem intramolecular amination and intermolecular Suzuki coupling (Scheme 2.174a).[407,408] In this methodology, *gem*-dihalovinyl substrates are compatible with a variety of aryl, alkenyl, and alkyl boron reagents, making it a versatile, modular approach. The tolerance for a variety of substitution patterns on the indole has been very thoroughly examined at positions 2–7. The mechanistic studies conducted show that two different pathways are likely to be in operation. One is *via* an initial alkynyl formation which then undergoes a 5-*endo-dig* cyclization to generate a C–N bond followed by Suzuki coupling. The other pathway is a direct Buchwald–Hartwig amination followed by Suzuki coupling. Soon after, the use of terminal alkenes and alkynes as coupling partners was published by the same group. 2-Vinylic indoles (Scheme 2.174b)[409] and 2-alkynyl indoles

(Scheme 2.174c)[410] were isolated in moderate to good yields. In 2011, Xi and co-workers reported that 2-alkynyl indoles can be prepared from *o*-bromo-(2,2-dibromovinyl)benzenes, terminal alkynes and anilines as well.[411] In the presence of $Pd(OAc)_2$/Xantphos with Cs_2CO_3 as a base in NMP at 120 °C, the corresponding indoles were prepared in moderate to good yields. Imidazoindolones are present as the key structural motif in the family of antifungals, fumiquinazolines, and the antagonist asperlicin. The first example of a CuI-catalyzed tandem intramolecular amidation forming substituted imidazoindolones from readily accessible *ortho gem*-dibromovinylanilines was described in 2006 (Scheme 2.174d).[412] This procedure represents an efficient one-pot procedure giving rapid access to a wide range of substituted imidazoindolones from readily prepared *gem*-dibromovinyl substrates *via* a novel tandem intramolecular C–N bond formation. This process is particularly attractive for industrial use since copper iodide is a very cheap, air-stable compound and can be used without purification, in contrast to the high cost of palladium and problematic removal of Pd residues from polar compounds. Additionally, this methodology could provide facile access to the natural products asperlicin and fumiquinazolines. More recently, an intermolecular version was reported.[413] 2-Amidoindoles and tetrahydroindolo[1,2-*a*]quinazolines were produced in a shorter time and in high yields. By using *gem*-dibromovinylanilides and sulfonamides as the substrates through the *in situ* formation of ynamides followed by a base-promoted intramolecular hydroamidation, the desired products were produced in good yields. In 2011, a one-pot synthesis of unsymmetrical 2,2′-biindolyl derivatives through a Cu-catalyzed *N*-arylation–Pd-catalyzed direct arylation sequence was described.[414] The reaction involved easily prepared *o*-*gem*-dibromovinyl substrates, and the desired biindolyls were obtained in moderate to good yields (Scheme 2.174e). Later on, the group reported the synthesis of 6*H*-isoindolo[2,1-*a*]indol-6-ones through one-pot sequential coupling reactions, which comprised a copper-catalyzed C–N coupling cyclization and palladium-catalyzed C–H activation.[415] General chemicals benzoyl chlorides and *o*-*gem*-dibromovinyl anilines were employed as the starting substrates.

In 2008, Alper and co-workers developed a tandem palladium-catalyzed carbonylative transformation of *gem*-dibromovinyl anilines.[416] By this tandem palladium-catalyzed N,C-coupling–carbonylation, 2-carboxyindoles were prepared in good yields under 10 atm of carbon monoxide and at 110 °C (Scheme 2.175a). The catalyst system tolerates a variety of functional groups, and the reactions were performed in a highly selectivity manner. In 2009, a palladium-catalyzed carbonylative Suzuki reaction of *gem*-dibromovinyl anilines was established.[417] By this newly developed protocol, 2-aroylindoles were prepared in moderate to good yields, through a one-pot palladium-catalyzed C,N-coupling–carbonylation–Suzuki coupling sequence (Scheme 2.175b). The reaction tolerates various functional groups, thus providing practical access to a wide range of 2-aroyl or 2-heteroaroylindoles from readily accessible starting materials.

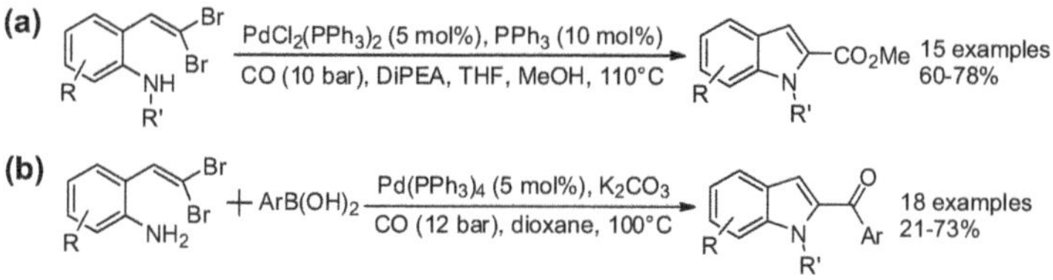

Scheme 2.175 Palladium-catalyzed carbonylation of *gem*-dibromovinyl anilines.

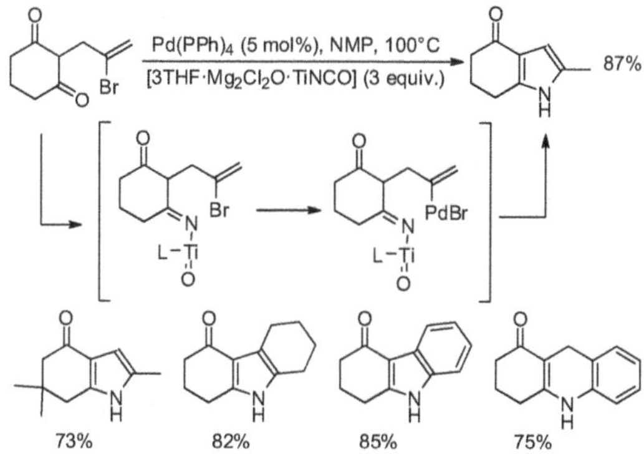

Scheme 2.176 Palladium-catalyzed nitrogenation–transmetallation process.

In 1991, a new C–N–C bond formation reaction using a nitrogenation–transmetalation process was described.[418] Ketones and aryl or vinyl halides couple to give divinyl or arylvinyl amines in the presence of the titanium isocyanate complex [$3THF\cdot Mg_2Cl_2O\cdot TiNCO$] and a palladium catalyst, *via* transmetallation of the titano imine complex with aryl or vinyl palladium bromide. Moderate to good yields of the desired products were observed (Scheme 2.176).

Lebel and Leogane developed a novel one-pot Curtius rearrangement–palladium-catalyzed indolization process in 2008.[419] The procedure allows the direct synthesis of 2,3-disubstituted and 3-substituted indoles starting from readily available 2-iodobenzoic acid. A synergistic effect between the two reactions of the process was observed, with a by-product of the first reaction serving as a reagent in the second synthetic step. In addition, the use of a one-pot procedure leads to higher yields while generating less by-products and chemical residues. This multicomponent process was also used to synthesize the first indole *N*-carboxamide derivatives through a heteroannulation procedure. 2-(2-Haloalkenyl)-aryl halides were applied as substrates for indole synthesis as well.[420,421] Willis's group achieved this type of transformation in 2006, and various indoles were formed in good yields (Scheme 2.177a). More recently, a practical and scalable Pd-catalyzed intramolecular C–N bond forming strategy for the construction of N-functionalized C2-/C3-substituted indole scaffolds was developed.[422] By this procedure, enamines derived from a broad variety of amines (alkyl,

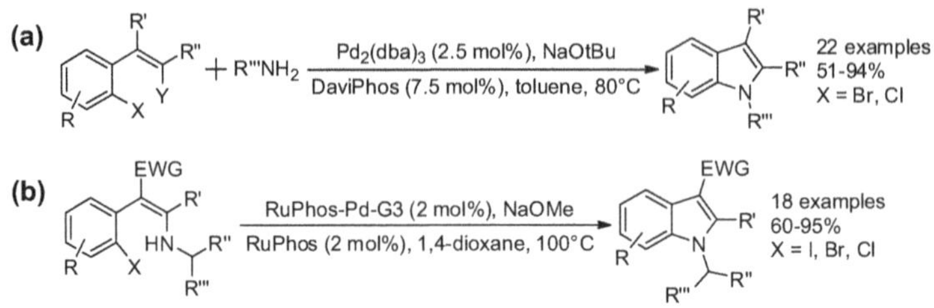

Scheme 2.177 Pd-catalyzed synthesis of indoles from 2-(2-haloalkenyl)-aryl halides.

Reaction Procedure (Scheme 2.177a): Sodium *tert*-butoxide (125 mg, 1.298×10^{-3} mol) was added to an oven-dried flask charged with $Pd_2(dba)_3$ (12 mg, 1.3×10^{-5} mol) and DaviPhos (16 mg, 3.968×10^{-5} mol) under nitrogen. The flask was flushed with nitrogen and the reagents suspended in anhydrous toluene (1.04 mL). To this, 1-bromo-2-(2-bromovinyl)benzene (136 mg, 5.192×10^{-4} mol) and aniline (58 mg, 57 mL, 6.230×10^{-4} mol) were added and the reaction mixture was heated at 80 °C for 5 hours under nitrogen. After cooling, the reaction mixture was diluted with diethyl ether (5 mL), filtered through a Celite pad and washed with diethyl ether (30 mL). The filtrate was reduced under vacuum. The product was separated *via* flash chromatography (eluent: 0.5% diethyl ether–hexane) to afford the indole.

(Scheme 2.177b): To a re-sealable vial containing a magnetic stir bar, methanesulfonato(2-dicyclohexylphosphino-2′,6′-di-i-propoxy-1,1′-biphenyl)(2-amino-1,1′-biphenyl-2-yl) palladium(II) (0.96 mmol, 0.02 equiv.) and 2-dicyclohexylphosphino-2′,6′-di-i-propoxy-1,1′-biphenyl (0.96 mmol, 0.02 equiv.) was added a solution of enamine (47.78 mmol, 1 equiv.) in 1,4-dioxane (0.3–0.5 M in enamine, ~8–10 mL g^{-1}). The vessel was subjected to three cycles of evacuation and purging with N_2 (g), followed by addition of NaOMe (67.72 mmol, ~1.5 equiv.). After an additional three cycles of evacuation and purging with N_2 (g), the heterogeneous brown/red reaction mixture was heated to 100–110 °C with stirring for ~1 h. After 1 h, the green heterogeneous reaction mixture was allowed to cool to room temperature and filtered over a bed of Celite/silica gel. The filter cake was washed with EtOAc (2×) and the filtrate was concentrated *in vacuo* to afford a yellow oil. This resultant oil was purified *via* silica gel chromatography typically using hexanes–ethyl acetate (4 : 1) as the eluent to afford the pure compound.

branched/chiral, or aromatic) can be assimilated into indole rings with ease. Additionally, indoles with diverse substitutions (around the aromatic ring and C2-positions) and electronic properties may also be synthesized (Scheme 2.177b). The use of immobilized enamines was reported by Kondo's group as early as 2002.[423,424] CuI can catalyze this reaction as well; the desired indoles were produced in excellent yields by using CsOAc as the base in DMSO at 30 °C.[425] In 2008, CuI was applied as a catalyst for the one-pot synthesis of substituted 2-trifluoromethylindoles from β-halo-β-(trifluoromethyl)-(*o*-halostyrene) and primary amines.[426] 2-Trifluoromethylindoles were prepared in moderate yields with variation of substituents at the benzene ring and nitrogen atom in DMSO with K_2CO_3 as a base at 100 °C. A palladium-catalyzed reaction between aryl iodides and 1-isocyano-2-(arylethynyl)benzene leads to 3-acyl-2-arylindoles selectively.[427] Tetracyclic indoles were formed when 2-iodoaniline, 2-iodophenol, or 2-iodobenzyl nitrile was applied as a coupling partner.

Several other copper-catalyst-based methodologies for indole synthesis have been reported recently. In 2014, a novel procedure for the synthesis of 2-aroylindole derivatives *via* one-pot CuI-catalyzed S_NAr reaction of *o*-bromochalcones with sodium azide and subsequent intramolecular cyclization through nitrene C–H insertion was developed by Ramana and co-workers (Scheme 2.178a).[428] This protocol was also applicable to 2′-bromocinnamates giving indole-2-carboxylates. This Cu-catalyzed process involves a set of three reactions: (i) S_NAr with azide and (ii) conversion of azide to nitrene; and (iii) intramolecular insertion of nitrene across the C–H bond with the net formation of two new C–N bonds. Koenig and co-workers developed a procedure for the synthesis of indole-2-carboxylic esters in 2010.[429] A variety of desired products are accessible in yields up to 61% through a ligand-free, copper-catalyzed reaction of a series of commercially available 2-halo aryl aldehydes with benign glycine amidoesters, including the common reagent ethyl acetamidoacetate (Scheme 2.178b). Ding's group developed an efficient ligand-free copper-catalyzed procedure for the synthesis of indoles (Scheme 2.178c).[430] The reaction undergoes a condensation–coupling–deformylation cascade process with 2-halo aryl aldehydes or ketones and ethyl isocyanoacetates as substrates. The reactions performed well at room temperature or 50 °C for iodo- and bromo-substituted substrates.

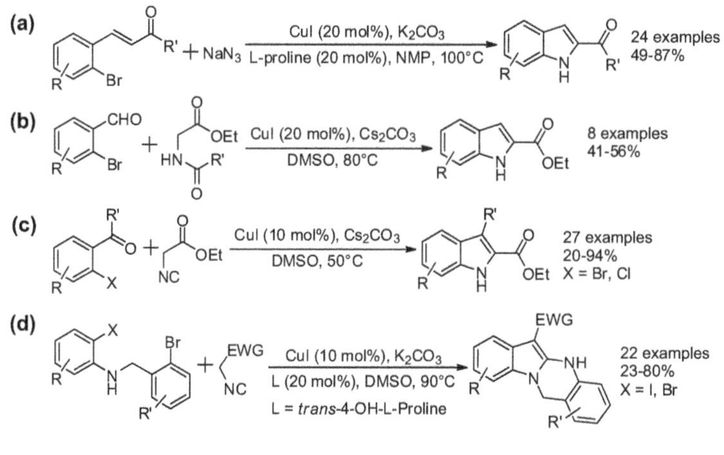

Scheme 2.178 Cu-catalyzed alternative procedure for indole synthesis.

Chloride substituted substrates also successfully yielded the desired indole-2-carboxylic acid esters with acceptable yields when the temperature was elevated to 80 °C. Furthermore, the reaction displayed excellent functional group compatibility and high chemical selectivity in the presence of a broad range of functional groups. This transformation can be performed in ionic liquid as well.[431] In 2012, a facile and efficient copper-catalyzed domino synthesis of 5,12-dihydroindolo[2,1-*b*]quinazoline derivatives was developed.[432] The reaction proceeded *via* copper-catalyzed Ullmann type intermolecular C–C and intramolecular C–N couplings under mild conditions. The protocol displays wide functional group compatibility and provides the desired products in moderate to good yields (Scheme 2.178d).

The isoindole nucleus is becoming increasingly important due to its photochemical properties and ability to participate in cycloaddition reactions. Moreover, it forms a component unit of some bioactive compounds ascribed with antiproliferative activity. In 2009, Solé and Serrano reported a Pd-catalyzed intramolecular α-arylation of α-amino acid esters to provide substituted isoindole derivatives that have been used in Diels–Alder reactions to access diverse skeletal frameworks.[433] The methodology consisted of α-arylation followed by dehydrogenation of the previously formed product to result in isoindoles in good yields (Scheme 2.179a). Later on, an efficient Cu-catalyzed condensation–α-arylation cascade process of 2-halobenzaldehydes with glycine esters to isoindoles was described.[434] The reactions proceeded well with 2-brombenzaldehydes and glycine alkyl esters and displayed good functional group compatibility and scalability. The desired alkyl 2*H*-isoindole-2-carboxylates were produced in good yields under mild conditions (Scheme 2.179b).

1*H*-Indazoles have been recognized in a number of biologically active molecules and have been reported as antidepressants and contraceptives as well. The reaction between 2-halobenzaldehydes and hydrazines offers the most direct pathway. In 2004, Cho and co-workers reported the synthesis of indazoles from 2-bromobenzaldehydes and arylhydrazines (Scheme 2.180a).[435] The reactions were performed in toluene at 100 °C in the presence of a catalytic amount of a palladium catalyst and phosphorus chelating ligands (DPPF or DPPP) along with NaO*t*Bu as the base. Meanwhile, a study by Voskoboynikov's group showed that *rac*-BINAP and DPEphos can be applied as effective ligands and Cs_2CO_3 or K_3PO_4 as an alternative base as well. Electron-rich, bulky ligands commonly employed

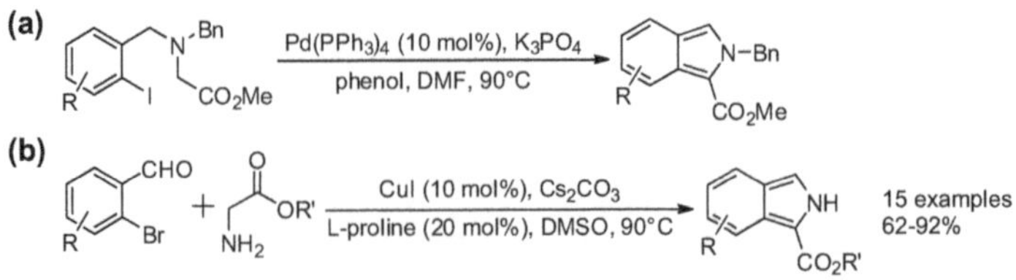

Scheme 2.179 Procedures for isoindole synthesis.

for intermolecular amination such as $PtBu_3$ and *o*-$PhC_6H_4PtBu_2$ were shown to be ineffective for cyclization and led instead to extensive oligomerization and tarring.[436] $TsNHNH_2$ was explored as a coupling partner with 2-bromo acetophenones too.[437] In this case, P(*o*-tolyl)$_3$ was found to be the best ligand with LiHMDS as the base. This method was successfully applied in the synthesis of nigellicine. Copper salts were also applied as effective catalysts for this transformation (Scheme 2.180b).[438–440] Recently, Lee's group reported that 2*H*-indazoles could be prepared from 2-bromobenzaldehydes, primary amines, and sodium azide.[441] Based on this copper-catalyzed, one-pot, three-component reaction, indazoles were formed in moderate to excellent yields with high functional group tolerance (Scheme 2.180c). Here, the copper catalyst plays a key role in the formation of C–N and N–N bonds. Cu^{II}-hydrotalcite[442] and CuO nanoparticles[443] were applied as heterogeneous catalysts for this reaction as well. In these cases, the catalysts were recovered and reused. In 2000, Song and Yee reported the synthesis of indazoles *via* palladium-catalyzed intramolecular amination.[444,445] A variety of 2-aryl-2*H*-indazoles were synthesized from the corresponding *N*-aryl-*N*-(*o*-bromobenzyl) hydrazines in good yields with the combination of $Pd(OAc)_2$/DPPF/*t*BuONa as the best system (Scheme 2.180d). However, the authors did not clarify how the oxidation occurred. Ma and co-workers developed a CuBr-catalyzed coupling reaction of 2-halobenzonitriles with hydrazine carboxylic esters and a CuBr/4-hydroxy-L-proline-catalyzed coupling reaction of 2-bromobenzonitriles with *N′*-arylbenzohydrazides to the corresponding 3-aminoindazoles in 2013 (Scheme 2.180e).[446] The reactions proceed

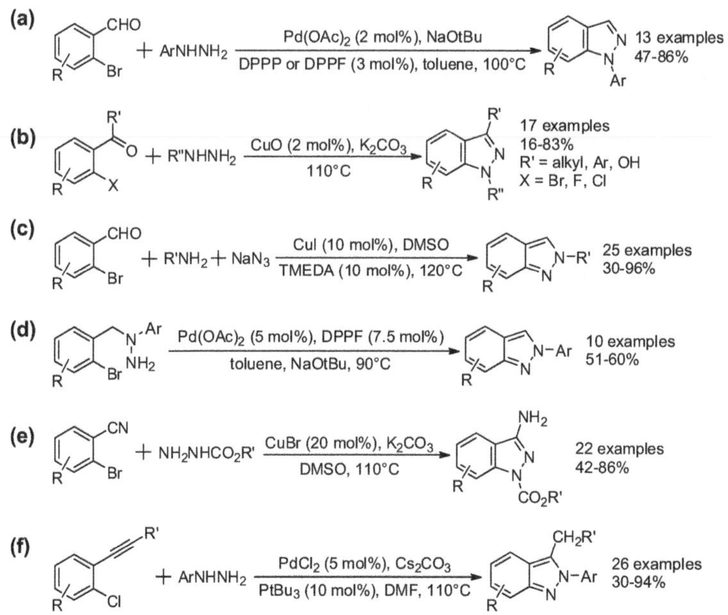

Scheme 2.180 Procedures for indazole synthesis.

smoothly at 60–90 °C to provide substituted 3-aminoindazoles in good yields through a cascade coupling–condensation (or coupling–deacylation–condensation). In the case of using 2-fluoro benzonitriles and methylhydrazine as substrates, the reactions proceeded in a catalyst-free manner.[447] Alternatively, 2*H*-indazoles could be produced from 2-halophenylacetylenes and hydrazines as well (Scheme 2.180f).[448] This method involves the first regioselective coupling of aryl halides with monosubstituted hydrazines, which undergo a reaction at the unsubstituted N atom, and the incorporation of this reaction into a domino coupling–cyclization–isomerization sequence. Moderate to excellent yields of the desired products were obtained.

Carbazole has been known as an important intermediate for artificial dyes and its photophysical properties. The reaction between primary amines or primary amides and 2,2′-dihalobiphenyls is the most straightforward pathway. Both palladium and copper catalysts were explored in this topic.[449–454] Indolines as an important class of heterocyclic compounds have been extensively studied during the past decades. As early as 1986, Grigg and coworkers reported the synthesis of indoline derivatives from various *N*-vinyl- and *N*-allyl-amides of 2-iodobenzoic acids. Afterwards, systematic studies on this topic have been carried out by them and other groups.[455–463] Regarding the substrates, the corresponding indole derivatives were reported as well. In 2012, a more detailed study was reported, which gave an efficient procedure for the construction of fused indolines with a 2-quaternary center through a palladium-catalyzed intramolecular Heck reaction of *N*-2(2-halobenzoxyl)-2,3-disubstituted indoles (Scheme 2.181).[464] This protocol provided straightforward access to diverse fused indolines with good functional group tolerance.

Transition metal-catalyzed C–H activation has been explored and applied in the preparation of heterocyclic compounds during the last few years. In 2010, a mechanistic study on palladium-catalyzed intramolecular C(sp^3)–H activation was reported.[465] The group found that diminishing the Lewis basicity of the nitrogen lone pair is crucial for this catalytic activity. A range of *N*-methylamides and sulfonamides react exclusively at primary C(sp^3)–H bonds to afford the products of alkane arylation in good yields (Scheme 2.182a). The isolation of a Pd(II) reaction intermediate has enabled an evaluation of the reaction mechanism with a focus on the role of the bases in the C(sp^3)–H bond cleaving step. The results of these stoichiometric studies, together with kinetic isotope effect experiments, provide rare experimental support for a concerted metalation–deprotonation (CMD) transition state, which has previously been proposed in alkane C(sp^3)–H arylation. Moreover, DFT calculations have uncovered the additional role of

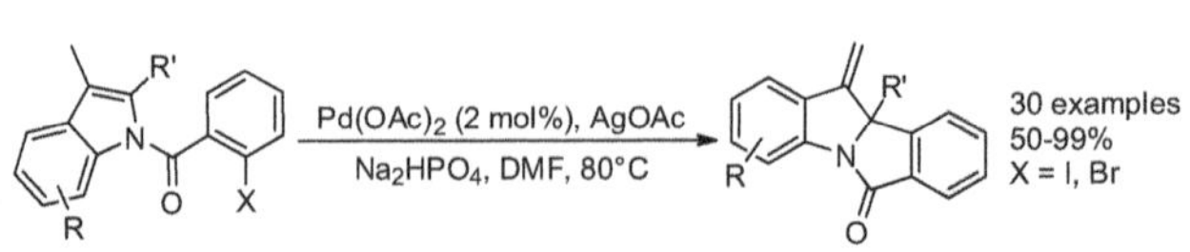

Scheme 2.181 Pd-catalyzed indoline synthesis.

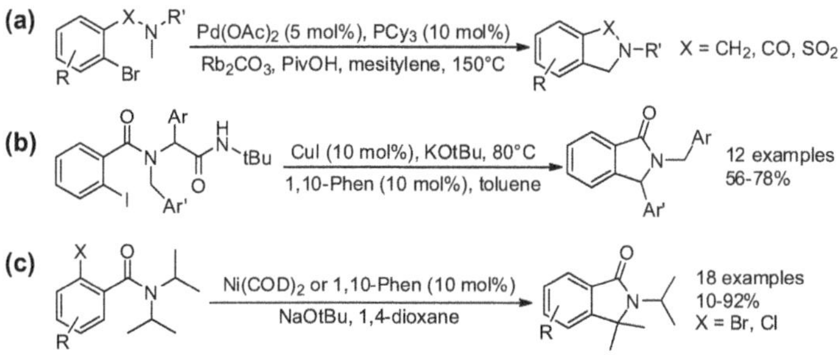

Scheme 2.182 Applying C–H activation in indoline synthesis.

the pivalate additive as a promoter of phosphine dissociation from the Pd(II) intermediate, enabling the CMD transition state. Finally, kinetic studies were performed, revealing the reaction rate expression and its relationship with the concentration of pivalate. Later on, the cyclization of Ugi-adducts was reported,[466,467] which was studied further with a copper catalyst (Scheme 2.182b).[468] More recently, $Ni(COD)_2$ or 1,10-phenanthroline was explored in this transformation (Scheme 2.182c).[469] Based on their study, aryl and alkyl radical intermediates were proposed.

The cyclization of *o*-halobenzamides with terminal alkynes gives another pathway for isoindolinone synthesis. In 1997, Kundu's group reported a palladium-catalyzed cyclization of *o*-iodobenzamides and its *N*-substituted derivatives with terminal alkynes.[470–472] The desired (*Z*)-3-aryl(alkyl)idene isoindolinones were formed in moderate to good yields (Scheme 2.183a). Recently, this methodology has been extended to *o*-bromobenzamides with the assistance of microwaves and without copper salt.[473] By adding cetyltrimethylammonium bromide (CTAB) as a surfactant with $Pd(CH_3CN)Cl_2$ as a catalyst and 1,4-bis(4-pyridyl)-2,3-diaza-1,3-butadiene as a ligand, this transformation can be realized under aerobic conditions in water.[474] Cossy and co-workers reported an efficient stereoselective access to (*E*)-3-(arylmethylene)isoindolin-1-ones in 2004.[475,476] By using ynamides and arylboronic acids as substrates, followed by Pd(0)-catalyzed Heck–Suzuki–Miyaura domino reaction sequences, the desired products were formed in moderate to good yields (Scheme 2.183b). 3-Methyleneisoindolin-1-ones can be produced when aryl boronic acids are replaced with ammonium formate, which can be produced by palladium-catalyzed carbonylation of 2-bromoacetophenones with primary amines alternatively (Scheme 2.183c).[477] In 1989, the use of a titanium–isocyanate complex as a nitrogen source, which was prepared from $TiCl_4$ under an atmospheric pressure of molecular nitrogen and carbon monoxide, for isoindolinone synthesis was developed.[478] In the presence of a palladium catalyst, by the combination of carbonylation and nitrogenation, isoindolinone and quinazolinone derivatives were synthesized from *o*-halophenyl alkyl ketone in one step in good yields (Scheme 2.183d). The reaction proceeds through the enol

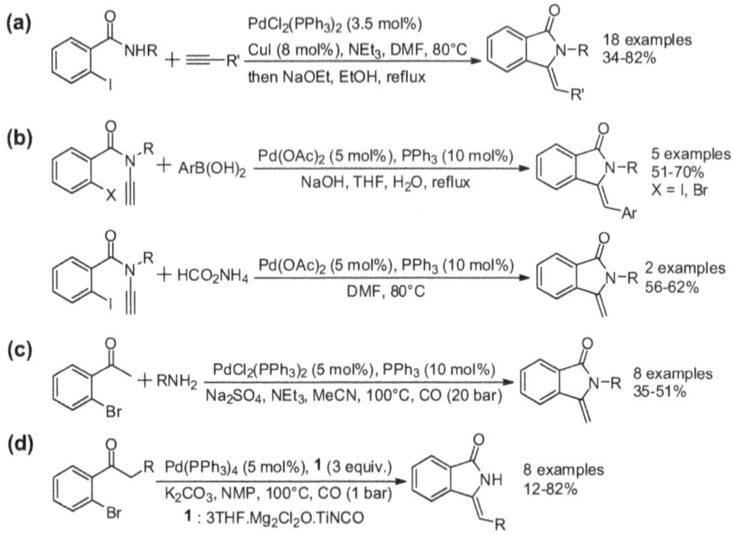

Scheme 2.183 Palladium-catalyzed isoindolinone synthesis.

lactone, generated by palladium-catalyzed carbonylation to *o*-halophenyl alkyl ketone, as the intermediate and then reacts with the titanium–isocyanate complex to give the final product.

In 2009, Ma, Jiang and their co-workers developed a CuI/L-proline-catalyzed coupling–additive cyclization domino reaction process for assembling substituted 3-methyleneisoindolin-1-ones from 2-bromobenzamides and terminal alkynes.[479] Variation of *N*-substituents, aromatic ring, and the methylene part was proven possible and provided the corresponding products in good yields (Scheme 2.184). Notably, this methodology was applied in the synthesis of lennoxamine precursor. Afterwards, this procedure was modified and new technologies and copper catalysts were tested. For example, $Cu(OAc)_2 \cdot H_2O$/DBU,[480] $Cu(OAc)_2 \cdot H_2O$/2,2′-biimidazole,[481] water as solvent,[482] and heterogeneous copper oxide.[483]

2-Bromobenzaldehydes have been applied as a class of promising substrates for palladium-catalyzed carbonylative synthesis of isoindolinones. In 1997, Shim and co-workers reported a palladium-catalyzed synthesis of 3-(alkylamino)isoindolinones by carbonylative cyclization of 2-bromobenzaldehyde with primary amines.[484] With NEt_3 as the base under solvent-free conditions, the desired products were formed in good yields (Scheme 2.185a). Interestingly, 3-position non-substituted isoindolinones can be produced by adding DMF as the solvent (Scheme 2.185b).[485] In addition to hydrazines,[486] tricyclic isoindolinones can be produced by using 2-aminoethanol or diamines as the coupling partners (Scheme 2.185c).[487,488] Based on a mechanistic study with 2-iodobenzoyl chloride and aldimines as starting materials, the group elucidated the reaction pathway and explored the reaction further (Scheme 2.185d).[489,490] Grigg and co-workers showed

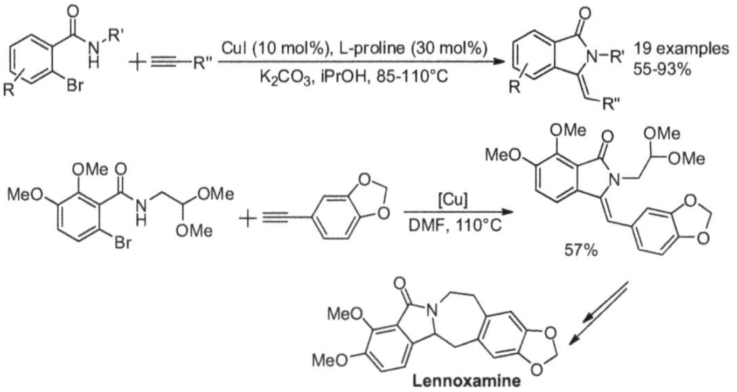

Scheme 2.184 Copper-catalyzed isoindolinone synthesis from 2-bromobenzamides.

Reaction Procedure (Scheme 2.184): An oven-dried Schlenk tube was charged with CuI (0.05 mmol), L-proline (0.15 mmol), potassium carbonate (1 mmol), and 2-bromobenzylamide (0.5 mmol). The tube was evacuated and backfilled with argon (3 times), and then phenylacetylene (83 μl, 0.75 mmol), and *i*-PrOH (1.0 mL) were added. The reaction mixture was stirred at 85 °C for 24–28 h. After cooling, 30 mL ethyl acetate was added. The organic layer was separated and washed with 1 N HCl (15 mL), brine (20 mL), and dried over Na_2SO_4. After removal of the solvent *in vacuo*, the residue was purified by silica gel chromatography to give the desired product.

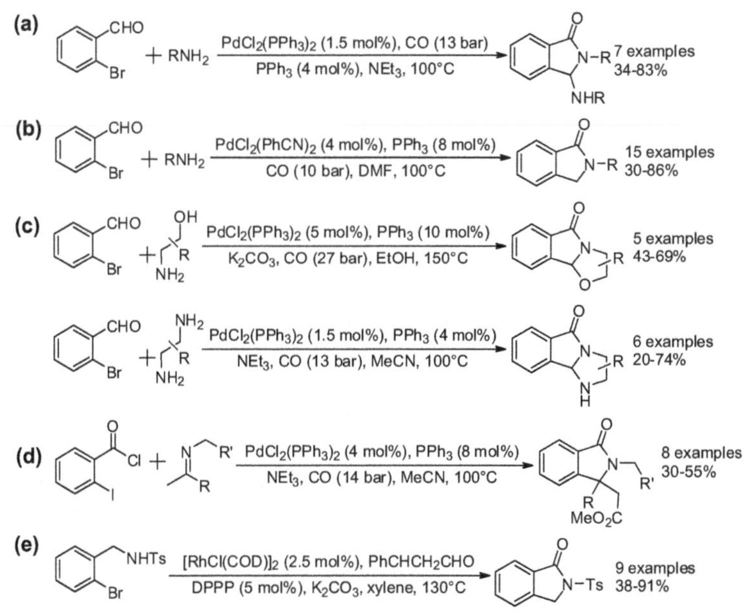

Scheme 2.185 Carbonylative synthesis of isoindolinones.

that palladium-catalyzed carbonylative intramolecular cyclization of 2-halo benzylamines is a powerful pathway for isoindolinone preparation.[491] Using mono-substituted benzylamines as substrates *via* C–H activation was proved possible.[492] A rhodium complex was applied in the carbonylative cyclization of 2-bromo benzylamines too (Scheme 2.185e).[493] In this case, no CO gas was needed, aldehyde was applied as a CO source. Aldehydes having electron-withdrawing groups showed a higher ability to donate the carbonyl moiety. Low to moderate yields of the five-, six-, and seven-membered heterocycles were obtained. Among the products, γ-butyrolactams can be produced from the corresponding alkyl halides. Grigg's group later reported the synthesis of isoindolinones from 2-iodobenzylbromides and primary amines *via* carbonylation.[494] A well-defined palladacycle was used as the catalyst. Recently, Kollár and co-workers restudied this transformation.[495] 1-Isoindolinone derivatives were synthesised in high yields by using 2-iodobenzyl bromide and 2-iodobenzylamine as bifunctional substrates *via* carbonylation with $Pd(OAc)_2$ as the catalyst precursor. Depending on the *N*-nucleophiles, two types of compounds were synthesised with 2-iodobenzyl bromide: the use of primary amines, including amino acid methylesters, resulted in the formation of *N*-substituted 1-isoindolinones, while secondary amines react both with the benzyl bromide and iodoarene moieties resulting in the corresponding *ortho*-(*N*-piperidino/morpholinomethyl)-benzamides. The parent 1-isoindolinone was obtained in a facile, highly chemoselective intramolecular aminocarbonylation of 2-iodobenzylamine. More recently, heterogeneous palladium catalysts were tested in the carbonylative cyclization of 2-bromobenzylamine from case to case.[496,497] Alper and co-workers reported the carbonylative cyclization of 1-halo-2-alkynylbenzenes with primary amines to 3-methyleneisoindolin-1-ones in good yields and high selectivities in favor of the *Z*-isomers.[34]

Isoindole-1,3-diones, better known as phthalimides, are key structural units of a variety of biologically important compounds, many of which are pharmaceutically significant. In 2008, Larock and co-worker developed a palladium-catalyzed aminocarbonylation of *o*-halobenzoates to produce 2-substituted isoindole-1,3-diones in good yields.[498] A range of isoindole-1,3-diones have been obtained by this one-step procedure (Scheme 2.186a). The reaction conditions are mild and the products are easy to isolate in good yields. A halogen moiety can also be introduced into the products, which provides a useful handle for further functionalization of the resulting heterocycles. This methodology tolerates a number of functional groups, including alcohol, ketone, methoxy, and nitro groups, and works well for both aliphatic and aromatic primary amines.[499] Later on, *o*-iodobenzoic acids were shown to be suitable substrates as well.[500] In this case, a supported Pd–NHC complex was applied as the catalyst, which makes the catalyst reusable. A nickel-catalyzed coupling of isocyanates with *o*-iodobenzoates to phthalimides was developed by Cheng's group (Scheme 2.186b).[501] A possible reaction mechanism was given—the catalytic reaction was probably initiated by the reduction of Ni(II) species to Ni(0) species by zinc powder. The iodo

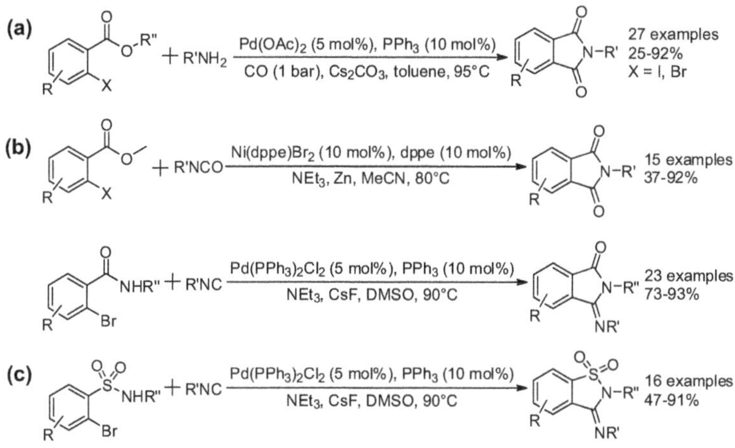

Scheme 2.186 Synthesis of isoindole-1,3-diones from *o*-iodobenzoates.

compound then undergoes oxidative addition to the Ni(0) complex to form the Ni(II) intermediate followed by insertion of an isocyanate molecule which will subsequently lead to the final product after imidation and regenerate the Ni(II) species. Jiang and co-workers developed the cyclization of *in situ* formed 2-bromobenzamides with isocyanides in 2012.[502] (3*E*)-(Imino)-isoindolin-1-ones and (3*E*)-(imino)thiaisoindoline 1,1-dioxides were afforded in good to excellent yields (up to 93%) with high stereoselectivity (Scheme 2.186c). Notably, this transformation was successfully extended to the synthesis of phenanthridines and dibenzooxazepines.

In 1999, Buchwald and Yang reported a palladium-catalyzed cyclization of secondary amides and carbamates. Indolone was produced in 82% yield with $Pd(OAc)_2$/MOP as the catalyst and K_2CO_3 as the base in toluene at 100 °C.[503] In 2005, Turner's group reported a procedure starting from 5-halophenyl acetic acids and primary amines (Scheme 2.187a).[504] Microwave heating was applied to complete the reaction in a short time. Various amines were applicable, not only aliphatic amines, but also benzyl amines and anilines. The cyclization of Ugi-products to give the corresponding indolones *via* C–N bond formation was reported as well.[505] The intramolecular cyclization of 2-bromophenyl pyrrol-2-yl ketone with a palladium or copper catalyst to the corresponding fluorazones was also described.[506,507] In 2011, a copper-catalyzed intramolecular *N*-arylation of hydroxamic acid derivatives was presented.[508] Based on this transformation, a series of *N*-alkoxyindol-2-ones was produced from the corresponding 2-(2-bromoaryl)acetylhydroxamates in moderate to excellent yields (Scheme 2.187b). The reaction conditions tolerate standard hydroxyl protecting groups on the hydroxylamine moiety and are also applicable for the synthesis of six-membered *N*-alkoxybenzolactams. Alternatively, Hsieh and co-workers discovered a method that can synthesize indolones from *o*-bromobenzyl cyanides.[509] By using copper as the catalyst, various oxindoles

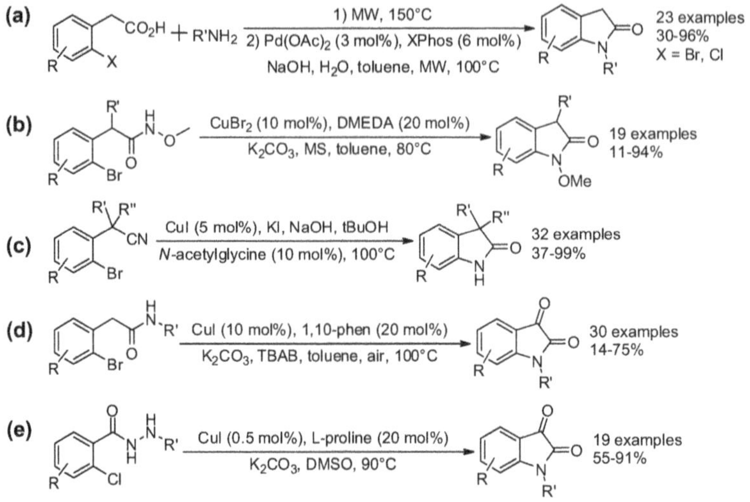

Scheme 2.187 Synthesis of indolones and indazolones.

were formed in good to excellent yields with tolerance of various substituents (Scheme 2.187c). In addition, this method could be applied to synthesize horsfiline and coerulescine in a few steps with high total yields. Interestingly, Xu and co-workers developed an efficient copper-catalyzed procedure for the synthesis of *N*-substituted isatins from arylacetamides.[510] A wide range of electronically and structurally varied nitrogen fragments could be assembled through this tandem C–O/C–N bond-forming process in good yields (Scheme 2.187d). Wang and co-workers developed a one-pot procedure for the synthesis of 2-(*N*-sulfonylimino)indolines.[511] The procedure combined a copper-catalyzed three-component reaction of sulfonyl azides, *o*-bromophenylacetylenes and amines with copper-catalyzed intramolecular C–N coupling in one sequence, which afforded the products in moderate to good yields. The resulting 2-(*N*-sulfonylimino)-indolines could be easily transformed to pharmaceutically valuable oxindoles (indolin-2-ones). A copper catalyst was also applied in the synthesis of indazolones.[512,513] By using 2-halobenzohydrazides as the substrates, a variety of 1-alkyl- and aryl-substituted indazolones were synthesized in good yields *via* a copper-catalyzed intramolecular C–N bond formation (Scheme 2.187e). Cyclization of hydrazine derived Ugi products with a copper or palladium catalyst to the corresponding indazolones was reported as well.[514]

The intramolecular amination of aryl halides could lead to the formation of saturated heterocycles. Depending on the length of the carbon chain, five-, six-, seven- and macroheterocycles can be produced in the presence of transition metal catalysts. Buchwald and co-workers developed a series of palladium-catalyzed catalytic systems with phosphines as the ligand.[515–517] Alternative systems have been developed during the past years as well, with

NHC (IMs · HCl) as the ligand or even under ligand-free conditions.[518–522] Copper salts were applied in these transformations as well.[523–533] In 2003, a nickel-catalyzed system was also developed.[534] With Ni(0) generated *in situ* as the catalyst and 2,2′-bipyridine or *N*,*N*′-bis(2,6-diisopropylphenyl)dihydroimidazol-2-ylidene (SIPr) as the ligand, the intramolecular coupling of aryl chlorides with amines was achieved with NaO*t*Bu as the base. Indoles, quinolines, benzazepines, benzoxazines, and benzoxazepines were produced in good yields from the corresponding substrates.

Procedures for the synthesis of fused isoindolines were developed as well. In 2005, a procedure based on using *o*-iodobenzyl azide and acetylenes as substrates with palladium and copper as the catalyst was developed.[535] Triazole fused isoindolines were produced in good yields (Scheme 2.188a). In this work, the authors proposed that the reaction started with Sonogashira coupling and then a Click reaction to give the final product. Later on, it was shown that this transformation can start with a copper-catalyzed Click reaction and then palladium-catalyzed C–H activation at the *ortho* position to give the same products.[536] A heterogeneous palladium and copper catalyst (Pd–Cu/C) was prepared and applied as well.[537] In 2006, Mori and co-workers reported a palladium-catalyzed intramolecular C–H arylation of imidazole derivatives.[538] The cyclization precursor was prepared in an excellent yield by the reaction of imidazole with 2-iodobenzyl bromide in the presence of NaH. The desired 5*H*-imidazo[5,1-*a*]isoindole was produced in good yield *via* palladium-catalyzed intramolecular C–H arylation at 100 °C (Scheme 2.188b). A procedure based on *in situ* generation of substrates was developed as well.[539] This transformation can be realized by a copper catalyst [CuI (20 mol%), 1,10-Phen (20 mol%), K_3PO_4, DMF, 145 °C] too.[540] In 2009, a protocol for the synthesis of imidazoisoindol-3-ones was described.[541] By employing a palladium-catalyzed intramolecular C–H insertion reaction of substituted 2-haloaryl

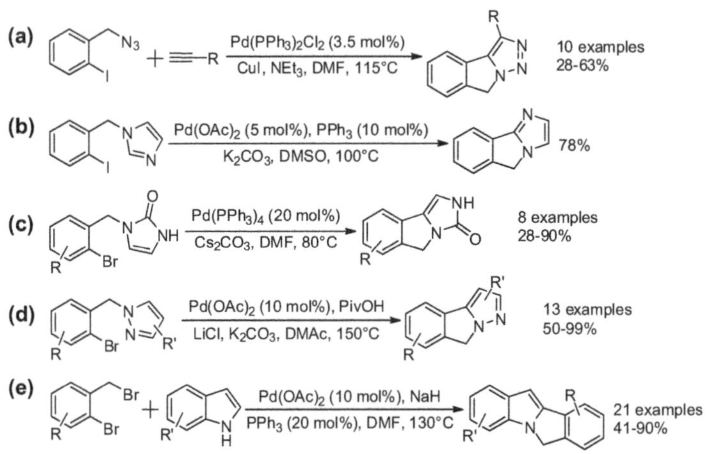

Scheme 2.188 Synthesis of fused isoindolines and related compounds.

imidazolin-2-ones, the desired products were isolated in good yields (Scheme 2.188c). In 2010, Heo and co-workers reported a procedure for the synthesis of pyrazolo-[5,1-*a*]isoindoles employing a palladium-catalyzed intramolecular C–H bond activation of 1-(2-halobenzyl)pyrazoles (Scheme 2.188d).[542] In this methodology, the use of lithium chloride (LiCl) was found to be essential in these reactions, to suppress further C–H bond activation at the C-3 position of pyrazolo[5,1-*a*]isoindole, when C-3 is unsubstituted. This protocol can also be applied to the synthesis of a pyrazolo[5,1-*a*]isoquinoline possessing a six-membered central ring system and a fully substituted pyrazolo[5,1-*a*]isoindole using sequential intra- and intermolecular C–H bond activation. More recently, a palladium-catalyzed domino *N*-benzylation/C–H arylation of nitrogen heterocycles with 2-bromobenzyl bromides was reported by Laha's group.[543] A variety of fused nitrogen heterocycles with diverse substitution patterns were produced in good yields (Scheme 2.188e). Here, PPh_3 was found to be the real catalyst for the reaction between benzyl bromide and N-H to produce the intermediate. The intramolecular cyclization with activated CH_2 was developed as well.[544–546]

Knochel and Ren reported a Pd-catalyzed intramolecular cyclization with a methyl group in 2006.[547] The key step of these ring closures is a chemoselective intramolecular C–H activation of a methyl group at position 2 of a pyrrole ring. Good yields of the desired fused products can be produced (Scheme 2.189).

Wu and Luo developed a novel route for the efficient assembly of indeno[1,2-*c*]pyrrole derivatives *via* a palladium-catalyzed tandem reaction of 2-alkynylbromobenzene with propargylic sulfonamide in 2012.[548] The starting materials are easily available, and the reaction proceeds smoothly with good functional group tolerance. This strategy involving double insertion of triple bonds shows high efficiency, and good yields of the corresponding products were isolated (Scheme 2.190).

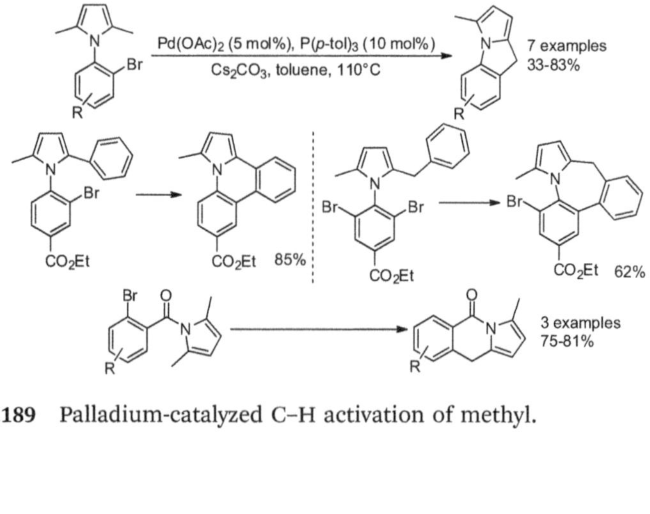

Scheme 2.189 Palladium-catalyzed C–H activation of methyl.

Reaction Procedure (Scheme 2.189): A mixture of the aryl bromide (1.0 mmol), $Pd(OAc)_2$ (5 mol%), tri(*p*-tolyl)phosphine (10 mol%), and Cs_2CO_3 (1.2 mmol) was heated at 110 °C in toluene (5 mL) under N_2 in a sealed tube for 12 h. After the reaction mixture had cooled to room temperature, water (10 mL) was added. The mixture was extracted with diethyl ether (3×30 mL), and the combined extracts were washed with brine, dried over Na_2SO_4, and concentrated *in vacuo*. Purification by flash chromatography (hexane–diethyl ether, 10 : 1) provided the desired product.

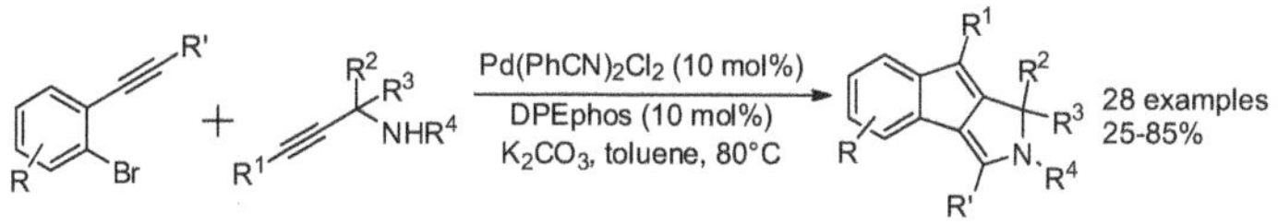

Scheme 2.190 Palladium-catalyzed indeno[1,2-*c*]pyrrole synthesis.

Reaction Procedure (Scheme 2.190): A mixture of 2-alkynylbromobenzene (0.2 mmol), propargylic sulfonamide (0.3 mmol), $Pd(PhCN)_2Cl_2$ (10 mol%), DPEPhos (10 mol%) and K_2CO_3 (0.4 mmol) in toluene (0.5 mL) was stirred at 80 °C. After completion of the reaction as indicated by TLC (usually 10 hours), the residue was purified immediately by flash chromatography on silica gel to provide indeno[1,2-*c*]pyrrole.

2.5 Miscellaneous

A procedure for the preparation of benzo[*b*]thiophenes by Pd(0)-catalyzed intramolecular cyclization of allyl (and propargyl) *o*-iodophenyl sulfides was developed in 1993.[549] Benzo[*b*]thiophenes were prepared in good yields by intramolecular Pd(0)-catalyzed Heck reaction of allyl *o*-iodophenyl sulfides. Intramolecular cyclization of *o*-iodophenyl propargyl sulfide in the presence of a hydride donor gave 3-methylene-2,3-dihydrobenzo[*b*]thiophene, which reacts with several enophiles in ene type reactions. However, allyl (and propargyl) aryl sulfides react with palladium(II) chloride to afford polymeric $[PdCl(SAr)]_2$. In 2004, a nickel-catalyzed electrochemical synthesis of dihydrobenzo[*b*]thiophene derivatives was developed.[550] The intramolecular electrochemical reductive cyclization of *ortho*-haloaryl allyl thioethers catalyzed by Ni(II) complexes associated to cyclam ligands afforded dihydrobenzo[*b*]thiophene derivatives in moderate to good yields. By a similar concept, benzosiloles, six- and eight-membered silacyclic skeletons can be produced as well. In 2012, Xi and co-workers developed an efficient Pd-catalyzed intramolecular Mizoroki–Heck reaction of vinylsilanes to construct benzosiloles, six- and eight-membered silacyclic skeletons.[551] The stability of the formed silacyclic rings controlled the reaction sites (α- or β-position) of the vinylsilane moieties. This method provides an efficient route to benzosilole derivatives, six- and eight-membered silacyclic compounds.

In summary, the main applications of double-functionalized arenes in the synthesis of five-membered heterocycles have been discussed. Various types of five-membered heterocycles were produced by the combination of different functional groups. In the following chapter, the preparation of six-membered heterocycles will be discussed.

References

1. R. J. Perry and S. R. Turner, *J. Org. Chem.*, 1991, **56**, 6573–6579.
2. H. Cao and H. Alper, *Org. Lett.*, 2010, **12**, 4126–4129.
3. D. Marosvölgyi-Haskó, A. Petz, A. Takács and L. Kollár, *Tetrahedron*, 2011, **67**, 9122–9128.
4. J. Chen, K. Natte, A. Spannenberg, H. Neumann, M. Beller and X.-F. Wu, *Org. Biomol. Chem.*, 2014, **12**, 5578–5581.
5. W. Li, M. Beller, X.-F. Wu, non-published results.
6. D. N. Sawant, Y. S. Wagh, K. D. Bhatte and B. M. Bhanage, *Eur. J. Org.*, 2011, 6719–6724.
7. M. V. Khedkar, S. R. Khan, D. N. Sawant, D. B. Bagal and B. M. Bhanage, *Adv. Synth. Catal.*, 2011, **353**, 3415–3422.
8. M. V. Khedkar, A. R. Shinde, T. Sasaki and B. M. Bhanage, *J. Mol. Catal. A: Chem.*, 2014, **385**, 91–97.
9. X.-F. Wu, S. Oschatz, M. Sharif, A. Flader, L. Krey, M. Beller and P. Langer, *Adv. Synth. Catal.*, 2013, **355**, 3581–3585.
10. Y. Terao, T. Satoh, M. Miura and M. Nomura, *Bull. Chem. Soc. Jpn.*, 1999, **72**, 2345–2350.
11. F. Churruca, R. SanMartin, I. Tellitu and E. Domínguez, *Eur. J. Org. Chem.*, 2005, 2481–2490.
12. M. C. Willis, D. Taylor and A. T. Gillmore, *Tetrahedron*, 2006, **62**, 11513–11520.
13. B. Lu, B. Wang, Y. Zhang and D. Ma, *J. Org. Chem.*, 2007, **72**, 5337–5341.
14. N. Aljaar, C. C. Malakar, J. Conrad, S. Strobel, T. Schleid and U. Beifuss, *J. Org. Chem.*, 2012, **77**, 7793–7803.
15. G. Altenhoff and F. Glorius, *Adv. Synth. Catal.*, 2004, **346**, 1661–1664.
16. X.-F. Wu, H. Neumann, S. Neumann and M. Beller, *Tetrahedron Lett.*, 2013, **54**, 3040–3042.
17. (a) S.-K. Xiang, D.-X. Zhang, H. Hu, J.-L. Shi, L.-G. Liao, C. Feng, B.-Q. Wang, K.-Q. Zhao, P. Hu, H. Yang and W.-H. Yu, *Adv. Synth. Catal.*, 2013, **355**, 1495–1499; (b) D.-X. Zhang, S.-K. Xiang, H. Hu, W. Tan, C. Feng, B.-Q. Wang, K.-Q. Zhao, P. Hu and H. Yang, *Tetrahedron*, 2013, **69**, 10022–10029.
18. (a) L. Ackermann and A. Althammer, *Angew. Chem., Int. Ed.*, 2007, **46**, 1627–1629; (b) L. Ackermann, A. Althammer and P. Mayer, *Synthesis*, 2009, 3493–3503.

19. Y. Kitamura, S. Yoshikaawa, T. Furuta and T. Kan, *Synlett*, 2008, 377–380.
20. S. D. Edmondson, A. Mastracchio and E. R. Parmee, *Org. Lett.*, 2000, **2**, 1109–1112.
21. T. Jensen, H. Pedersen, B. Bang-Andersen, R. Madsen and M. Joergensen, *Angew. Chem., Int. Ed.*, 2008, **47**, 888–890.
22. J. M. Knapp, J. S. Zhu, D. J. Tantillo and M. J. Kurth, *Angew. Chem., Int. Ed.*, 2012, **51**, 10588–10591.
23. W. Hao, W. Geng, W.-X. Zhang and Z. Xi, *Chem. Eur. J.*, 2014, **20**, 2605–2612.
24. (a) J. Barluenga, A. Jiménez-Aquino, C. Valdés and F. Aznar, *Angew. Chem., Int. Ed.*, 2007, **46**, 1529–1532; (b) J. Barluenga, A. Jiménez-Aquino, F. Aznar and C. Valdés, *J. Am. Chem. Soc.*, 2009, **131**, 4031–4041; (c) J. Barluenga, A. Jiménez-Aquino, F. Aznar and C. Valdés, *Chem. Eur. J.*, 2010, **16**, 11707–11711.
25. X. Deng, H. McAlister and N. S. Mani, *J. Org. Chem.*, 2009, **74**, 5742–5745.
26. H. Hammoud, M. Schmitt, F. Bihel, C. Antheaume and J.-J. Bourguignon, *J. Org. Chem.*, 2012, **77**, 417–423.
27. D. Zhao, J. Hu, N. Wu, X. Huang, X. Qin, J. Lan and J. You, *Org. Lett.*, 2011, **13**, 6516–6519.
28. N. T. Jui and S. L. Buchwald, *Angew. Chem., Int. Ed.*, 2013, **52**, 11624–11627.
29. K. T. J. Loones, B. U. W. Maes, C. Meyers and J. Deruytter, *J. Org. Chem.*, 2006, **71**, 260–264.
30. K. T. J. Loones, B. U. W. Maes and R. A. Dommisse, *Tetrahedron*, 2007, **63**, 8954–8961.
31. T. Beresneva, J. Popelis and E. Abele, *Chem. Hetero. Comp.*, 2013, **49**, 345–347.
32. J. H. Kim, Y. S. Chun, H. Shin and S. Lee, *Synthesis*, 2012, 1809–1817.
33. J. H. Kim and S. Lee, *Synthesis*, 2012, 1464–1476.
34. H. Cao, L. McNamee and H. Alper, *Org. Lett.*, 2008, **10**, 5281–5284.
35. Z.-W. An, M. Catellani and G. P. Chiusoli, *J. Organomet. Chem.*, 1989, **371**, C51–C52.
36. P. G. Ciattini, E. Morera, G. Ortar and S. S. Rossi, *Tetrahedron*, 1991, **47**, 6449–6456.
37. E. Rizzi, S. Dallavalle, L. Merlini, G. L. Beretta, G. Partesi and F. Zunino, *Bioorg. Med. Chem. Lett.*, 2005, **15**, 4313–4316.
38. X. Lv, Y. Liu, W. Qian and W. Bao, *Adv. Synth. Catal.*, 2008, **350**, 2507–2512.
39. Q. Ding, X. Liu, B. Cao, Z. Zong and Y. Peng, *Tetrahedron Lett.*, 2011, **52**, 1964–1967.
40. F. Yao, W. Hao and M.-Z. Cai, *J. Organomet. Chem.*, 2013, **723**, 137–142.
41. D. E. Ames and A. Opalko, *Synthesis*, 1983, 234–235.
42. D. E. Ames and A. Opalko, *Tetrahedron*, 1984, **40**, 1919–1925.
43. B. Dajka-Halász, K. Monsieurs, O. Éliás, L. Károlyházy, P. Tapolcsányi, B. U. W. Maes, Z. Riedl, G. Hajós, R. A. Dommisse, G. L. F. Lemière, J. Kosmrlj and P. Mátyus, *Tetrahedron*, 2004, **60**, 2283–2291.

44. W. S. Yoon, S. J. Lee, S. K. Kang, D.-C. Ha and J. D. Ha, *Tetrahedron Lett.*, 2009, **50**, 4492–4494.
45. H. Xu and L.-L. Fan, *Chem. Pharm. Bull.*, 2008, **56**, 1496–1498.
46. H. Lusic, R. Uprety and A. Deiters, *Org. Lett.*, 2010, **12**, 916–919.
47. Z. Liu and R. C. Larock, *Org. Lett.*, 2004, **6**, 3739–3741.
48. M. Parisien, D. Valette and K. Fagnou, *J. Org. Chem.*, 2005, **70**, 7578–7584.
49. W. S. Yue and J. J. Li, *Org. Lett.*, 2002, **4**, 2201–2203.
50. (a) B. Xiao, T.-J. Gong, Z.-J. Liu, J.-H. Liu, D.-F. Luo, J. Xu and L. Liu, *J. Am. Chem. Soc.*, 2011, **133**, 9250–9253; (b) Y. Wei and N. Yoshikai, *Org. Lett.*, 2011, **13**, 5504–5507.
51. E. Motti, N. D. Ca, D. Xu, S. Armani, B. M. Aresta and M. Catellani, *Tetrahedron*, 2013, **69**, 4421–4428.
52. M. Tobisu, Y. Kita, Y. Ano and N. Chatani, *J. Am. Chem. Soc.*, 2008, **130**, 15982–15989.
53. D. M. Ferguson, S. R. Rubolph and D. Kalyani, *ACS Catal.*, 2014, **4**, 2395–2401.
54. R. Grigg, J. Redpath, V. Sridharan and D. Wilson, *Tetrahedron Lett.*, 1994, **35**, 4429–4432.
55. R. Grigg, J. P. Major, F. M. Martin and M. Whittaker, *Tetrahedron Lett.*, 1999, **40**, 7709–7711.
56. R. Grigg, J. M. Sansano, V. Santhakumar, V. Sridharan, R. Thangavelanthum, M. Thornton-Pett and D. Wilson, *Tetrahedron*, 1997, **53**, 11803–11826.
57. A. Casaschi, R. Grigg, J. M. Sansano, D. Wilson and J. Redpath, *Tetrahedron*, 2000, **56**, 7541–7551.
58. P. Fretwell, R. Grigg, J. M. Sansano, V. Sridharan, S. Sukirthalingam, D. Wilson and J. Redpath, *Tetrahedron*, 2000, **56**, 7525–7539.
59. U. Anwar, A. Casaschi, R. Grigg and J. M. Sansano, *Tetrahedron*, 2001, **57**, 1361–1367.
60. U. Anwar, M. P. Fielding, R. Grigg, V. Sridharan and C. J. Urch, *J. Organomet. Chem.*, 2006, **691**, 1476–1487.
61. R. Grigg, V. Savic, V. Sridharan and C. Terrier, *Tetrahedron*, 2002, **58**, 8613–8620.
62. P. Liu, L. Huang, Y. Lu, M. Dilmeghani, J. Baum, T. Xiang, J. Adams, A. Tasker, R. Larsen and M. M. Faul, *Tetrahedron Lett.*, 2007, **48**, 2307–2310.
63. M. Szlosek-Pinaud, P. Diaz, J. Martinez and F. Lamaty, *Tetrahedron*, 2007, **63**, 3340–3349.
64. J.-Q. Chen, J.-H. Xie, D.-H. Bao, S. Liu and Q.-L. Zhou, *Org. Lett.*, 2012, **14**, 2714–2717.
65. O. René, D. Lapointe and K. Fagnou, *Org. Lett.*, 2009, **11**, 4560–4563.
66. T. Piou, L. Neuville and J. Zhu, *Org. Lett.*, 2012, **14**, 3760–3763.
67. H. S. Kim, S. Gowrisankar, S. H. Kim and J. N. Kim, *Tetrahedron Lett.*, 2008, **49**, 3858–3861.
68. S. G. Neuman and M. Lautens, *J. Am. Chem. Soc.*, 2011, **133**, 1778–1780.

69. S. G. Newman, J. K. Howell, N. Nicolaus and M. Lautens, *J. Am. Chem. Soc.*, 2011, **133**, 14916–14919.
70. S. Olivero and E. Duñach, *J. Chem. Soc., Chem. Commun.*, 1995, 2497–2498.
71. R. Lhermet, M. Durandetti and J. Maddaluno, *Beilstein J. Org. Chem.*, 2013, **9**, 710–716.
72. C. Uyeda, Y. Tan, G. C. Fu and J. C. Peters, *J. Am. Chem. Soc.*, 2013, **135**, 9548–9552.
73. (a) T. Fujioka, T. Nakamura, H. Yorimitsu and K. Oshima, *Org. Lett.*, 2002, **4**, 2257–2259; (b) W. Affo, H. Ohmiya, T. Fujioka, Y. Ikeda, T. Nakamura, H. Yorimitsu, K. Oshima, Y. Imamura, T. Mizuta and K. Miyoshi, *J. Am. Chem. Soc.*, 2006, **128**, 8068–8077.
74. (a) K. Miura, M. Tomita, Y. Yamada and A. Hosomi, *J. Org. Chem.*, 2007, **72**, 787–792; (b) C. G. Martin, J. A. Murphy and C. R. Smith, *Tetrahedron Lett.*, 2000, **41**, 1833–1836; (c) N. Hayashi, I. Shibata and A. Baba, *Org. Lett.*, 2005, **7**, 3093–3096; (d) K. C. Majumdar and S. Alam, *Org. Lett.*, 2006, **8**, 4059–4062; (e) T. Fujita, S. Sanada, Y. Chiba, K. Sugiyama and J. Ichikawa, *Org. Lett.*, 2014, **16**, 1398–1401; (f) C. Morice, M. Domostoj, K. Briner, A. Mann, J. Suffer and C.-G. Wermuth, *Tetrahedron Lett.*, 2001, **42**, 6499–6502; (g) R. Rai and D. B. Collum, *Tetrahedron Lett.*, 1994, **35**, 6221–6224; (h) S. Wolff and H. M. R. Hoffmann, *Synthesis*, 1988, 760–763; (i) H. M. R. Hoffmann, B. Schmidt and S. Wolff, *Tetrahedron*, 1989, **45**, 6133–6126; (j) A. Y. Fedorov and J.-P. Finet, *Eur. J. Org. Chem.*, 2004, 2040–2045; (k) D. P. Curran and M. J. Totleben, *J. Am. Chem. Soc.*, 1992, **114**, 6050–6058; (l) A. Ogawa, Y. Sumino, T. Nanke, S. Ohya, N. Sonoda and T. Hirao, *J. Am. Chem. Soc.*, 1997, **119**, 2745–2746; (m) H. Bhandal, V. F. Patel, G. Pattenden and J. J. Russell, *J. Chem. Soc., Perkin Trans.*, 1990, **1**, 2691–2701.
75. R. Grigg, V. Sridharan and L.-H. Xu, *J. Chem. Soc., Chem. Commun.*, 1995, 1903–1904.
76. R. Grigg and V. Savic, *Tetrahedron Lett.*, 1996, **37**, 6565–6568.
77. L. F. Tietze and R. Schimpf, *Chem. Ber.*, 1994, **127**, 2235–2240.
78. D. M. Hodgson and C. Wells, *Tetrahedron Lett.*, 1994, **35**, 1601–1604.
79. M. Durandetti, L. Hardou, R. Lhermet, M. Rouen and J. Maddaluno, *Chem. Eur. J.*, 2011, **17**, 12773–12783.
80. R. C. Larock, N. Berrios-Peña and K. Narayanan, *J. Org. Chem.*, 1990, **55**, 3447–3450.
81. R. V. Rozhkov and R. C. Larock, *J. Org. Chem.*, 2010, **75**, 4131–4134.
82. H.-P. Bi, X.-Y. Liu, F.-R. Gou, L.-N. Guo, X.-H. Duan, X.-Z. Shu and Y.-M. Liang, *Angew. Chem., Int. Ed.*, 2007, **46**, 7068–7071.
83. W. Li and M. Shi, *Eur. J. Org. Chem.*, 2009, 270–274.
84. T. Boi, A. Deagostino, C. Prandi, S. Tabasso, A. Toppino and P. Venturello, *Org. Biomol. Chem.*, 2010, **8**, 2020–2027.
85. D. E. Emrich and R. C. Larock, *J. Organomet. Chem.*, 2004, **689**, 3756–3766.
86. (a) C. D. Netto, A. J. M. da Silva, E. J. S. Salustiano, T. S. Bacelar, I. G. Rica, M. C. M. Cavalcante, V. M. Rumjanek and P. R. R. Costa,

Bioorg. Med. Chem., 2010, **18**, 1610–1616; (b) R. A. C. Leão, V. D. Pinho, A. S. Coelho, C. D. Buarque, P. F. Moraes, D. A. Alonso, C. Nájera and P. R. R. Costa, *Eur. J. Org. Chem.*, 2011, 3313–3316.
87. J. John, R. Rajan, S. S. Chand, P. Prakash, N. Joseph, E. Suresh and K. V. Radhakrishnan, *Tetrahedron*, 2013, **69**, 152–159.
88. M. Lafrance, S. I. Gorelsky and K. Fagnou, *J. Am. Chem. Soc.*, 2007, **129**, 14570–14571.
89. R. C. Larock, E. K. Yum, M. J. Doty and K. K. C. Sham, *J. Org. Chem.*, 1995, **60**, 3270–3271.
90. N. G. Kundu, M. Pal, J. S. Mahanty and M. De, *J. Chem. Soc., Perkin Trans. 1*, 1997, 2815–2820.
91. A. Arcadi, F. Marinelli and S. Cacchi, *Synthesis*, 1986, 749–751.
92. T. Konno, J. Chae, T. Ishihara and H. Yamanaka, *Tetrahedron*, 2004, **60**, 11695–11700.
93. R. Grigg, V. Sridharan and D. A. Sykes, *Tetrahedron*, 2008, **64**, 8952–8962.
94. M. Yamaguchi, H. Katsumata and K. Manbe, *J. Org. Chem.*, 2013, **78**, 9270–9281.
95. G. W. Kabalka, L. Wang and R. M. Pagni, *Tetrahedron*, 2001, **57**, 8017–8028.
96. M. Shigeta, J. Watanabe and G.-i. Konishi, *Tetrahedron Lett.*, 2013, **54**, 1761–1764.
97. S. Ghosh, J. Das and F. Saikh, *Tetrahedron Lett.*, 2012, **53**, 5883–5886.
98. E. Genin, R. Amengual, V. Michelet, M. Savignac, A. Jutand, L. Neuville and J.-P. Genet, *Adv. Synth. Catal.*, 2004, **346**, 1733–1741.
99. M. Pal, V. Subramanian and K. R. Yeleswarapu, *Tetrahedron Lett.*, 2003, **44**, 8221–8225.
100. C. Rossy, E. Fouquet and F.-X. Felpin, *Beilstein J. Org. Chem.*, 2013, **9**, 1426–1431.
101. D. Saha, R. Dey and B. C. Ranu, *Eur. J. Org. Chem.*, 2010, 6067–6071.
102. A. Ohtaka, T. Teratani, R. Fujii, K. Ikeshita, T. Kawashima, K. Tatsumi, O. Shimomura and R. Nomura, *J. Org. Chem.*, 2011, **76**, 4052–4060.
103. E. K. Yum, O.-K. Yang, J.-E. Kim and H.-J. Park, *Bull. Korean Chem. Soc.*, 2013, **34**, 2645–2649.
104. C. Li, Y. Zhang, P. Li and L. Wang, *J. Org. Chem.*, 2011, **76**, 4692–4696.
105. M. Thévenin, S. Thoret, P. Grellier and J. Dubois, *Bioorg. Med. Chem.*, 2013, **21**, 4885–4892.
106. Y. Kommagalla, K. Srinivas and C. V. Ramana, *Tetrahedron Lett.*, 2013, **54**, 1824–1827.
107. K. Goswami, I. Duttagupta and S. Sinha, *J. Org. Chem.*, 2012, 77, 7081–7085.
108. M. W. Khan, M. J. Alam, M. A. Rashid and R. Chowdhury, *Bioorg. Med. Chem.*, 2005, **13**, 4796–4805.
109. P. M. Byers, J. I. Rashid, R. K. Mohamed and I. V. Alabugin, *Org. Lett.*, 2012, **14**, 6032–6035.
110. K. Okuro, M. Furuune, M. Enna, M. Miura and M. Nomura, *J. Org. Chem.*, 1993, **58**, 4716–4721.

111. (a) C. E. Castro, E. J. Gauchan and D. C. Owsley, *J. Org. Chem.*, 1966, **31**, 4071–4078; (b) G. J. S. Doad, J. A. Barltrop, C. M. Petty and T. C. Owen, *Tetrahedron Lett.*, 1989, **30**, 1597–1598.
112. (a) C. G. Bates, P. Saejueng, J. M. Murphy and D. Venkataraman, *Org. Lett.*, 2002, **4**, 4727–4729; (b) P. Saejueng, C. G. Bates and D. Venkataraman, *Synthesis*, 2005, 1706–1712.
113. J.-H. Li, J.-L. Li, D.-P. Wang, S.-F. Pi, Y.-X. Xie, M.-B. Zhang and X.-C. Hu, *J. Org. Chem.*, 2007, **72**, 2053–2057.
114. E. A. Jaseer, D. J. C. Prasad and G. Sekar, *Tetrahedron*, 2010, **66**, 2077–2082.
115. R. Wang, S. Mo, Y. Lu and Z. Shen, *Adv. Synth. Catal.*, 2011, **353**, 713–718.
116. R. Cano, M. Yus and D. J. Ramón, *Tetrahedron*, 2012, **68**, 1393–1400.
117. V. F. Patel, G. Pattenden and J. J. Russell, *Tetrahedron Lett.*, 1986, **27**, 2303–2306.
118. R. C. Larock and D. E. Stinn, *Tetrahedron Lett.*, 1988, **29**, 4687–4690.
119. X. Xie, B. Chen, J. Lu, J. Han, X. She and X. Pan, *Tetrahedron Lett.*, 2004, **45**, 6235–6237.
120. M. Gardiner, R. Grigg, V. Sridharan and N. Vicker, *Tetrahedron Lett.*, 1998, **39**, 435–438.
121. M. Gardiner, R. Grigg, M. Kordes, V. Sridharan and N. Vicker, *Tetrahedron*, 2001, **57**, 7729–7735.
122. K. Parthasarathy, M. Jeganmohan and C.-H. Cheng, *Org. Lett.*, 2006, **8**, 621–623.
123. M. Durandetti, L. Hardou, M. Clément and J. Maddaluno, *Chem. Commun.*, 2009, 4753–4755.
124. Z. Liu, Y. Xia, S. Zhou, L. Wang, Y. Zhang and J. Wang, *Org. Lett.*, 2013, **15**, 5032–5035.
125. M. Leibeling, M. Pawliczek, D. Kratzer, D. Stalke and D. B. Werz, *Org. Lett.*, 2012, **14**, 346–349.
126. D. Ma, Q. Cai and X. Xie, *Synlett*, 2005, 1767–1770.
127. D. P. Sant'Ana, V. D. Pinho, M. C. L. S. Maior and P. R. R. Costa, *Tetrahedron Lett.*, 2009, **50**, 3753–3755.
128. H. Yuan, K. Bi, B. Li, R. Yue, J. Ye, Y. Shen, L. Shan, H. Jin, Q. Sun and W. Zhang, *Org. Lett.*, 2013, **15**, 4742–4745.
129. A. R. Kapdi, A. Karbelkar, M. Naik, S. Pednekar, C. Fisher, C. Schulzke and M. Tromp, *RSC Adv.*, 2013, **3**, 20905–20912.
130. P. Shah, M. D. Santana, J. García, J. L. Serrano, M. Naik, S. Pednekar and A. R. Kapdi, *Tetrahedron*, 2013, **69**, 1446–1453.
131. D. E. Ames and D. Bull, *Tetrahedron*, 1982, **38**, 383–387.
132. B. Yin, C. Cai, G. Zeng, R. Zhang, X. Li and H. Jiang, *Org. Lett.*, 2012, **14**, 1098–1101.
133. J. Setsune, T. Ueda, K. Shikata, K. Matsukawa, T. Iida and T. Kitao, *Tetrahedron*, 1986, **42**, 2647–2656.
134. J. M. O'Connor, B. J. Stallman, W. G. Clark, A. Y. L. Shu, R. E. Spada, T. M. Stevenson and H. A. Dieck, *J. Org. Chem.*, 1983, **48**, 807–809.

135. R. C. Larock and E. K. Yum, *Tetrahedron*, 1996, **52**, 2743–2758.
136. J. John, I. U. E. Suresh and K. V. Radhakrishnan, *J. Am. Chem. Soc.*, 2009, **131**, 5042–5043.
137. R. Grigg, J. Redpath, V. Sridharan and D. Wilson, *Tetrahedron Lett.*, 1994, **35**, 7661–7664.
138. M. Lautens, E. Tayama and C. Herse, *J. Am. Chem. Soc.*, 2005, **127**, 72–73.
139. R. C. Larock, N. Berrios-Peña and C. A. Fried, *J. Org. Chem.*, 1991, **56**, 2615–2617.
140. W. Shu, Q. Yu and S. Ma, *Adv. Synth. Catal.*, 2009, **351**, 2807–2810.
141. C. Zhu and S. Ma, *Org. Lett.*, 2013, **15**, 2782–2785.
142. M.-G. Braun, M. H. Katcher and A. G. Doyle, *Chem. Sci.*, 2013, **4**, 1216–1220.
143. X. Wang, J. Liu, H. Guo, C. Ma, X. Gao, K. Zhou and G. Huang, *Synthesis*, 2012, 1037–1042.
144. R. C. Larock and E. K. Yum, *J. Am. Chem. Soc.*, 1991, **113**, 6689–6690.
145. D. Wensbo, A. Erlksson, T. Jeschke, U. Annby and S. Gronowitz, *Tetrahedron Lett.*, 1993, **34**, 2823–2826.
146. C. Amatore, E. Blart, J. P. Genet, A. Jutand, S. Lemaire-Audoire and M. Savignac, *J. Org. Chem.*, 1995, **60**, 6829–6839.
147. H. A. Oskooie, M. M. Heravi and F. K. Behbahani, *Molecules*, 2007, **12**, 1438–1446.
148. S. S. Palimkar, P. H. Kumar, R. J. Lahoti and K. V. Srinivasan, *Tetrahedron*, 2006, **62**, 5109–5115.
149. B. Z. Lu, W. Zhao, H.-X. Wei, M. Dufour, V. Farina and C. H. Senanayake, *Org. Lett.*, 2006, **8**, 3271–3274.
150. K. B. Hong, C. W. Lee and E. K. Yum, *Tetrahedron Lett.*, 2004, **45**, 693–697.
151. L. Djakovitch, V. Dufaud and R. Zaidi, *Adv. Synth. Catal.*, 2006, **348**, 715–724.
152. Y. Monguchi, S. Mori, S. Aoyagi, A. Tsutsui, T. Maegawa and H. Sajiki, *Org. Biomol. Chem.*, 2010, **8**, 3338–3342.
153. A. Ahmed, M. Ghosh, P. Sarkar and J. K. Ray, *Tetrahedron Lett.*, 2013, **54**, 6691–6694.
154. S. Cacchi, G. Fabrizi and L. M. Parisi, *Org. Lett.*, 2003, **5**, 3843–3846.
155. H. Jiang, H. Fu, R. Qiao, Y. Jiang and Y. Zhao, *Synthesis*, 2008, 2417–2426.
156. F. liu and D. Ma, *J. Org. Chem.*, 2007, **72**, 4844–4850.
157. T. Ponpandian and S. Muthusubramanian, *Tetrahedron Lett.*, 2012, **53**, 4248–4252.
158. R. C. Larock, E. K. Yum and M. D. Refvik, *J. Org. Chem.*, 1998, **63**, 7652–7662.
159. M. Shen, G. Li, B. Z. Lu, A. Hossain, F. Roschangar, V. Farina and C. H. Senanayake, *Org. Lett.*, 2004, **6**, 4129–4132.
160. L. Ackermann, R. Sandmann, A. Villar and L. T. Kaspar, *Tetrahedron*, 2008, **64**, 769–777.

161. J. Liu, M. Shen, Y. Zhang, G. Li, A. Khodabocus, S. Rodriguez, B. Qu, V. Farina, C. H. Senanayake and B. Z. Lu, *Org. Lett.*, 2006, **8**, 3573–3575.
162. A. Keivanloo, M. Bakherad, H. Nasr-Isfahani and S. Esmaily, *Tetrahedron Lett.*, 2012, **53**, 3126–3130.
163. T. Konno, J. Chae, T. Ishihara and H. Yamanaka, *J. Org. Chem.*, 2004, **69**, 8258–8265.
164. F. Roschangar, J. Liu, E. Estanove, M. Dufour, S. Rodríguez, V. Farina, E. Hickey, A. Hossain, P. Jones, H. Lee, B. Z. Lu, R. Varsolona, J. Schröder, P. Beaulieu, J. Gillard and C. H. Senanayake, *Tetrahedron Lett.*, 2008, **49**, 363–366.
165. X. Cui, J. Li, Y. Fu, L. Liu and Q.-X. Guo, *Tetrahedron Lett.*, 2008, **49**, 3458–3462.
166. P. He, Y. Du, G. Liu, C. Cao, Y. Shi, J. Zhang and G. Pang, *RSC Adv.*, 2013, **3**, 18345–18350.
167. K. Srinivas, P. Saiprathima, K. Balaswamy and M. M. Rao, *J. Organomet. Chem.*, 2013, **741-742**, 162–167.
168. W. J. Ang, C. Tai, L. Lo and Y. Lam, *RSC Adv.*, 2014, **4**, 4921–4929.
169. L. Yu, X. Jiang, L. Wang, Z. Li, D. Wu and X. Zhou, *Eur. J. Org. Chem.*, 2010, 5560–5562.
170. P. Li, L. Wang, M. Wang and F. You, *Eur. J. Org. Chem.*, 2008, 5946–5951.
171. X. Yu, E. Park, T. P. Kondratyuk, J. M. Pezzuto and D. Sun, *Org. Biomol. Chem.*, 2012, **10**, 8835–8847.
172. G. K. Jana and S. Sinha, *Tetrahedron Lett.*, 2012, **53**, 1671–1674.
173. A. L. Smith, G. I. Stevenson, C. J. Swain and J. L. Castro, *Tetrahedron Lett.*, 1998, **39**, 8317–8320.
174. F. Messina, M. Botta, F. Corelli and C. Villani, *Tetrahedron: Asymmetry*, 2000, **11**, 1681–1685.
175. N. Olivi, P. Spruyt, J. Peyrat, M. Alami and J. Brion, *Tetrahedron Lett.*, 2004, **45**, 2607–2610.
176. O. Russo, S. Messaoudi, A. Hamez, N. Olivi, J. Peyrat, J. Brion, S. Sicsic, I. Berque-Bestel and M. Alami, *Tetrahedron*, 2007, **63**, 10671–10683.
177. K. Dooleweerdt, T. Ruhland and T. Skrydstrup, *Org. Lett.*, 2009, **11**, 221–224.
178. S. Kramer, K. Dooleweerdt, A. T. Lindhardt, M. Rottländer and T. Skrydstrup, *Org. Lett.*, 2009, **11**, 4208–4211.
179. J. Cao, Y. Xu, Y. Kong, Y. Cui, Z. Hu, G. Wang, Y. Deng and G. Lai, *Org. Lett.*, 2012, **14**, 38–41.
180. A. Kondoh, H. Yorimitsu and K. Oshima, *Org. Lett.*, 2010, **12**, 1476–1479.
181. M. Layek, V. Gajare, D. Kalita, A. Islam, K. Mukkanti and M. Pal, *Tetrahedron Lett.*, 2009, **50**, 3867–3871.
182. T. Nanjo, S. Yamamoto, C. Tsukano and Y. Takemoto, *Org. Lett.*, 2013, **15**, 3754–3757.
183. V. F. Patel, G. Pattenden and J. J. Russell, *Tetrahedron Lett.*, 1986, **27**, 2303–2306.

184. (a) M. Mori and Y. Ban, *Tetrahedron Lett.*, 1976, 1803–1806; (b) M. Mori, Y. Hashimoto and Y. Ban, *Tetrahedron Lett.*, 1980, 631–634.
185. (a) R. Odle, B. Blevins, M. Ratcliff and L. S. Hegedus, *J. Org. Chem.*, 1980, **45**, 2709–2710; (b) L. S. Hegedus, T. A. Mulhern and A. Mori, *J. Org. Chem.*, 1985, **50**, 4282–4288; (c) M. Mori, K. Chiba and Y. Ban, *Tetrahedron Lett.*, 1977, 1037–1040.
186. R. C. Larock and S. Babu, *Tetrahedron Lett.*, 1987, **28**, 5291–5294.
187. T. Sakamoto, Y. Kondo, A. Yasuhara and H. Yamanaka, *Tetrahedron*, 1991, **47**, 1877–1886.
188. S. Caddick and W. Kofie, *Tetrahedron Lett.*, 2002, **43**, 9347–9350.
189. M. L. Weinrich and H. P. Beck, *Tetrahedron Lett.*, 2009, **50**, 6968–6972.
190. Y. Liu, B. Yao, C.-L. Deng, R.-Y. Tang, X.-G Zhang and J.-H. Li, *Org. Lett.*, 2011, **13**, 1126–1129.
191. A. Osuka, Y. Mori and H. Suzuki, *Chem. Lett.*, 1982, 2031–2034.
192. (a) T. Watanabe, S. Arai and A. Nishida, *Synlett*, 2004, 907–909; (b) J. Maruyama, H. Yamashita, T. Watanabe, S. Arai and A. Nishida, *Tetrahedron*, 2009, **65**, 1327–1335.
193. H. H. Nguyen and M. J. Kurth, *Org. Lett.*, 2013, **15**, 362–365.
194. (a) J. M. Harris and A. Padwa, *Org. Lett.*, 2003, **5**, 4195–4197; (b) E. M. Mmutlane, J. M. Harris and A. Padwa, *J. Org. Chem.*, 2005, **70**, 8055–8063.
195. C. Wang and J. Sperry, *Tetrahedron*, 2013, **69**, 4563–4577.
196. J. Barluenga, M. A. Fernández, F. Aznar and C. Valdés, *Chem. Eur. J.*, 2005, **11**, 2276–2283.
197. H. Suzuki, S. V. Thiruvikraman and A. Osuka, *Synthesis*, 1984, 616–617.
198. M. T. Baumgartner, M. A. Nazareno, M. C. Murguía, A. B. Pierini and R. A. Rossi, *Synthesis*, 1999, 2053–2056.
199. C. Chen, D. R. Lieberman, R. D. Larsen, T. R. Verhoeven and P. J. Reider, *J. Org. Chem.*, 1997, **62**, 2676–2677.
200. Y. Jia and J. Zhu, *J. Org. Chem.*, 2006, **71**, 7826–7834.
201. U. S. Sørensen and E. Pombo-Villar, *Helv. Chim. Acta*, 2004, **87**, 82–89.
202. L. M. Blair and J. Sperry, *Synlett*, 2013, 1931–1936.
203. R. P. Karuvalam, K. R. Haridas, A. M. Sajith and A. Muralidharan, *Tetrahedron Lett.*, 2013, **54**, 5126–5129.
204. Z. Jin, S. Guo, L. Qiu, G. Wu and J. Fang, *Appl. Organomet. Chem.*, 2011, **25**, 502–507.
205. M. Nazaré, C. Schneider, A. Lindenschmidt and D. W. Will, *Angew. Chem., Int. Ed.*, 2004, **43**, 4526–4528.
206. S. Tanimori, H. Ura and M. Kirihata, *Eur. J. Org. Chem.*, 2007, 3977–3980.
207. M. A. Ali and T. Punniyamurthy, *Synlett*, 2011, 623–626.
208. Y. Chen, Y. Wang, Z. Sun and D. Ma, *Org. Lett.*, 2008, **10**, 625–628.
209. X. Yang, H. Fu, R. Qiao, Y. Jiang and Y. Zhao, *Adv. Synth. Catal.*, 2010, **352**, 1033–1038.
210. M. Jiang, J. Li, F. Wang, Y. Zhao, X. Dong and W. Zhao, *Org. Lett.*, 2012, **14**, 1420–1423.

211. R. B. Bedford and C. S. J. Cazin, *Chem. Commun.*, 2002, 2310–2311.
212. R. B. Bedford and M. Betham, *J. Org. Chem.*, 2006, **71**, 9403–9410.
213. R. B. Bedford, C. P. Butts, M. F. Haddow, R. Osborne and R. F. Sankey, *Chem. Commun.*, 2009, 4832–4834.
214. R. B. Bedford, J. G. Bowen and A. L. Weeks, *Tetrahedron*, 2013, **69**, 4389–4394.
215. P. Franck, S. Hostyn, B. Dajka-Halász, Á. Polonka-Bálint, K. Monsieurs, P. Mátyus and B. U. W. Maes, *Tetrahedron*, 2008, **64**, 6030–6034.
216. (a) T. H. M. Jonckers, B. U. W. Maes, G. L. F. Lemière, G. Rombouts, L. Pieters, A. Haemers and R. A. Dommisse, *Synlett*, 2003, 615–618; (b) C. Meyers, G. Rombouts, K. T. J. Loones, A. Coelho and B. U. W. Maes, *Adv. Synth. Catal.*, 2008, **350**, 465–470.
217. G. V. Baelen, S. Hostyn, L. Dhooghe, P. Tapolcsányi, P. Mátyus, G. Lemière, R. Dommisse, M. Kaiser, R. Brun, P. Cos, L. Maes, G. Hajós, Z. Riedl, I. Nagy, B. U. W. Maes, G. Hajós, Z. Riedl, I. Nagy, B. U. W. Maes and L. Pieters, *Bioorg. Med. Chem.*, 2009, **17**, 7209–7217.
218. N. D. Cá, G. Sassi and M. Catellani, *Adv. Synth. Catal.*, 2008, **350**, 2179–2182.
219. T. Iwaki, A. Yasuhara and T. Sakamoto, *J. Chem. Soc., Perkin Trans. 1*, 1999, 1505–1510.
220. (a) Z. Liu and R. C. Larock, *Org. Lett.*, 2004, **6**, 3739–3741; (b) Z. Liu and R. C. Larock, *Tetrahedron*, 2007, **63**, 347–355.
221. C. Lu, N. A. Markina and R. C. Larock, *J. Org. Chem.*, 2012, **77**, 11153–11160.
222. J. García-Fortanet, F. Kessler and S. L. Buchwald, *J. Am. Chem. Soc.*, 2009, **131**, 6676–6677.
223. Y. Kitamura, S. Yoshikawa, T. Furuta and T. Kan, *Synlett*, 2008, 377–380.
224. B. Zou, Q. Yuan and D. Ma, *Org. Lett.*, 2007, **9**, 4291–4294.
225. Z. Li, H. Sun, H. Jiang and H. Liu, *Org. Lett.*, 2008, **10**, 3263–3266.
226. W. S. Cheung, R. J. Patch and M. R. Player, *J. Org. Chem.*, 2005, **70**, 3741–3744.
227. R. Yanada, S. Obika, T. Inokuma, K. Yanada, M. Yamashita, S. Ohta and Y. Takemoto, *J. Org. Chem.*, 2005, **70**, 6972–6975.
228. G. R. Dong, S. Park, D. Lee, K. J. Shin and J. H. Seo, *Synlett*, 2013, 1993–1997.
229. S. Tang, P. Peng, Z. Wang, B. Tang, C. Deng, J. Li, P. Zhong, P. Zhong and N. Wang, *Org. Lett.*, 2008, **10**, 1875–1878.
230. R. Shintani, T. Yamagami and T. Hayashi, *Org. Lett.*, 2006, **8**, 4799–4801.
231. J. Schönhaber, W. Frank and T. J. J. Müller, *Org. Lett.*, 2010, **12**, 4122–4125.
232. D. M. D'Souza, C. Muschelknautz, F. Rominger and T. J. J. Müller, *Org. Lett.*, 2010, **12**, 3364–3367.
233. M. C. McDermott, G. R. Stephenson, D. L. Hughes and A. J. Walkington, *Org. Lett.*, 2006, **8**, 2917–2920.

234. J. C. de Mendonca, M. O. F. Goulart, E. Léonel and J. Nédélec, *Tetrahedron Lett.*, 2002, **43**, 6343–6345.
235. R. Grigg, B. Putnikovic and C. J. Urch, *Tetrahedron Lett.*, 1996, **37**, 695–698.
236. S. Lee and J. F. Hartwig, *J. Org. Chem.*, 2001, **66**, 3402–3415.
237. M. Bararjanian, S. Balalaie, F. Rominger, B. Movassagh and H. R. Bijanzadeh, *J. Org. Chem.*, 2010, **75**, 2806–2812.
238. M. Bararjanian, S. Hosseinzadeh, S. Balalaie and H. R. Bijanazdeh, *Tetrahedron*, 2011, **67**, 2644–2650.
239. M. Bararjanian, S. Hosseinzadeh, S. Balalaie, H. R. Bijanzadeh and E. Wolf, *Tetrahedron Lett.*, 2011, **52**, 3329–3332.
240. Z. An, M. Catellani and G. P. Chiusoli, *J. Organomet. Chem.*, 1990, **397**, C31–C32.
241. M. Genelot, V. Dufaud and L. Djakovitch, *Tetrahedron*, 2011, **67**, 976–981.
242. M. von Seebach, R. Grigg and A. de Meijere, *Eur. J. Org. Chem.*, 2002, 3268–3275.
243. F. Ozawa, H. Yanagihara and A. Yamamoto, *J. Org. Chem.*, 1986, **51**, 415–417.
244. C. Mukhopadhyay, P. K. Tapaswi and R. J. Butcher, *Org. Biomol. Chem.*, 2010, **8**, 4720–4729.
245. C. T. Brain and S. A. Brunton, *Tetrahedron Lett.*, 2002, **43**, 1893–1895.
246. J. E. R. Sadig, R. Foster, F. Wakenhut and M. C. Willis, *J. Org. Chem.*, 2012, **77**, 9473–9486.
247. J. Alonso, N. Halland, M. Nazaré, O. R'kyek, M. Urmann and A. Lindenschmidt, *Eur. J. Org. Chem.*, 2011, 234–237.
248. N. Zheng, K. W. Anderson, X. Huang, H. N. Nguyen and S. L. Buchwald, *Angew. Chem., Int. Ed.*, 2007, **46**, 7509–7512.
249. C. Venkatesh, G. S. M. Sundaram, H. Ila and H. Junjappa, *J. Org. Chem.*, 2006, **71**, 1280–1283.
250. S. Kumar, H. Ila and H. Junjappa, *J. Org. Chem.*, 2009, **74**, 7046–7051.
251. Z. Wu, Q. Huang, X. Zhou, L. Yu, Z. Li and D. Wu, *Eur. J. Org. Chem.*, 2011, 5242–5245.
252. B. Zou, Q. Yuan and D. Ma, *Angew. Chem., Int. Ed.*, 2007, **46**, 2598–2601.
253. Z. Hu, W. Ye, H. Zou and Y. Yu, *Synth. Commun.*, 2010, **40**, 222–228.
254. Z. Wu, L. Zou, Z. Jiang, D. Wu, Z. Li and X. Zhou, *Eur. J. Org. Chem.*, 2010, 4971–4975.
255. D. Yang, H. Fu, L. Hu, Y. Jiang and Y. Zhao, *J. Org. Chem.*, 2008, **73**, 7841–7844.
256. Y. Qu, L. Pan, Z. Wu and X. Zhou, *Tetrahedron*, 2013, **69**, 1717–1719.
257. K. Hirano, A. T. Biju and F. Glorius, *J. Org. Chem.*, 2009, **74**, 9570–9572.
258. P. Saha, T. Ramana, N. Purkait, M. A. Ali, R. Paul and T. Punniyamurthy, *J. Org. Chem.*, 2009, **74**, 8719–8725.
259. H. Yu, M. S. Zhang and L. R. Cui, *Chin. Chem. Lett.*, 2012, **23**, 573–575.
260. J. Peng, M. Ye, C. Zong, F. Hu, L. Feng, X. Wang, Y. Wang and C. Chen, *J. Org. Chem.*, 2011, **76**, 716–719.

261. N. Zheng and S. L. Buchwald, *Org. Lett.*, 2007, **9**, 4749–4751.
262. H. Yuan, Y. Chen, C. Chen and B. Chen, *Chin. J. Chem.*, 2013, **31**, 1247–1249.
263. Y. Kim, M. R. Kumar, N. Park, Y. Heo and S. Lee, *J. Org. Chem.*, 2011, **76**, 9577–9583.
264. H. Jin, X. Xu, J. Gao, J. Zhong and Y. Wang, *Adv. Synth. Catal.*, 2010, **352**, 347–350.
265. X. Lv and W. Bao, *J. Org. Chem.*, 2009, **74**, 5618–5621.
266. F. Wang, S. Cai, Q. Liao and C. Xi, *J. Org. Chem.*, 2011, **76**, 3174–3180.
267. G. Qiu and J. Wu, *Chem. Commun.*, 2012, **48**, 6046–6048.
268. G. Yuan, H. Liu, J. Gao, K. Yang, Q. Niu, H. Mao, X. Wang and X. Lv, *J. Org. Chem.*, 2014, **79**, 1749–1757.
269. (a) C. Chen, C. Chen, B. Li, J. Tao and J. Peng, *Molecules*, 2012, **17**, 12506–12520; (b) P. Roy and A. Pramanik, *Tetrahedron Lett.*, 2013, **54**, 5243–5245; (c) S. M. Barolo, Y. Wang, R. A. Rossi and G. D. Cuny, *Tetrahedron*, 2013, **69**, 5487–5494.
270. G. Shen and W. Bao, *Adv. Synth. Catal.*, 2010, **352**, 981–986.
271. (a) G. Evindar and R. A. Batey, *J. Org. Chem.*, 2006, **71**, 1802–1808; (b) R. D. Viirre, G. Evindar and R. A. Batey, *J. Org. Chem.*, 2008, **73**, 3452–3459.
272. J. H. Spatz, T. Bath, M. Umkehrer, J. Bardin, G. Ross, C. Burdack and J. Kolb, *Tetrahedron Lett.*, 2007, **48**, 9030–9034.
273. D. Xu, L. Sun and Q. You, *Tetrahedron*, 2012, **68**, 4248–4251.
274. N. Barbero, M. Carril, R. SanMartin and E. Domínguez, *Tetrahedron*, 2007, **63**, 10425–10432.
275. M. L. Kantam, G. T. Venkana, K. B. S. Kumar, V. Balasubrahmanyam and S. Bhargava, *Synlett*, 2009, 1753–1756.
276. N. Khatun, S. Guin, S. K. Rout and B. K. Patel, *RSC Adv.*, 2014, **4**, 10770–10778.
277. K. Takagi, T. Iwachido and N. Hayama, *Chem. Lett.*, 1987, 839–840.
278. S. Pi, C. Zhang, R. Tang and J. Li, *Synlett*, 2009, 3032–3036.
279. C. Benedi, F. Bravo, P. Uriz, E. Fernández, C. Claver and S. Castillón, *Tetrahedron Lett.*, 2003, **44**, 6073–6077.
280. (a) L. L. Joyce, G. Evindar and R. A. Batey, *Chem. Commun.*, 2004, 446–447; (b) L. L. Joyce and R. A. Batey, *Org. Lett.*, 2009, **11**, 2792–2795.
281. Y. Cheng, Q. Peng, W. Fan and P. Li, *J. Org. Chem.*, 2014, **79**, 5812–5819.
282. T. Itoh and T. Mase, *Org. Lett.*, 2007, **9**, 3687–3689.
283. J. Qiu, X. Zhang, R. Tang, P. Zhong and J. Li, *Adv. Synth. Catal.*, 2009, **351**, 2319–2323.
284. Q. Ding, B. Cao, X. Liu, Z. Zong and Y. Peng, *Green Chem.*, 2010, **12**, 1607–1610.
285. (a) W. R. Bowman, H. Heaney and P. H. G. Smith, *Tetrahedron Lett.*, 1982, **23**, 5093–5096; (b) A. Osuka, Y. Uno, H. Horiuchi and H. Suzuki, *Synthesis*, 1984, 145–146; (c) K. Konishi, I. Nishiguchi and T. Hirashima, *Synthesis*, 1984, 254–255; (d) R. Wang, Z. Chen, L. Yue, W. Pan and J. Zhao, *Tetrahedron Lett.*, 2012, **53**, 4529–4531;

(e) T. B. Nguyen, L. Ermolenko and A. Al-Mourabit, *Org. Lett.*, 2013, **15**, 4218–4221; (f) R. Wang, W. Yang, L. Yue, W. Pan and H. Zeng, *Synlett*, 2012, 1643–1648.

286. S. Murru, H. Ghosh, S. K. Sahoo and B. K. Patel, *Org. Lett.*, 2009, **11**, 4254–4257.
287. E. A. Jaseer, D. J. C. Prasad, A. Dandapat and G. Sekar, *Tetrahedron Lett.*, 2010, **51**, 5009–5012.
288. (a) J. Wang, F. Peng, J. Jiang, Z. Lu, L. Wang, J. Bai and Y. Pan, *Tetrahedron Lett.*, 2008, **49**, 467–470; (b) S. Okamoto, Y. Sakai, S. Watanabe, S. Nishi, A. Yoneyama, H. Katsumata, Y. Kosaki, R. Sato, M. Shiratori, M. Shibuno and T. Shishido, *Tetraheron Lett.*, 2014, **55**, 1909–1912.
289. G. Shen, X. Lv and W. Bao, *Eur. J. Org. Chem.*, 2009, 5897–5901.
290. (a) S. Murru, P. Mondal, R. Yella and B. K. Patel, *Eur. J. Org. Chem.*, 2009, 5406–5413; (b) S. K. Sahoo, N. Khatun, A. Gogoi, A. Deb and B. K. Patel, *RSC Adv.*, 2013, **3**, 438–446.
291. S. K. Rout, S. Guin, J. Nath and B. K. Patel, *Green Chem.*, 2012, **14**, 2491–2498.
292. Y. Guo, R. Tang, P. Zhong and J. Li, *Tetrahedron Lett.*, 2010, **51**, 649–652.
293. Y. Sun, Y. Zhang, X. Cui and W. Wang, *Adv. Synth. Catal.*, 2011, **353**, 1174–1178.
294. W. Zhang, Y. Yue, D. Yu, L. Song, Y. Xu, Y. Tian and Y. Guo, *Adv. Synth. Catal.*, 2012, **354**, 2283–2287.
295. R. Yao, H. Liu, Y. Wu and M. Cai, *Appl. Organomet. Chem.*, 2013, **27**, 109–113.
296. N. Zhao, L. Liu, F. Wang, J. Li and W. Zhang, *Adv. Synth. Catal.*, 2014, **356**, 2575–2579.
297. J. Liu, X. Zhang, J. Yang and L. Wang, *Appl. Organometal. Chem.*, 2014, **28**, 198–203.
298. J. Yang, P. Li and L. Wang, *Tetrahedron*, 2011, **67**, 5543–5549.
299. R. Xiao, W. Hao, J. Ai and M. Cai, *J. Organometal. Chem.*, 2012, **705**, 44–50.
300. M. Kaname, M. Minoura and H. Sashida, *Tetrahedron Lett.*, 2011, **52**, 505–508.
301. (a) D. Ma, S. Xie, P. Xue, X. Zhang, J. Dong and Y. Jiang, *Angew. Chem., Int. Ed.*, 2009, **48**, 4222–4225; (b) J. Li, Y. Zhang, Y. Jiang and D. Ma, *Tetrahedron Lett.*, 2012, **53**, 2511–2513.
302. J. Liu, Q. Gui, Z. Yang, Z. Tan, R. Guo and J. Shi, *Synthesis*, 2013, 943–951.
303. X. Zhang, W. Zeng, Y. Yang, H. Huang and Y. Liang, *Org. Lett.*, 2014, **16**, 876–879.
304. N. Park, Y. Heo, M. R. Kumar, Y. Kim, K. H. Song and S. Lee, *Eur. J. Org. Chem.*, 2012, 1984–1993.
305. W. Xue, Y. Guo, F. Gao, H. Li and A. Wu, *Org. Lett.*, 2013, **15**, 890–893.
306. (a) D. Ma, X. Lu, L. Shi, H. Zhang, Y. Jiang and X. Liu, *Angew. Chem., Int. Ed.*, 2011, **50**, 1118–1121; (b) L. Shi, X. Liu, H. Zhang, Y. Jiang and D. Ma, *J. Org. Chem.*, 2011, **76**, 4200–4204.

307. G. Satish, K. H. V. Reddy, K. Ramesh, K. Karnakar and Y. V. D. Nageswar, *Tetrahedron Lett.*, 2012, **53**, 2518–2521.
308. H. Xiang, J. Qi, Q. He, M. Jiang, C. Yang and L. Deng, *Org. Biomol. Chem.*, 2014, **12**, 4633–4636.
309. D. J. C. Prasad and G. Sekar, *Org. Biomol. Chem.*, 2013, **11**, 1659–1665.
310. S. M. Soria-Castro and A. B. Peñéñory, *Beilstein J. Org. Chem.*, 2013, **9**, 467–475.
311. H. Deng, Z. Li, F. Ke and X. Zhou, *Chem. Eur. J.*, 2012, **18**, 4840–4843.
312. R. Wang, Y. Ding, H. Liu, S. Peng, J. Ren and L. Li, *Tetrahedron Lett.*, 2014, **55**, 945–949.
313. Y. Xu and J. Zhang, *Tetrahedron Lett.*, 1985, **26**, 4771–4774.
314. T. J. Brunker, B. J. Anderson, N. F. Blank, D. S. Glueck and A. L. Rheingold, *Org. Lett.*, 2007, **9**, 1109–1112.
315. L. Bonnafoux, R. Gramage-Doria, F. Colobert and F. R. Leroux, *Chem. Eur. J.*, 2011, **17**, 11008–11016.
316. F. Wang, C. Chen, G. Deng and C. Xi, *J. Org. Chem.*, 2012, **77**, 4148–4151.
317. B. S. Bhakuni, S. J. Balkrishna, A. Kumar and S. Kumar, *Tetrahedron Lett.*, 2012, **53**, 1354–1357.
318. V. Krasikova and M. Katkevics, *Chem. Hetero. Comp.*, 2013, **48**, 1684–1690.
319. (a) S. J. Balkrishna, B. S. Bhakuni, D. Chopra and S. Kumar, *Org. Lett.*, 2010, **12**, 5394–5397; (b) S. J. Balkrishna, B. S. Bhakuni and S. Kumar, *Tetrahedron*, 2011, **67**, 9565–9575.
320. T. Kashiki, S. Shinamura, M. Kohara, E. Miyazaki, K. Takimiya, M. Ikeda and H. Kuwabara, *Org. Lett.*, 2009, **11**, 2473–2475.
321. Y. Li, C. Nie, H. Wang, X. Li, F. Verpoort and C. Duan, *Eur. J. Org. Chem.*, 2011, 7331–7338.
322. L. Sun, C. Deng, R. Tang and X. Zhang, *J. Org. Chem.*, 2011, **76**, 7546–7550.
323. V. Guilarte, M. A. Fernández-Rodríguez, P. García-García, E. Hernando and R. Sanz, *Org. Lett.*, 2011, **13**, 5100–5103.
324. M. Kuhn, F. C. Falk and J. Paradies, *Org. Lett.*, 2011, **13**, 4100–4103.
325. (a) W. You, X. Yan, Q. Liao and C. Xi, *Org. Lett.*, 2010, **12**, 3930–3933; (b) Q. Liao, W. You, Z. Lou, L. Wen and C. Xi, *Tetrahedron Lett.*, 2013, **54**, 1475–1477; (c) P. Zhao, H. Yin, H. Gao and C. Xi, *J. Org. Chem.*, 2013, **78**, 5001–5006.
326. C. S. Bryan, J. A. Braunger and M. Lautens, *Angew. Chem., Int. Ed.*, 2009, **48**, 7064–7068.
327. F. Zeng and H. Alper, *Org. Lett.*, 2011, **13**, 2868–2871.
328. (a) J. Debray, M. Lemaire and F. Popowycz, *Synlett*, 2013, 37–40; (b) M. V. Patel, J. J. Rohde, V. Gracias and T. Kolasa, *Tetrahedron Lett.*, 2003, **44**, 6665–6667; (c) Z. Liao, X. Lv and M. Tao, *Res. Chem. Intermed.*, 2013, **39**, 4021–4024; (d) Z. Duan, S. Ranjit and X. Liu, *Org. Lett.*, 2010, **12**, 2430–2433; (e) M. Saito, I. Osaka, E. Miyazaki, K. Takimiya, H. Kuwabara and M. Ikeda, *Tetrahedron Lett.*, 2011, **52**, 285–288.

329. M. Palucki, J. P. Wolfe and S. L. Buchwald, *J. Am. Chem. Soc.*, 1996, **118**, 10333–10334.
330. G. Mann, C. Incarvito, A. L. Rheingold and J. F. Hartwig, *J. Am. Chem. Soc.*, 1999, **121**, 3224–3225.
331. A. N. Butkevich, A. Corbu, L. Meerpoel, I. Stansfield, P. Angibaud, P. Bonnet and J. Cossy, *Org. Lett.*, 2012, **14**, 4998–5001.
332. J. Niu, P. Guo, J. Kang, Z. Li, J. Xu and S. Hu, *J. Org. Chem.*, 2009, **74**, 5075–5078.
333. J. Huang, Y. Chen, J. Chan, M. L. Ronk, R. D. Larsen and M. M. Faul, *Synlett*, 2011, 1419–1422.
334. A. Svennebring, P. Nilsson and M. Larhed, *J. Org. Chem.*, 2007, **72**, 5851–5854.
335. X. Fang and X. Tong, *Tetrahedron Lett.*, 2010, **51**, 317–320.
336. J. Jin, Y. Luo, C. Zhou, X. Chen, Q. Wen, P. Lu and Y. Wang, *J. Org. Chem.*, 2012, **77**, 11368–11371.
337. D. L. Coffen, B. Schaer, F. T. Bizzarro and J. B. Cheung, *J. Org. Chem.*, 1984, **49**, 296–300.
338. P. De, Nonappa, K. Pandurangan, U. Maitra and S. Wailes, *Org. Lett.*, 2007, **9**, 3767–2770.
339. S. Udd, R. Jokela, R. Franzén and J. Tois, *Tetrahedron Lett.*, 2010, **51**, 1030–1033.
340. N. G. Kundu and M. Pal, *J. Chem. Soc., Chem. Commun.*, 1993, 86–88.
341. N. G. Kundu, M. Pal and B. Nandi, *J. Chem. Soc., Perkin Trans. 1*, 1998, 561–568.
342. L. Zhou and H. Jiang, *Tetrahedron Lett.*, 2007, **48**, 8449–8452.
343. D. Rambabu, G. P. Kumar, B. D. Kumar, R. Kapavarapu, M. V. B. Rao and M. Pal, *Tetrahedron Lett.*, 2013, **54**, 2989–2995.
344. K. Kobayashi, K. Hashimoto, S. Fukamachi and H. Konishi, *Synthesis*, 2008, 1094–1098.
345. A. Zanardi, J. A. Mata and E. Peris, *Organometallics*, 2009, **28**, 4335–4339.
346. Y. C. Fan and O. Kwon, *Org. Lett.*, 2012, **14**, 3264–3267.
347. C. G. Saluste, S. Crumpler, M. Furber and R. J. Whitby, *Tetrahedron Lett.*, 2004, **45**, 6995–6996.
348. M. Dell'Acqua, D. Facoetti, G. Abbiati and E. Rossi, *Tetrahedron*, 2011, **67**, 1552–1556.
349. M. R. Kumar, F. M. Irudayanathan, J. H. Moon and S. Lee, *Adv. Synth. Catal.*, 2013, **355**, 3221–3230.
350. E.-i. Negishi and J. M. Tour, *Tetrahedron Lett.*, 1986, **27**, 4869–4872.
351. C. Copéret, S. Ma, T. Sugihara and E.-i. Negishi, *Tetrahedron*, 1996, **52**, 11529–11544.
352. C. Copéret, T. Sugihara, G. Wu, I. Shimoyama and E.-i. Negishi, *J. Am. Chem. Soc.*, 1995, **117**, 3422–3431.
353. E.-i. Negishi, C. Copéret, S. Ma, T. Mita, T. Sugihara and J. M. Tour, *J. Am. Chem. Soc.*, 1996, **118**, 5904–5918.
354. E.-i. Negishi, C. Copéret, T. Sugihara, I. Shimoyama, Y. Zhang, G. Wu and J. M. Tour, *Tetrahedron*, 1994, **50**, 425–436.

355. A. Cowell and J. K. Stille, *J. Am. Chem. Soc.*, 1980, **102**, 4193–4198.
356. V. P. Baillargeon and J. K. Stille, *J. Am. Chem. Soc.*, 1986, **108**, 452–461.
357. X. Gong, P. W. Miller, A. D. Gee, N. J. Long, A. J. de Mello and R. Vilar, *Chem. Eur. J.*, 2012, **18**, 2768–2772.
358. H. Audrain, L. Martarello, A. Gee and D. Bender, *Chem. Commun.*, 2004, 558–559.
359. X. Wu, A. K. Mahalingam, Y. Wan and M. Alterman, *Tetrahedron Lett.*, 2004, **45**, 4635–4638.
360. J.-J. Brunet, C. Sidot and P. Caubere, *J. Org. Chem.*, 1983, **48**, 1166–1171.
361. M. Foa, F. Francalanci, E. Bencini and A. Gardano, *J. Organomet. Chem.*, 1985, **285**, 293–303.
362. D. Y. Lee, C. S. Cho, L. H. Jiang, X. Wu, S. C. Shim and D. H. Oh, *Synth. Commun.*, 1997, **27**, 3449–3455.
363. C. S. Cho, D. Y. Baek, H. Y. Kim, S. C. Shim and D. H. Oh, *Synth. Commun.*, 2000, **30**, 1139–1145.
364. C. S. Cho, D. Y. Baek and S. C. Shim, *J. Heterocycl. Chem.*, 1999, **36**, 1101–1104.
365. D. V. Kadnikov and R. C. Larock, *Mendeleev Commun.*, 2007, **17**, 74–76.
366. D. K. Rayabarapu, H.-T. Chang and C.-H. Cheng, *Chem. Eur. J.*, 2004, **10**, 2991–2996.
367. H.-T. Chang, M. Jeganmohan and C.-H. Cheng, *Chem. Eur. J.*, 2007, **13**, 4356–4363.
368. R. Shang, Y. Fu, J.-B. Li, S.-L. Zhang, Q.-X. Guo and L. Liu, *J. Am. Chem. Soc.*, 2009, **131**, 5738–5739.
369. L. Ackermann and J. Pospech, *Org. Lett.*, 2011, **13**, 4153–4155.
370. C.-H. Lin, Y.-J. Wang and C.-F. Lee, *Eur. J. Org. Chem.*, 2010, 4368–4371.
371. D. Zhao, N. Wu, S. Zhang, P. Xi, X. Su, J. Lan and J. You, *Angew. Chem., Int. Ed.*, 2009, **48**, 8729–8732.
372. V. Guilarte, M. P. Castroviejo, E. Álvarez and R. Sanz, *Beilstein J. Org. Chem.*, 2011, **7**, 1255–1260.
373. C. B. Lavery, N. L. Rotta-Loria, R. McDonald and M. Stradiotto, *Adv. Synth. Catal.*, 2013, **355**, 981–987.
374. (a) M. C. Willis, D. Taylor and A. T. Gillmore, *Org. Lett.*, 2004, **6**, 4755–4757; (b) M. C. Willis, D. Taylor and A. T. Gillmore, *Tetrahedron*, 2006, **62**, 11513–11520.
375. J. Faragó and A. Kotschy, *Synthesis*, 2009, 85–90.
376. C. Chen and P. G. Dormer, *J. Org. Chem.*, 2005, **70**, 6964–6967.
377. L. Ackermann and L. T. Kaspar, *J. Org. Chem.*, 2007, **72**, 6149–6153.
378. M. Carril, R. SanMartin, I. Tellitu and E. Domínguez, *Org. Lett.*, 2006, **8**, 1467–1470.
379. J. Bonnamour, M. Piedrafita and C. Bolm, *Adv. Synth. Catal.*, 2010, **352**, 1577–1581.
380. G. Chelucci, *Chem. Rev.*, 2012, **112**, 1344–1462.
381. S. Thielges, E. Meddah, P. Bisseret and J. Eustache, *Tetrahedron Lett.*, 2004, **45**, 907–910.

382. S. G. Newman, V. Aureggi, C. S. Bryan and M. Lautens, *Chem. Commun.*, 2009, 5236–5238.
383. W. Chen, Y. Zhang, L. Zhang, M. Wang and L. Wang, *Chem. Commun.*, 2011, **47**, 10476–10478.
384. Y. Ji, P. Li, X. Zhang and L. Wang, *Org. Biomol. Chem.*, 2013, **11**, 4095–4101.
385. W.-G. Lee, R. Gallardo-Macias, K. M. Frey, K. A. Spasov, M. Bollini, K. S. Anderson and W. L. Jorgensen, *J. Am. Chem. Soc.*, 2013, **135**, 16705–16713.
386. J. Liu, W. Chen, Y. Ji and L. Wang, *Adv. Synth. Catal.*, 2012, **354**, 1585–1592.
387. W. Chen, P. Li, T. Miao, L.-G. Meng and L. Wang, *Org. Biomol. Chem.*, 2013, **11**, 420–424.
388. W. Zhou, W. Chen and L. Wang, *Org. Biomol. Chem.*, 2012, **10**, 4172–4178.
389. M. L. N. Rao, D. N. Jadhav and P. Dasgupta, *Eur. J. Org. Chem.*, 2013, 781–788.
390. J. Liu, N. Zhang, Y. Yue, D. Wang, Y. Zhang, X. Zhang and K. Zhuo, *RSC Adv.*, 2013, **3**, 3865–3868.
391. X. Qin, X. Cong, D. Zhao, J. You and J. Lan, *Chem. Commun.*, 2011, **47**, 5611–5613.
392. W. Chen, M. Wang, P. Li and L. Wang, *Tetrahedron*, 2011, **67**, 5913–5919.
393. S. Ye, G. Liu, S. Pu and J. Wu, *Org. Lett.*, 2012, **14**, 70–73.
394. J. Liu, W. Chen and L. Wang, *RSC Adv.*, 2013, **3**, 4723–4730.
395. L. Ackermann, *Org. Lett.*, 2005, **7**, 439–442.
396. L. Ackermann, S. Barfüßer and H. K. Potukuchi, *Adv. Synth. Catal.*, 2009, **351**, 1064–1072.
397. Z.-Y. Tang and Q.-S. Hu, *Adv. Synth. Catal.*, 2006, **348**, 846–850.
398. C. B. Lavery, R. McDonald and M. Stradiotto, *Chem. Commun.*, 2012, **48**, 7277–7279.
399. P. G. Alsabeh, R. J. Lundgren, L. E. Longobardi and M. Stradiotto, *Chem. Commun.*, 2011, **47**, 6936–6938.
400. N. Halland, M. Nazaré, J. Alonso, O. R'kyek and A. Lindenschmidt, *Chem. Commun.*, 2011, **47**, 1042–1044.
401. A. Prakash, M. Dibakar, K. Selvakumar, K. Ruckmani and M. Sivakumar, *Tetrahedron Lett.*, 2011, **52**, 5625–5628.
402. H. Wang, Y. Li, L. Jiang, R. Zhang, K. Jin, D. Zhao and C. Duan, *Org. Biomol. Chem.*, 2011, **9**, 4983–4986.
403. L. Ackermann, W. Song and R. Sandmann, *J. Organomet. Chem.*, 2011, **696**, 195–201.
404. K.-S. Masters, M. Wallesch and S. Bräse, *J. Org. Chem.*, 2011, **76**, 9060–9067.
405. S. G. Newman and M. Lautens, *J. Am. Chem. Soc.*, 2010, **132**, 11416–11417.
406. A. R. Kunzer and M. D. Wendt, *Tetrahedron Lett.*, 2011, **52**, 1815–1818.

407. Y.-Q. Fang and M. Lautens, *Org. Lett.*, 2005, **7**, 3549–3552.
408. Y.-Q. Fang and M. Lautens, *J. Org. Chem.*, 2008, **73**, 538–549.
409. A. Fayol, Y.-Q. Fang and M. Lautens, *Org. Lett.*, 2006, **8**, 4203–4206.
410. M. Nagamochi, Y.-Q. Fang and M. Lautens, *Org. Lett.*, 2007, **9**, 2955–2958.
411. Y. Liang, T. Meng, H.-J. Zhang and Z. Xi, *Synlett*, 2011, 911–914.
412. J. Yuen, Y.-Q. Fang and M. Lautens, *Org. Lett.*, 2006, **8**, 653–656.
413. S. E. Kiruthika and P. T. Perumal, *Org. Lett.*, 2014, **16**, 484–487.
414. Z. J. Wang, F. Yang, X. Lv and W. Bao, *J. Org. Chem.*, 2011, **76**, 967–970.
415. H. F. He, S. Dong, Y. Chen, Y. Yang, Y. Le and W. Bao, *Tetrahedron*, 2012, **68**, 3112–3116.
416. T. O. Vieira, L. A. Meaney, Y. L. Shi and H. Alper, *Org. Lett.*, 2008, **10**, 4899–4901.
417. M. Arthuis, R. Pontikis and J. C. Florent, *Org. Lett.*, 2009, **11**, 4608–4611.
418. Y. Uozumi, M. Mori and M. Shibasaki, *J. Chem. Soc., Chem. Commun.*, 1991, 81–83.
419. O. Leogane and H. Lebel, *Angew. Chem., Int. Ed.*, 2008, **47**, 350–352.
420. M. C. Willis, G. N. Brace, T. J. K. Findlay and I. P. Holmes, *Adv. Synth. Catal.*, 2006, **348**, 851–856.
421. L. C. Henderson, M. J. Lindon and M. C. Willis, *Tetrahedron*, 2010, **66**, 6632–6638.
422. R. G. Vaswani, B. K. Albrecht, J. E. Audia, A. Coté, L. A. Dakin, M. Duplessis, V. S. Gehling, J. C. Harmange, M. C. Hewitt, Y. Leblanc, C. G. Nasveschuk and A. M. Taylor, *Org. Lett.*, 2014, **16**, 4114–4117.
423. K. Yamazaki, Y. Nakamura and Y. Kondo, *J. Chem. Soc., Perkin Trans. 1*, 2002, 2137–2138.
424. K. Yamazaki, Y. Nakamura and Y. Kondo, *J. Org. Chem.*, 2003, **68**, 6011–6019.
425. T. Noji, K. Okano, T. Fukuyama and H. Tokuyama, *Org. Synth.*, 2011, **88**, 388–397.
426. M. G. Mokrushin, A. V. Shastin, V. M. Muzalevskiy, E. S. Balenkova and V. G. Nenajenko, *Mendeleev Commun.*, 2008, **18**, 327–328.
427. T. Nanjo, S. Yamamoto, C. Tsukano and Y. Takemoto, *Org. Lett.*, 2013, **15**, 3754–3757.
428. Y. Goriya and C. V. Ramana, *Chem. Commun.*, 2014, **50**, 7790–7792.
429. S. G. Koenig, J. W. Dankwardt, Y. Liu, H. Zhao and S. P. Singh, *Tetrahedron Lett.*, 2010, **51**, 6549–6551.
430. Q. Cai, Z. Li, J. Wie, C. Ha, D. Pei and K. Ding, *Chem. Commun.*, 2009, 7581–7583.
431. L. Gu and X. Li, *J. Braz. Chem. Soc.*, 2011, **22**, 2036–2039.
432. M. Jiang, J. Li, F. Wang, Y. Zhao, F. Zhao, X. Dong and W. Zhao, *Org. Lett.*, 2012, **14**, 1420–1423.
433. D. Solé and O. Serrano, *Org. Biomol. Chem.*, 2009, **7**, 3382–3384.
434. S. Biswas, M. Nayak, S. Kanojiya and S. Batra, *Adv. Synth. Catal.*, 2011, **353**, 3330–3334.

435. C. S. Cho, D. K. Lim, N. H. Heo, T. J. Kim and S. C. Shim, *Chem. Commun.*, 2004, 104–105.
436. A. Y. Lebedev, A. S. Khartulyari and A. Z. Voskoboynikov, *J. Org. Chem.*, 2005, **70**, 596–602.
437. K. Inamoto, M. Katsuno, T. Yoshino, Y. Arai, K. Hiroya and T. Sakamoto, *Tetrahedron*, 2007, **63**, 2695–2711.
438. D. Viña, E. del Olmo, J. L. López-Pérez and A. S. Feliciano, *Org. Lett.*, 2007, **9**, 525–528.
439. X. Xiong, Y. Jiang and D. Ma, *Org. Lett.*, 2012, **14**, 2552–2555.
440. Y. K. Bae and C. S. Cho, *Appl. Organomet. Chem.*, 2013, **27**, 224–227.
441. M. R. Kumar, A. Park, N. Park and S. Lee, *Org. Lett.*, 2011, **13**, 3542–3545.
442. A. N. Prasad, R. Srinivas and B. M. Reddy, *Catal. Sci. Technol.*, 2013, **3**, 654–658.
443. N. Khatun, A. Gogoi, P. Basu, P. Das and B. K. Patel, *RSC Adv.*, 2014, **4**, 4080–4084.
444. J. J. Song and N. K. Yee, *Org. Lett.*, 2000, **2**, 519–521.
445. J. J. Song and N. K. Yee, *Tetrahedron Lett.*, 2001, **42**, 2937–2940.
446. L. Xu, Y. Peng, Q. Pan, Y. Jiang and D. Ma, *J. Org. Chem.*, 2013, **78**, 3400–3404.
447. H. J. Liu, S. F. Huang, C. L. Chen and M. H. Lin, *Tetrahedron*, 2013, **69**, 3907–3912.
448. N. Halland, M. Nazaré, O. R'kyek, J. Alonso, M. Urmann and A. Lindenschmidt, *Angew. Chem., Int. Ed.*, 2009, **48**, 6879–6882.
449. K. Nozaki, K. Takahashi, K. Nakano, T. Hiyama, H. Z. Tang, M. Fujiki, S. Yamaguchi and K. Tamao, *Angew. Chem., Int. Ed.*, 2003, **42**, 2051–2053.
450. T. Kitawaki, Y. Hayashi, A. Ueno and N. Chida, *Tetrahedron*, 2006, **62**, 6792–6801.
451. Y. Zhou and J. G. Verkade, *Adv. Synth. Catal.*, 2010, **352**, 616–620.
452. W. Geng, W. X. Zhang, W. Hao and Z. Xi, *J. Am. Chem. Soc.*, 2012, **134**, 20230–20233.
453. E. Li, X. Xu, h. Li, H. Zhang, X. Xu, X. Yuan and Y. Li, *Tetrahedron*, 2009, **65**, 8961–8968.
454. Q. Liao, L. Zhang, F. Wang, S. Li and C. Xi, *Eur. J. Org. Chem.*, 2010, 5426–5431.
455. R. Grigg, V. Sridharan, P. Stevenson and T. Worakun, *J. Chem. Soc., Chem. Commun.*, 1986, 1697–1699.
456. R. Grigg, V. Sridharan, P. Stevenson and S. Sukirthalingam, *Tetrahedron*, 1989, **45**, 3557–3568.
457. R. Grigg, V. Loganathan, V. Santhakumar, V. Sridharan and A. Teasdale, *Tetrahedron Lett.*, 1991, **32**, 687–690.
458. B. Burns, R. Grigg, V. Santhakumar, V. Sridharan, P. Stevenson and T. Worakun, *Tetrahedron*, 1992, **48**, 7297–7320.
459. D. Brown, R. Grigg, V. Sridharan and V. Tambyrajah, *Tetrahedron Lett.*, 1995, **36**, 8137–8140.

460. A. García, D. Rodríguez, L. Castedo, C. Saá and D. Domínguez, *Tetrahedron Lett.*, 2001, **42**, 1903–1905.
461. R. Grigg, V. Sridharan, P. Stevenson, S. Sukirthalingam and T. Worakun, *Tetrahedron*, 1990, **46**, 4003–4018.
462. A. P. Kozikowski and D. Ma, *Tetrahedron Lett.*, 1991, **32**, 3317–3320.
463. G. Satyanarayana and M. E. Maier, *Tetrahedron*, 2012, **68**, 1745–1749.
464. L. Zhao, Z. Li, L. Chang, J. Xu, H. Yao and X. Wu, *Org. Lett.*, 2012, **14**, 2066–2069.
465. S. Rousseaux, S. I. Gorelsky, B. K. W. Chung and K. Fagnou, *J. Am. Chem. Soc.*, 2010, **132**, 10692–10705.
466. L. Zhang, F. Zhao, M. Zheng, Y. Zhai and H. Liu, *Chem. Commun.*, 2013, **49**, 2894–2896.
467. W. Chen, L. Jin, Y. Zhu, X. Cao, L. Zheng and W. Mo, *Synlett*, 2013, 1856–1860.
468. V. Tyagi, S. Khan and P. M. S. Chauhan, *Tetrahedron Lett.*, 2013, **54**, 1279–1284.
469. W. C. Wertjes, L. C. Wolfe, P. J. Waller and D. Kalyani, *Org. Lett.*, 2013, **15**, 5986–5989.
470. M. W. Khan and N. G. Kundu, *Synlett*, 1997, 1435–1437.
471. N. G. Kundu, M. W. Khan and R. Mukhopadhyay, *Tatrahedron*, 1999, **55**, 12361–12376.
472. N. G. Kundu and M. W. Khan, *Tetrahedron*, 2000, **56**, 4777–4792.
473. M. Hellal and G. D. Cuny, *Tetrahedron Lett.*, 2011, **52**, 5508–5511.
474. S. Sarkar, S. Dutta, R. Dey and S. Naskar, *Tetrahedron Lett.*, 2012, **53**, 6789–6792.
475. S. Couty, B. Liégault, C. Meyer and J. Cossy, *Org. Lett.*, 2004, **6**, 2511–2514.
476. S. Couty, B. Liégault, C. Meyer and J. Cossy, *Tetrahedron*, 2006, **62**, 3882–3895.
477. C. S. Cho, H. S. Shim, H. J. Choi, T. J. Kim and S. C. Shim, *Synth. Commun.*, 2002, **32**, 1821–1827.
478. Y. Uozumi, N. Kawasaki, E. Mori, M. Mori and M. Shibasaki, *J. Am. Chem. Soc.*, 1989, **111**, 3725–3727.
479. L. Li, M. Wang, X. Zhang, Y. Jiang and D. Ma, *Org. Lett.*, 2009, **11**, 1309–1312.
480. L. Zhang, Y. Zhang, X. Wang and J. Shen, *Molecules*, 2013, **18**, 654–665.
481. J. Pan, Z. Xu, R. Zeng and J. Zou, *Chin. J. Chem.*, 2013, **31**, 1022–1026.
482. S. Sarkar, R. Pal and A. K. Sen, *Tetrahedron Lett.*, 2013, **54**, 4273–4276.
483. S. Sarkar, N. Chatterjee, M. Roy, R. Pal, S. Sarkar and A. K. Sen, *RSC Adv.*, 2014, **4**, 7024–7029.
484. C. S. Cho, L. H. Jiang, D. Y. Lee, S. C. Shim, H. S. Lee and S. D. Cho, *J. Heterocycl. Chem.*, 1997, **34**, 1371–1374.
485. C. S. Cho and W. X. Ren, *Tetrahedron Lett.*, 2009, **50**, 2097–2099.
486. C. Han, Y. Shen, P. Lu and Y. Wang, *Chin. J. Chem.*, 2013, **31**, 182–186.
487. C. C. Cho, D. Y. Chu, D. Y. Lee, S. C. Shim, T. J. Kim, W. T. Lim and N. H. Heo, *Synth. Commun.*, 1997, **27**, 4141–4158.

488. C. S. Cho, L. H. Jiang and S. C. Shim, *Synth. Commun.*, 1998, **28**, 849–857.
489. C. S. Cho, X. Wu, L. H. Jiang, S. C. Shim, H. J. Choi and T. J. Kim, *J. Heterocycl. Chem.*, 1998, **35**, 265–268.
490. C. S. Cho, H. S. Shim, H. J. Choi, T. J. Kim, S. C. Shim and M. C. Kim, *Tetrahedron Lett.*, 2000, **41**, 3891–3893.
491. R. Grigg, V. Sridharan and A. Thayaparan, *Tetrahedron Lett.*, 2003, **44**, 9017–9019.
492. K. Orito, A. Horibata, T. Nakamura, H. Ushito, H. Nagasaki, M. Yuguchi, S. Yamashita and M. Tokuda, *J. Am. Chem. Soc.*, 2004, **126**, 14342–14343.
493. T. Morimoto, M. Fujioka, K. Fuji and K. Kakiuchi, *J. Organomet. Chem.*, 2007, **692**, 625–634.
494. R. Grigg, L. Zhang, S. Collard and A. Keep, *Tetrahedron Lett.*, 2003, **44**, 6979–6982.
495. D. Marosvölgyo, A. Takács, Z. Riedl and L. Kollár, *Tetrahedron*, 2011, **67**, 1036–1040.
496. Y. Zhu, L. Chuanzhao, A. O. Biying, M. Sudarmadji, A. Chen, D. T. Tuan and A. M. Seayad, *Dalton Trans.*, 2011, **40**, 9320–9325.
497. T. T. Dang, Y. Zhu, J. S. Y. Ngiam, S. C. Ghosh, A. Chen and A. M. Seayad, *ACS Catal.*, 2013, **3**, 1406–1410.
498. S. A. Worlikar and R. C. Larock, *J. Org. Chem.*, 2008, **73**, 7175–7180.
499. J. R. Martinelli, D. A. Waston, D. M. M. Freckmann, T. E. Barder and S. L. Buchwald, *J. Org. Chem.*, 2008, **73**, 7102–7107.
500. M. V. Khedkar, S. R. Khan, K. P. Dhake and B. M. Bhanage, *Synthesis*, 2012, 2623–2629.
501. J. C. Hsieh and C. H. Cheng, *Chem. Commun.*, 2005, 4554–4556.
502. B. Liu, Y. Li, H. Jiang, M. Yin and H. Huang, *Adv. Synth. Catal.*, 2012, **354**, 2288–2300.
503. B. H. Yang and S. L. Buchwald, *Org. Lett.*, 1999, **1**, 35–37.
504. R. R. Poondra and N. J. Turner, *Org. Lett.*, 2005, **7**, 863–866.
505. C. Kalinski, M. Umkehrer, G. Ross, J. Kolb, C. Burdack and W. Hiller, *Tetrahedron Lett.*, 2006, **47**, 3423–3426.
506. M. B. Calvert and J. Sperry, *J. Heterocycl. Chem.*, 2014, **51**, 282–284.
507. S. Wang, Q. Yang, J. Dong, C. Li, L. Sun, C. Song and J. Chang, *Eur. J. Org. Chem.*, 2013, 7631–7634.
508. T. Kukosha, N. Trufilkina and M. Katkevics, *Synlett*, 2011, 2525–2528.
509. J. C. Hsieh, A. Y. Cheng, J. H. Fu and T. W. Kang, *Org. Biomol. Chem.*, 2012, **10**, 6404–6409.
510. J. Sun, B. Liu and B. Xu, *RSC Adv.*, 2013, **3**, 5824–5827.
511. H. Jin, B. Zhou, Z. Wu, Y. Shen and Y. Wang, *Tetrahedron*, 2011, **67**, 1178–1182.
512. S. Tanimori, Y. Ozaki, Y. Iesaki and M. Kirihata, *Synlett*, 2008, 1973–1976.
513. S. Tanimori, Y. Kobayashi, Y. Iesaki, Y. Ozaki and M. Kirihata, *Org. Biomol. Chem.*, 2012, **10**, 1381–1387.

514. S. J. Wlsch, C. Kalinski, M. Umkehrer, G. Ross, J. Kolb, C. Burdack and L. A. Wessjohann, *Tetrahedron Lett.*, 2012, **53**, 2298–2301.
515. A. S. Guram, R. A. Rennels and S. L. Buchwald, *Angew. Chem.*, 1995, **107**, 1456–1458.
516. J. P. Wolfe, R. A. Rennels and S. L. Buchwlad, *Tetrahedron*, 1996, **52**, 7525–7546.
517. S. Wagaw, R. A. Rennels and S. L. Buchwald, *J. Am. Chem. Soc.*, 1997, **119**, 8451–8458.
518. I. Bytschkov, H. Siebeneicher and S. Doye, *Eur. J. Org. Chem.*, 2003, 2888–2902.
519. D. L. Priebbenow, S. G. Stewart and F. M. Pfeffer, *Org. Biomol. Chem.*, 2011, **9**, 1508–1515.
520. K. S. Kumar, P. M. Kumar, V. S. Rao, A. A. Jafar, C. L. T. Meda, R. Kapavarapu, K. V. L. Parsa and M. Pal, *Org. Biomol. Chem.*, 2012, **10**, 3098–3103.
521. R. Harada, N. Nishida, S. Uchiito, Y. Onozaki, N. Kurono, H. Senboku, T. Masao, T. Ohkuma and K. Orito, *Eur. J. Org. Chem.*, 2012, 366–379.
522. J. S. Nakhla, J. W. Kampf and J. P. Wolfe, *J. Am. Chem. Soc.*, 2006, **128**, 2893–2901.
523. W. R. Bowman, H. Heaney and P. H. G. Smith, *Tetrahedron Lett.*, 1984, **25**, 5821–5824.
524. A. Klapars, X. Huang and S. L. Buchwald, *J. Am. Chem. Soc.*, 2002, **124**, 7421–7428.
525. F. Y. Kwong and S. L. Buchwald, *Org. Lett.*, 2003, **5**, 793–796.
526. A. Minatti and S. L. Buchwald, *Org. Lett.*, 2008, **10**, 2721–2724.
527. Y. M. Zhu, L. N. Qin, R. Liu, S. J. Ji and H. Katayama, *Tetrahedron Lett.*, 2007, **48**, 6262–6266.
528. T. Kubo, C. Katoh, K. Yamada, K. Okano, H. Tokuyama and T. Fukuyama, *Tetrahedron*, 2008, **64**, 11230–11236.
529. (a) F. Zhou, J. Guo, J. Liu, K. Ding, S. Yu and Q. Cai, *J. Am. Chem. Soc.*, 2012, **134**, 14326–14329; (b) Q. Cai and F. Zhou, *Synlett*, 2013, 408–411.
530. L. D. S. Yadav, B. S. Yadav and V. K. Rai, *Synthesis*, 2006, 1868–1872.
531. W. Yang, L. Ye, D. Huang, M. Liu, J. Ding, J. Chen and H. Wu, *Tetrahedron*, 2013, **69**, 9852–9856.
532. Y. B. Huang, C. T. Yang, J. Yi, X. J. Deng, Y. Fu and L. Liu, *J. Org. Chem.*, 2011, **76**, 800–810.
533. A. Arnanz, M. Pintado-Sierra, A. Corma, M. Iglesias and F. Sánchez, *Adv. Synth. Catal.*, 2012, **354**, 1347–1355.
534. R. Omar-Amrani, A. Thomas, E. Brenner, R. Schneider and Y. Fort, *Org. Lett.*, 2003, **5**, 2311–2314.
535. (a) C. Chowdhury, S. B. Mandal and B. Achari, *Tetrahedron Lett.*, 2005, **46**, 8531–8534; (b) K. Brahma, B. Achari and C. Chowdhury, *Synthesis*, 2013, 545–555.
536. V. Fiandanese, G. Marchese, A. Punzi, F. Lannone and G. G. Rafaschieri, *Tetrahedron*, 2010, **66**, 8846–8853.

537. C. Rossy, J. Majimel, M. T. Delapierre, E. Fouquet and F. X. Felpin, *J. Organomet. Chem.*, 2014, **755**, 78–85.
538. N. Arai, M. Takahashi, M. Mitani and A. Mori, *Synlett*, 2006, 3170–3172.
539. V. Gracias, A. F. Gasiecki, T. G. Pagano and S. W. Djuric, *Tetrahedron Lett.*, 2006, **47**, 8873–8876.
540. Y. Huang, W. Chen, D. Zhao, C. Chen, H. Yin, L. Zheng, M. Jin and S. Han, *Chin. J. Chem.*, 2013, **31**, 1007–1010.
541. S. R. Dandepally and A. L. Williams, *Tetrahedron Lett.*, 2009, **50**, 1395–1398.
542. Y. L. Choi, H. Lee, B. T. Kim, K. Choi and J. N. Heo, *Adv. Synth. Catal.*, 2010, **352**, 2041–2049.
543. J. K. Laha, N. Dayal, S. Singh and R. Bhimpuria, *Eur. J. Org. Chem.*, 2014, 5469–5475.
544. S. Berteina and A. De Mesmaeker, *Synlett*, 1998, 1227–1230.
545. O. Gaertzen and S. L. Buchwald, *J. Org. Chem.*, 2002, **67**, 465–475.
546. D. Solé and O. Serrano, *J. Org. Chem.*, 2010, **75**, 6267–6270.
547. H. Ren and P. Knochel, *Angew. Chem., Int. Ed.*, 2006, **45**, 3462–3465.
548. Y. Luo and J. Wu, *Org. Lett.*, 2012, **14**, 1592–1595.
549. N. Arnau, M. Moreno-Mañas and R. Pleixats, *Tetrahedron*, 1993, **49**, 11019–11028.
550. J. Pelletier, S. Olivero and E. Duñach, *Synth. Commun.*, 2004, **34**, 3343–3348.
551. K. Ouyang, Y. Liang and Z. Xi, *Org. Lett.*, 2012, **14**, 4572–4575.

CHAPTER 3

Six-membered Heterocycle Synthesis

In this chapter, the applications of double-functionalized arenes in the preparation of six-membered heterocycles will be discussed. The contents are sub-divided into the classes of substrates applied.

3.1 1,2-Dihaloarenes

In 2004, Jing and co-workers reported a palladium-catalyzed coupling of 1,2-dibromobenzenes with 1,2-dialcohols.[1] By the combination of palladium and phosphine, a number of 1,4-benzodioxanes were produced in moderate yields (Scheme 3.1). This method was used as a key step in the synthesis of two natural products as well: isoamericanol A and isoamericanin A. The copper-catalyzed *N*,*O*-arylation of 1-bromo-2-iodobenzene with (*S*)-prolinols was also reported.[2] In the presence of CuI (2.5 mol%) as a catalyst with NaOH as a base in *i*PrOH, the corresponding products were formed in good yields.

In 2009, a one-step protocol for the synthesis of xanthones *via* Pd-catalyzed coupling between 1,2-dibromoarenes and salicylaldehydes was developed.[3] The success of the reaction relies heavily on the careful selection of a proper palladium catalyst, solvent and base; moderate yields of the desired products were formed (Scheme 3.2). In this communication, a reaction mechanism was proposed by the authors. The Pd(0) catalyst first underwent oxidative insertion into one of the carbon–halide bonds to generate the aryl–Pd(II) intermediate, which reacted with the phenolate to displace the halide, then underwent C–H activation of the C–H bond of the aldehyde to form intermediate **A**. After abstraction of the hydrogen atom by the base and reductive elimination to exclude the Pd(0) catalyst, the biaryl ketone intermediate was

RSC Catalysis Series No. 24
Heterocycles from Double-Functionalized Arenes: Transition Metal Catalyzed Coupling Reactions
By Xiao-Feng Wu and Matthias Beller

Published by the Royal Society of Chemistry, www.rsc.org

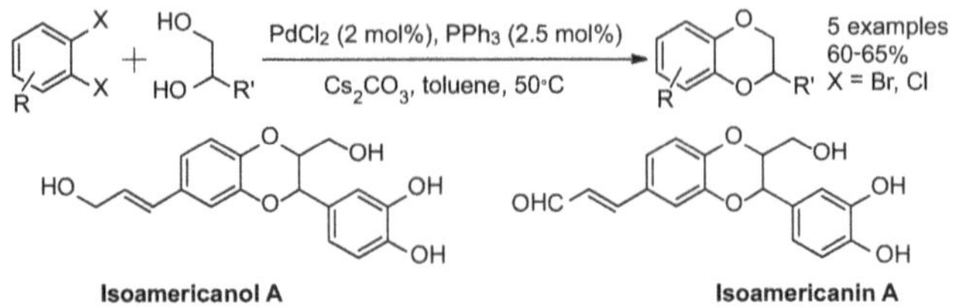

Scheme 3.1 Palladium-catalyzed synthesis of 1,4-benzodioxanes.

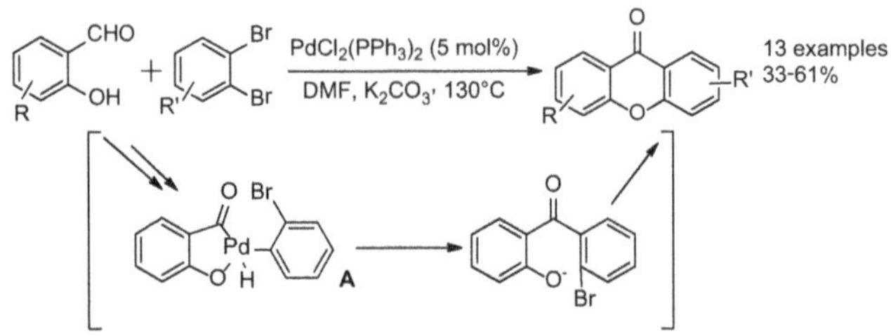

Scheme 3.2 Palladium-catalyzed synthesis of xanthones from salicylaldehydes.

Reaction Procedure (Scheme 3.2): In a 25 ml round-bottomed flask equipped with a reflux condenser under N_2 were placed salicylaldehyde (1 mmol) and 1,2-dibromobenzene (2 mmol), K_2CO_3 (2 mmol), $(PPh_3)_2PdCl_2$ (35 mg, 0.05 mmol) and 5 ml DMF. The mixture was heated to 130 °C for 12 hours before it was cooled to room temperature. The reaction mixture was diluted with diethyl ether (25 mL) and washed with water and brine. The aqueous layer was extracted with diethyl ether (3×25 mL). The organic layers were combined and dried with $MgSO_4$, filtered, and the solvent was removed under reduced pressure. The residue was purified by flash chromatography on silica gel (petroleum ether–ethyl acetate, 50 : 1) to afford the desired product.

produced. The Pd(0) catalyst reinserted into the other carbon halide bond on the ketone and the final product was produced after another round of the intramolecular displacement of the halide and reductive elimination.

Phenothiazines are a class of important heterocycles and are extensively used as drugs (such as the promazine series), insecticides, inhibitors of polymerization, optoelectronic materials, antioxidants, paints, *etc.* Abele's group found that phenothiazines can be prepared from the reaction between 2-aminobenzothiazoles and 1,2-dihaloarenes (Scheme 3.3a).[4] They proposed that the reaction mechanism for this transformation includes *N*-3 arylation of 2-amino-benzothiazoles leading to benzothiazoline imine. The ring opening of imine in the presence of Cs_2CO_3 afforded the caesium salt of thiol. The final products were formed after copper-catalyzed C–S bond formation. Meanwhile, Zeng and co-workers reported another CuI-catalyzed system.[5] With ligand-free CuI-catalyzed cascade C–S and C–N cross-coupling

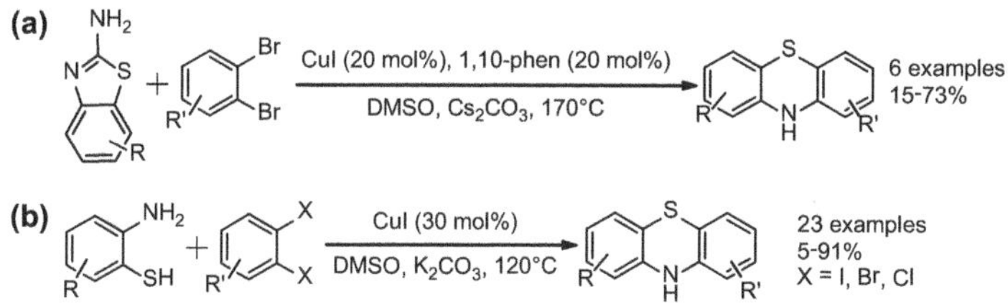

Scheme 3.3 Copper-catalyzed synthesis of phenothiazines.

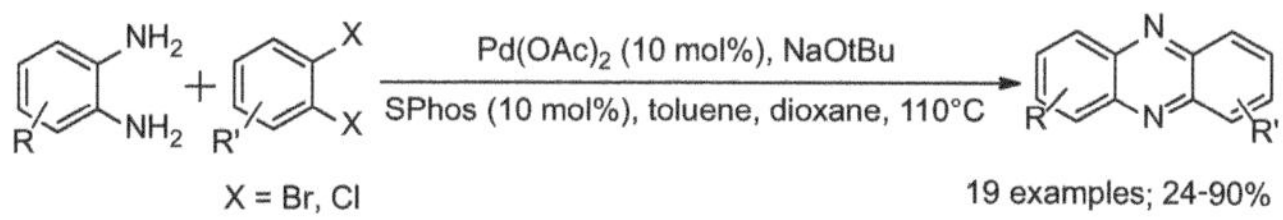

Scheme 3.4 Palladium-catalyzed synthesis of phenazines.

of (hetero)aryl *ortho*-dihalides and *ortho*-aminobenzenethiols, various phenothiazines were synthesized with excellent regioselectivity in moderate to excellent yields (Scheme 3.3b). Later on, copper oxide nanoparticles supported on graphene oxide[6] and $CuFe_2O_4$ nanoparticles[7] were applied in this type of coupling as well.

In 2013, a palladium-catalyzed double *N*-arylation of 1,2-diamino(hetero)-arenes with 1,2-dihalo(hetero)arenes for the synthesis of phenazines and pyridoquinoxalines was developed by Laha and co-workers.[8] A variety of functional groups, which included base-sensitive groups, were tolerated under the optimized conditions to afford phenazines or pyridoquinoxalines in good to excellent yields (Scheme 3.4).

A new palladium-catalyzed cross-coupling of the tosylhydrazone of a Mannich adduct with a 1,2-dihalogenated aromatic system followed by an intramolecular C–N bond forming reaction was developed in 2011.[9] In this methodology, the same palladium catalyst promotes both independent steps. Moderate to good yields of the desired products were isolated (Scheme 3.5a). Wang's group developed a facile and convergent palladium-catalyzed coupling of *o*-dihalobenzenes and *N*-tosylhydrazones to acridines.[10] A single palladium species was used to catalyze two independent reaction steps in a one-pot fashion too. This protocol is broad in scope with respect to both coupling partners, providing an efficient method to access heavily-functionalized acridine moieties in good yields in general (Scheme 3.5b). In addition, regioselective reactions have been achieved for unsymmetrical *o*-dihalobenzenes. A preliminary mechanistic investigation has revealed a favourable pathway consisting of a faster carbene migratory insertion and β-hydride elimination sequence followed by intramolecular C–N cross-coupling.

A new palladium-catalyzed direct annulation of benzophenone imines with *o*-dihalobenzenes proceeding through C–H and N–H bond activation to produce 6-arylphenanthridine derivatives was developed in 2012.[11]

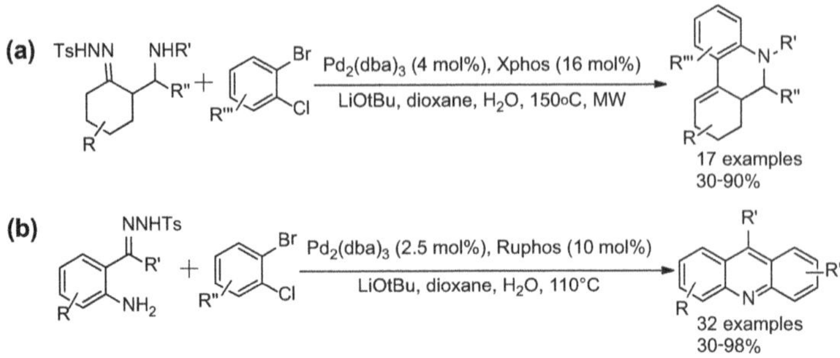

Scheme 3.5 Palladium-catalyzed cyclization of hydrazones.

Reaction Procedure (Scheme 3.5a): A 0.5–2 mL microwave vial was charged with *N*-tosylhydrazone (0.27 mmol, 1.08 equiv.), Xphos (16 mol%), tris(dibenzylideneacetone)dipalladium(0) (4 mol%), lithium *tert*-butoxide (1.025 mmol, 4.1 equiv.), dioxane (1.5 mL), distilled water (10 μl), the corresponding aryl halide (0.25 mmol, 1 equiv.) and a triangular magnetic stir bar. The vessel was sealed with a septum, placed into the microwave cavity and irradiated to maintain the reaction mixture at the desired temperature (150 °C) during the programmed time (15–120 min) in a Biotage Initiator microwave apparatus. When the reaction was completed, the vial was cooled down to room temperature using a propelled air flow. It was then opened and poured into a separating funnel. The vial was rinsed with a saturated solution of $NaHCO_3$ (10 mL) and dichloromethane (20 mL). The layers were separated and the aqueous phase was extracted three times with dichloromethane (20 ml). The combined organic phases were washed with brine, dried over $MgSO_4$ and filtered. The solvent was removed under reduced pressure and the residue was purified by flash chromatography on silica gel.

(Scheme 3.5b): $Pd_2(dba)_3$ (7 mg, 0.025 equiv.), RuPhos (14 mg, 0.1 equiv.), and 1,4-dioxane (1 mL) were added to an oven-dried microwave tube equipped with a Teflon®-coated magnetic stir bar. The tube was evacuated and backfilled with N_2 three times. The mixture was heated and stirred at 90 °C for 5 min, and then allowed to cool to room temperature. LiO*t*Bu (96 mg, 4.0 equiv.) and *N*-tosylhydrazone (136 mg, 1.5 equiv.) were added to the tube. The tube was evacuated and backfilled with N_2 three times. *o*-Dibromobenzene (71 mg, 0.3 mmol) and water (27 mg, 5.0 equiv.) were added to the tube by syringe. The reaction mixture was heated and stirred at 110 °C for 12 h, and then it was allowed to cool to room temperature. The mixture was filtered through a silica plug (washing with petroleum ether–ethyl acetate, 1 : 1). The filtrate was concentrated *in vacuo*, and the residue was purified by flash column chromatography (petroleum ether–ethyl acetate, 15 : 1) to yield 9-methylacridine.

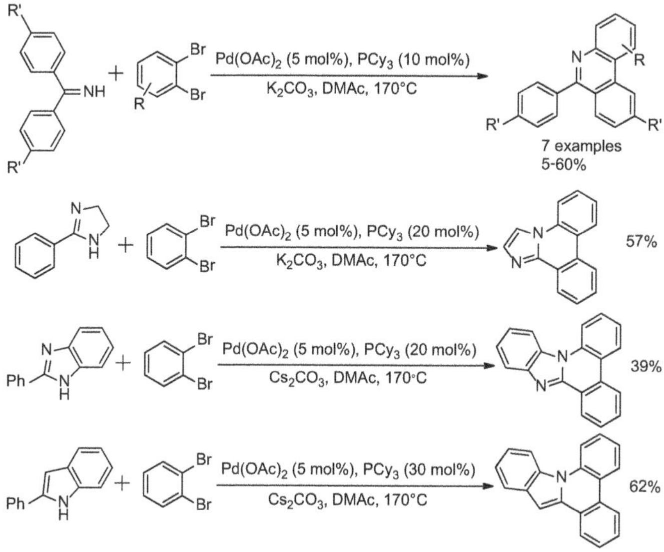

Scheme 3.6 Palladium-catalyzed synthesis of phenanthridines.

2-Phenyl-imidazole, -benzimidazole, and -indole were tested as substrates as well, and all underwent annulation to form the corresponding tetra- and pentacyclic compounds in low to moderate yields (Scheme 3.6). Soon after, more systematic and specific studies were performed by different groups and the application of the products as fluorescent materials was also demonstrated.[12–14]

In 2010, Hocek and co-workers employed intramolecular C–H arylations as a key step in the synthesis of fused purine systems: 13-substituted purino[8,9-*f*]phenanthridines and 11-substituted 5,6-dihydropurino[8,9-*a*]isoquinolines.[15] Moderate yields of the products can be achieved by double C–H arylations of 9-phenylpurines with 1,2-diiodobenzene. The consecutive Suzuki coupling of 9-(2-bromophenyl)purines with 2-bromophenylboronic acid followed by intramolecular C–H arylation was more efficient. 5,6-Dihydropurino[8,9-*a*]isoquinolines were prepared in quantitative yields by intramolecular C–H arylations of 9-(2-chlorophenethyl)purines. The novel fused heterocyclic systems of purino[8,9-*f*]phenanthridines and 5,6-dihydropurino[8,9-*a*]isoquinolines showed interesting absorption spectra and luminescence.

An efficient and simple tandem protocol for the synthesis of 1,2,3-triazole-fused imidazo-[1,2-*a*]pyridines was reported in 2013.[16] The reactions followed a copper-catalyzed tandem azide–alkyne cycloaddition (CuAAC), Ullmann-type C–N coupling, and intramolecular direct arylation sequence and showed high generality and functional group tolerance. Good yields (59–77%) of 1,2,3-triazole-fused imidazo[1,2-*a*]pyridines were produced in a single step (Scheme 3.7).

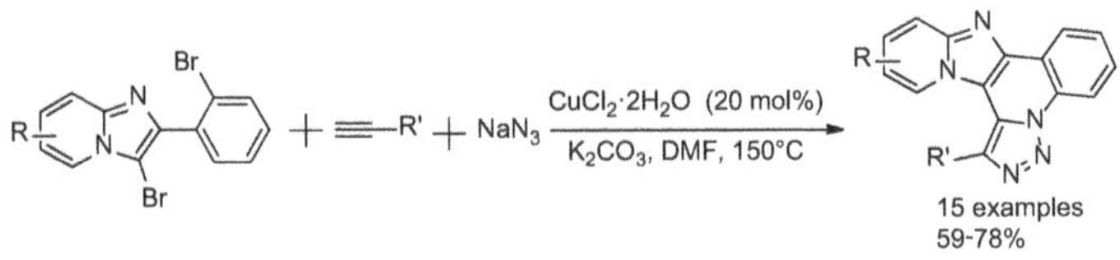

Scheme 3.7 Copper-catalyzed synthesis of fused imidazo-[1,2-*a*]pyridines.

Reaction Procedure (Scheme 3.7): A mixture of imidazole (0.5 mmol), phenylacetylene (0.6 mmol), sodium azide (0.6 mmol), $CuCl_2 \cdot 2H_2O$ (17 mg, 0.1 mmol) and K_2CO_3 (1.25 mmol) were vigorously stirred in DMF (3 mL) at 150 °C for 24 h. After cooling to room temperature, the reaction mixture was diluted with water and extracted into ethyl acetate. The organic layer was dried with anhydrous sodium sulfate and the solvent was evaporated. The crude compound was purified by flash chromatography on a short silica gel (ethyl acetate–hexanes) to afford the pure compound.

3.2 2-Halophenol Derivatives

Among the numerous oxygen-containing six-membered heterocycles, flavones and its derivatives are a representative class of biologically active compounds. In 1990, Kalinin and co-workers reported a palladium-catalyzed procedure for flavone synthesis *via* carbonylative cyclization of 2-iodophenols with terminal alkynes.[17] The reactions were performed in diethylamine at 120 °C, and the desired flavones were formed in good yields (Scheme 3.8a). To the best of our knowledge, this is the first example of the carbonylative synthesis of flavones from 2-iodophenols and acetylenes. Afterwards, modified procedures were developed, such as using water[18] or phosphonium salt ionic liquids[19] as a solvent, and applying microwave heating.[20] In these cases, only an atmospheric pressure of CO was required and good yields could be achieved. In 2000, Yang and Miao found that in addition to 2-iodophenols, iodophenol acetates can be applied as substrates as well.[21] Regiospecific carbonylative annulation of *o*-iodophenol acetates and acetylenes mediated by a palladium–thiourea–dppp complex in the presence of a base at 40 °C under a balloon pressure of CO generates diversified flavones in high yields (Scheme 3.8b). This newly developed report extended carbonylative annulation and provide a new method for the syntheses of flavonoid compounds. More recently, a palladium-catalyzed procedure based on using 2-bromophenols as substrates was developed by Zhuo and co-workers.[22] By using benzimidazolium ligand and $PdCl_2(PPh_3)_2$ as the catalyst system, the desired flavones were formed in moderate to good yields in an amine solvent (Scheme 3.8c). Interestingly, aurones can be selectively produced by slightly changing the amine solvent applied.

In 1997, Alper and Okuro studied the cyclization of 2-iodophenols with allenes.[23] By using $Pd(OAc)_2$ as a catalyst, DPPB as a ligand and DiPEA as a

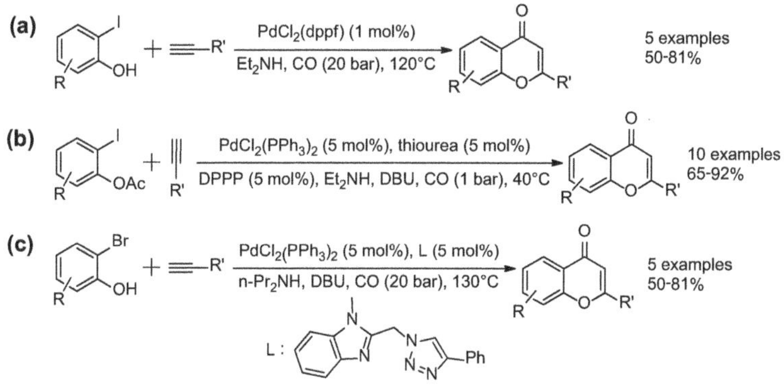

Scheme 3.8 Palladium-catalyzed carbonylative synthesis of flavones.

Reaction Procedure (Scheme 3.8a): Diethylamine (3 ml), *o*-iodophenol (0.91 mmol), phenylacetylene (1.82 mmol) and $PdCl_2$(dppf) (6.0 mg, 0.0091 mmol) were placed into an autoclave. The reaction was carried out under CO (20 bar) at 120 °C with magnetic stirring for 2 hours. The yield of flavone formed was determined by TIC and UV spectroscopy (81%). The product was isolated by TLC on silica gel (hexane–chloroform, 1 : 2 as eluent) and recrystallized from hexane.

(Scheme 3.8b): Diethylamine (5 ml), *o*-iodophenylacetate (1 mmol), phenylacetylene (2 mmol) and $PdCl_2(PPh_3)_2$ (5 mol%), thiourea (5 mol%), DPPP (5 mol%) and DBU (5 mmol) were placed into an autoclave and degased with CO for 10 min and then stirred at 40 °C for 48 h. The product was isolated by chromatography on silica gel (ethyl acetate–hexane, 2 : 3 as eluent).

(Scheme 3.8c): A 50 mL autoclave equipped with a magnetic stir bar was charged with 2-bromophenols (1.0 mmol), alkynes (2.0 mmol), Pd catalyst (0.05 mmol), benzimidazolium ligands (0.05 mmol), and solvent (*n*-Pr_2NH, 4.0 mL). Then the autoclave was pressurized with carbon monoxide to 20 bar (CO purity, 99.9%). The autoclave was placed in an oil bath pre-heated at 130 °C, and the reaction mixture was stirred for 24 h. After the reaction, the autoclave was cooled, and excess gas was purged slowly. After removal of the solvent *in vacuo*, the residual mixture was chromatographed by TLC using hexane–ethyl acetate to give the desired product.

base, the reactions proceeded in a regioselective manner in benzene under 20 bar of CO at 100 °C. The corresponding 3-methylene-2,3-dihydro- or 3-vinyl-4*H*-1-benzopyran-4-one derivatives were produced in good yields. Grigg's group studied this type of cyclization in 2000 (Scheme 3.9).[24,25] By changing the catalyst system, the selectivity of this transformation can be dramatically improved. They applied the products formed in cycloaddition and Diels–Alder reactions with azomethine ylide and imines.

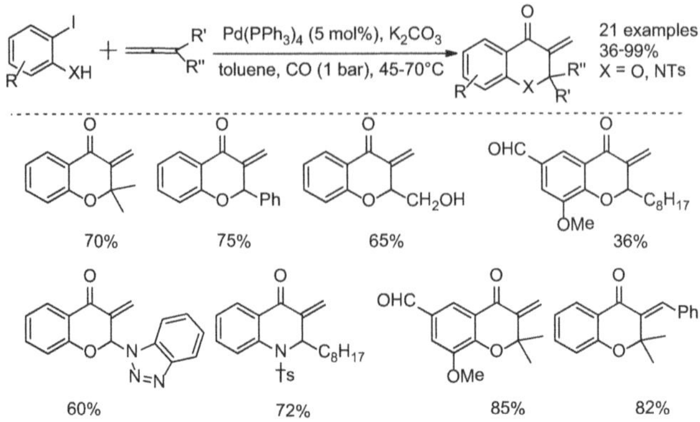

Scheme 3.9 Palladium-catalyzed carbonylative synthesis of benzopyranones.

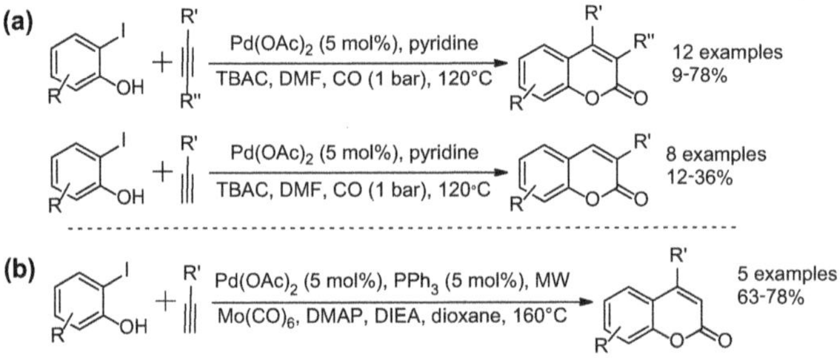

Scheme 3.10 Palladium-catalyzed carbonylative synthesis of coumarins.

The carbonylative cyclization of alkynes with 2-iodophenols to give coumarins was achieved as well. In 2000, Larock's group reported the first study on this selectivity.[26] A variety of 3,4-disubstituted coumarins were prepared in good yields from readily available starting materials (Scheme 3.10a). The reaction mechanism for this carbonylative annulation was proposed and involved (1) reduction of $Pd(OAc)_2$ to the actual Pd(0) catalyst, (2) oxidative addition of *o*-iodophenol to Pd(0), (3) insertion of the internal alkyne, (4) insertion of carbon monoxide, and (5) attack of the phenolic oxygen on the resulting acyl–palladium complex leading to formation of the coumarin and regeneration of the Pd(0) catalyst. As well as internal alkynes and 2-iodophenols, terminal alkynes and 2-iodoanilines can also be applied as substrates. In the case of using 2-iodoanilines as starting materials, their carbamate derivatives are required. In 2005, Xiao and Cao studied this transformation by using molybdenum hexacarbonyl as a convenient and solid carbon monoxide source.[27] The reactions were conducted at 160 °C for 30 min with microwave heating in the presence of diethylamine (DIEA) and DMAP in 1,4-dioxane. The corresponding chromen-2-one derivatives were afforded in good yields (Scheme 3.10b). This concept was applied in the synthesis of anthrapyran-2-ones in 2013.[28] Heterobimetallic Co_2Rh_2 nanoparticles were tested as heterogeneous catalysts for this transformation in 2004.[29] Moderate to good yields of the corresponding coumarins were produced under the same conditions. However, reusability, the advantage of heterogeneous catalysts, was not shown.

The carbonylative cyclization of 2-iodophenols with norbornadiene or norbornene was also carried out. In 1989, Catellani's group studied the annulation of 2-iodophenol with norbornadiene to coumarin (Scheme 3.11a).[30,31] In this method, the elimination of cyclopentadiene *via* retro-Diels–Alder reaction was involved in the formation of the terminal product. Fiaud and co-workers reported the palladium-catalyzed carbonylative cyclization of 2-iodophenol with norbornene in 1997 (Scheme 3.11b).[32] They reported that the selectivity for the production of the two regioisomers can be controlled. Catellani and co-workers studied the intramolecular cyclization of *ortho*-iodophenyl 3-butenoate to 4-methylcoumarin with a palladium catalyst.[33,34] 85% of 4-methylcoumarin was produced in the presence of benzonitrile and carbon monoxide, which is compulsory (Scheme 3.12a). Silva, Costa and their co-workers reported a tandem process for the synthesis of coumarins from 2-iodophenols and enoates with the assistance of a palladium catalyst (Scheme 3.12b).[35,36] The general scope of this Heck-lactonization involves *E*- and *Z*-enoates as substrates. It was shown that this reaction is sensitive to steric hindrance around the double bound in

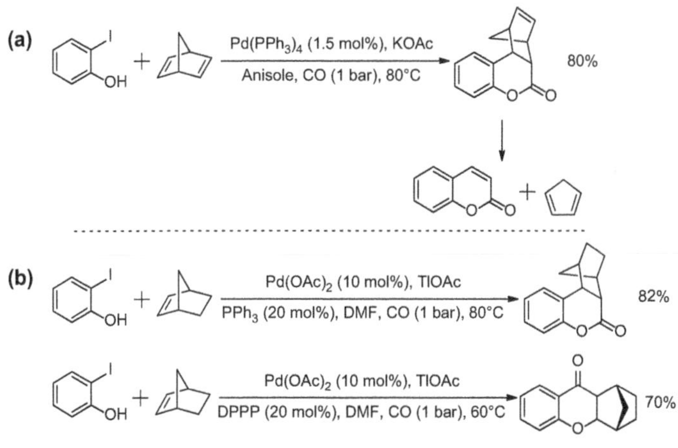

Scheme 3.11 Palladium-catalyzed carbonylative cyclization of alkenes.

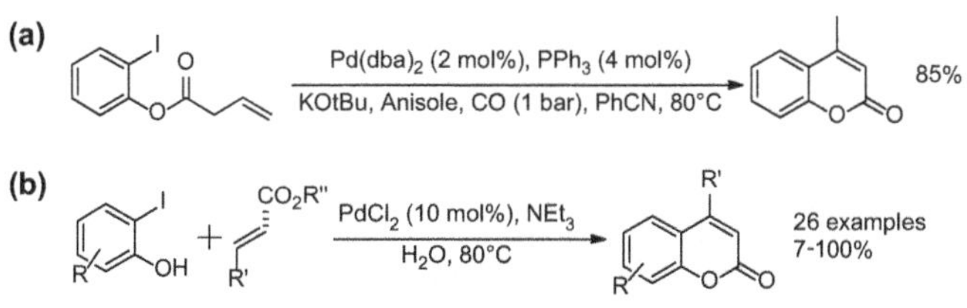

Scheme 3.12 Palladium-catalyzed cyclization of functionalized alkenes.

the enoates. The intervention of cationic palladium intermediates was evidenced by ESI(+)-MS(/MS) for reactions in the presence of Ag_2CO_3 or with H_2O as a solvent. A series of coumarins disubstituted at the 6- and 4-positions could be obtained in one step from enoates *via* a tandem Heck-lactonization reaction under eco-friendly conditions employing water and avoiding the use of silver salts and phosphines.

2*H*-1,4-Benzoxazin-3-(4*H*)-one is an important heterocycle scaffold which can be found in natural and designed synthetic bioactive compounds, ranging from herbicides and fungicides to therapeutic usable drugs. More specific examples, including blepharin and other glycosides with the 2-hydroxy-2*H*-1,4-benzoxazine-3(4*H*)-one skeleton isolated from gramineous plants, such as maize, wheat, rye and rice, have been suggested to act as plant resistance factors against microbial diseases and insects; **C-1027-chr 2** is an active ingredient of enediyne antitumor antibiotics. In 2009, Liu and co-workers reported an efficient and convenient method for preparing *N*-substituted 2*H*-1,4-benzoxazin-3-(4*H*)-ones from 2-halophenols *via* nucleophilic substitution with 2-chloroacetamides followed by a CuI-catalyzed coupling cyclization.[37] A broad spectrum of substrates can be effectively employed to afford the desired products in good yields (Scheme 3.13a). Meanwhile, Bao and co-workers developed a similar procedure with thermal heating.[38] Good to excellent yields of the desired products were isolated (Scheme 3.13b). In 2013, a palladium-catalyzed procedure was developed for

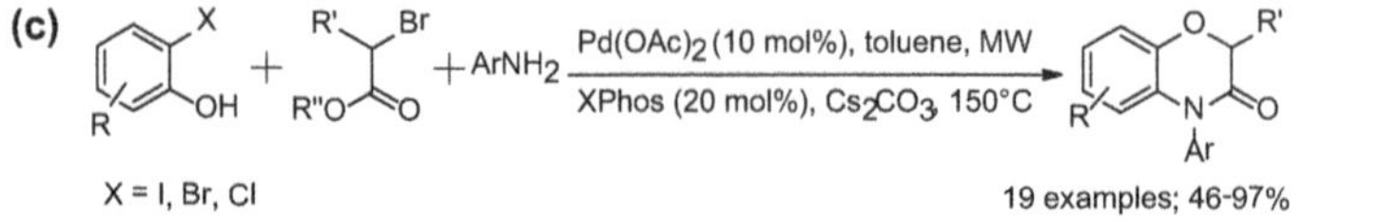

Scheme 3.13 Synthesis of 2*H*-1,4-benzoxazin-3-(4*H*)-ones.

Reaction Procedure (Scheme 3.13a): To a solution of 2-halophenol (0.5 mmol) and chloroacetamide (0.75 mmol) in DMSO (2 mL) were added CuI (0.1 mmol), Cs_2CO_3 (0.9 mmol), and DBU (0.75 mmol). The vial was sealed and the mixture was then irradiated (MW) for 10 min at 130 °C. The cold mixture was diluted with CH_2Cl_2 (20 mL), washed with brine, dried over anhydrous Na_2SO_4, and concentrated *in vacuo*. The residue was purified by flash column chromatography (petroleum ether–ethyl acetate, 10 : 1) to yield the expected product.

(Scheme 3.13b): An oven-dried Schlenk tube equipped with a Teflon valve was charged with a magnetic stir bar, CuI (0.05 mmol, 10 mol%), Cs_2CO_3 (1.1 mmol), 2-halophenol (0.60 mmol), 2-haloamides (0.50 mmol), and 1,10-phenanthroline (0.10 mmol, 20 mol%). The tube was evacuated and backfilled with N_2 (this procedure was repeated 3 times). Under a counter-flow of N_2, 1,4-dioxane (2.0 mL) was added by syringe and the mixture was stirred for about 24 h at 90 °C. The reaction mixture was cooled to room temperature, ethyl acetate (20 mL) was added, and the resulting suspension was filtered. The filtrate was concentrated, and the residue was purified by column chromatography on silica gel (PE–EtOAc, 10 : 1–2 : 1, v/v) to provide the desired product.

(Scheme 3.13c): To a pressure-safe vial (10 mL) containing a magnetic stir bar were added a 2-halophenol (0.3 mmol), an ethyl 2-(2-halophenoxy)alkanoate (0.45 mmol), an aryl amine (0.54 mmol), Cs_2CO_3 (1.05 mmol), $Pd(OAc)_2$ (6.7 mg, 0.03 mmol, 10 mol%), and XPhos (28.5 mg, 0.06 mmol, 20 mol%). The vial was then sealed with a silicon cap, evacuated and back-filled with N_2 through the cap (this procedure was repeated several times), and anhyd. degassed toluene (4 mL) was added *via* a syringe (through the cap). The resulting mixture was heated at 150 °C for 3 h under microwave irradiation. After cooling, the mixture was filtered through Celite and the filter-bed was rinsed with EtOAc. The filtrate was concentrated under reduced pressure and the residue purified by column chromatography on silica gel (EtOAc–PE) to afford the corresponding 3,4-dihydro-3-oxo-2*H*-1,4-benzoxazine.

accessing 3,4-dihydro-3-oxo-2*H*-1,4-benzoxazines from less reactive ethyl 2-(2-chlorophenyl) alkanoates and amines.[39] The use of XPhos as the ligand was found to be crucial for enhanced catalytic activity. In addition, a one-pot, three-component approach was established for the efficient synthesis of the target heterocycles from 2-halophenols, ethyl 2-bromoalkanoates, and aryl amines. It was found that microwave heating at high temperature had significant advantages in terms of improving the synthetic efficiency and achieving high yields compared to conventional heating, particularly with 2-chlorophenol as one of the substrates. The reported protocol tolerates a broad range of substrates, and provides high yields of the desired products (Scheme 3.13c).

In 2001, Grigg and co-workers reported a communication on the palladium-catalyzed cascade intramolecular cyclization of 2-iodobenzamides.[40] The corresponding spiro- or linear fused heterocycles were produced in good yields (Scheme 3.14). The transformation involved the generation of a (π-allyl)palladium intermediate from the allene moiety.

An efficient synthetic approach to obtain new quinoxalinones and oxazinones based on Cu-catalyzed Ullmann-type reactions was developed in 2011.[41] By intramolecular *N*-arylation of pyrrole and indole carboxamides

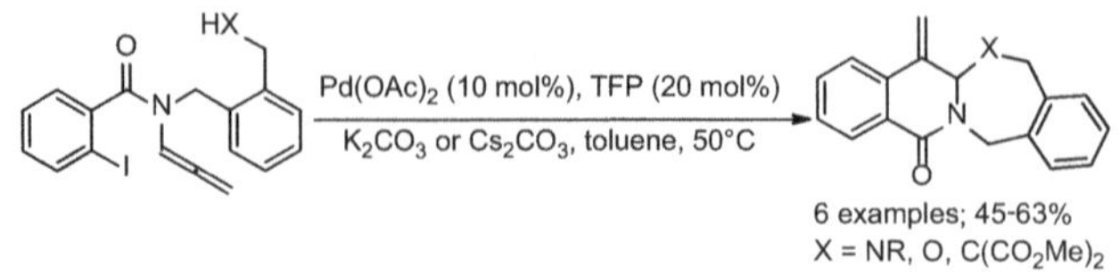

Scheme 3.14 Palladium-catalyzed cyclization of allenes.

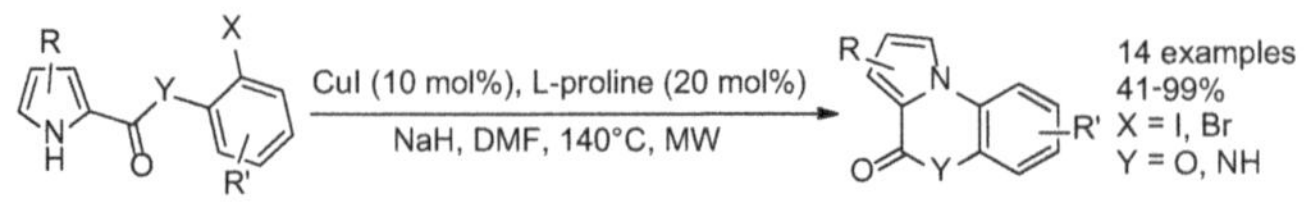

Scheme 3.15 Copper-catalyzed synthesis of fused quinoxalinones and oxazinones.

and carboxylates linked with a pendant haloarene, pyrrole- and indole-fused quinoxalinone and oxazinone derivatives were prepared in good yields (Scheme 3.15). The ring closure reactions were carried out by conventional heating and MW irradiation. The use of conventional heating afforded moderate to good yields of the quinoxalinone and oxazinone derivatives (34–72%), while the best results were obtained (41–99%) by using MW heating.

In 1999, Alper's group developed a novel method for the synthesis of benzo[*e*]-1,3-oxazin-4-ones and derivatives.[42] By cyclocarbonylation of *o*-iodophenols with carbodiimides with the assistance of a palladium complex and bidentate phosphine (DPPB) under CO pressure, the corresponding products can be produced in good to excellent yields with complete regioselectivity. Product yields are dependent on the nature of the substrate, catalyst, solvent, and base as well as the phosphine ligand. The reaction of *o*-iodophenols with unsymmetrical carbodiimides affords benzo[*e*]-1,3-oxazin-2-imine-4-ones in good yields and usually in a completely regioselective manner. Benzo[*e*]-1,3-oxazin-2,4-diones were obtained in good to excellent yields using the same procedure and a 1 : 2 ratio of *o*-iodophenol–isocyanate. Pyrido-1,3-oxazin-4-ones were isolated in fine yields using 2-hydroxy-3-iodopyridine instead of a 2-iodophenol as a reactant. The reaction mechanism was believed to involve *in situ* formation of a carbamate ester followed by palladium-catalyzed carbonylative amidation.

Bao's group developed several interesting procedures for the synthesis of 2,3-dihydro-1,4-benzodioxins. In 2008, they reported a procedure for the preparation of 2,3-dihydro-1,4-benzodioxins from *o*-iodophenols and epoxides in a one-pot manner in moderate to good yields (Scheme 3.16a).[43] Cu_2O/1,10-phenanthroline/Cs_2CO_3 was applied as the catalyst system and the reactions were suggested to occur *via* ring-opening–coupling sequences. Both aryl and aliphatic epoxides can be tolerated under these conditions. Later on, they developed a new tandem synthetic route to 1,4-benzodioxanes (Scheme 3.16b).[44] The reaction involves a cascade process of nucleophilic ring-opening and a CuBr-catalyzed intramolecular coupling transformation. Importantly, through a modification on the reactant structure, the problem

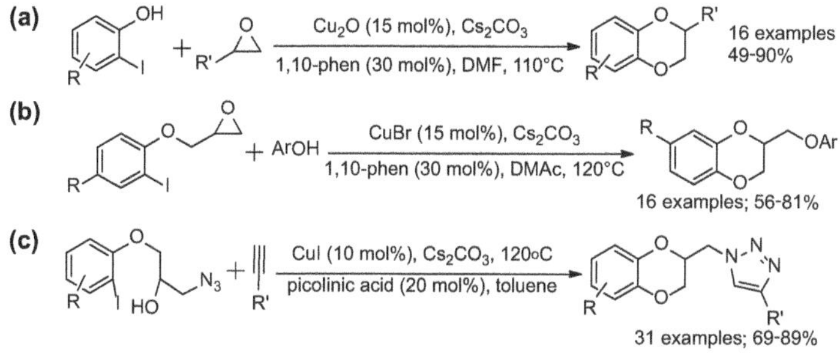

Scheme 3.16 Copper-catalyzed synthesis of 1,4-benzodioxins from epoxides.

of producing isomers *via* nucleophilic attack to different sites on the epoxide has been successfully avoided in this study. Hence, this methodology is a highly useful route for the synthesis of 1,4-benzodioxane derivatives. This procedure was applied in the synthesis of coumarinyl 1,4-benzodioxanes, which are potential anti-oxidants.[45] In 2013, Swamy and co-workers successfully developed a simple one-pot procedure for the synthesis of 1,2,3-triazole appended benzodioxines (benzodioxanes), benzoxazines, benzoxathiine and benzodioxepines (Scheme 3.16c).[46] The methodology involves two different kinds of reaction: a Click and intramolecular carbon–oxygen bond formation using a Cu(I) dual catalytic system. The wide applicability was amply proven by the diverse range of substrates used. Both aryl and vinyl iodide can be used as substrates and all the key products were characterized by single crystal X-ray crystallography. Alternatively, Buchwald's group found that 1,4-benzodioxines can be produced by palladium-catalyzed intramolecular C–O coupling reactions.[47,48] Sekar and co-workers developed a copper-catalyzed procedure for the synthesis of 1,4-benzodioxines through an Ullmann-type intramolecular C_{aryl}–O coupling cyclization reaction.[49] With a catalytic amount of a BINOL–CuI complex, the desired products were formed in moderate yields. Additionally, this methodology was successfully utilized as the key step in the total synthesis of isovanillyl sweetening agents 5-(2,3-dihydro-1,4-benzodioxin-2-yl)-2-methoxyphenol and 5-(2,3-dihydro-1,4-benzoxathiin-2-yl)-2-methoxyphenol in 15.8% and 14.85% overall yields in five steps from isovanillin.

In 2007, Gallagher and co-workers demonstrated an efficient and modular two-step palladium-catalyzed protocol for the coupling of 1,2-cyclic sulfamidates with readily available 2-bromophenols.[50] This method could give substituted and enantioenriched 1,4-benzoxazines in good yields (Scheme 3.17a). The value of this methodology was proven by the short and efficient entry to (3*S*)-3-methyl-1,4-benzoxazine, a late stage intermediate in the synthesis of levofloxacin. In 2011, a procedure for the synthesis of *N*-substituted 1,4-benzoxazine- and 1,4-benzothiazine-2-carboxylates *via* copper-catalyzed intramolecular amination of arylbromides was reported.[51] Starting from *ortho*-bromosubstituted phenoxyacetates or

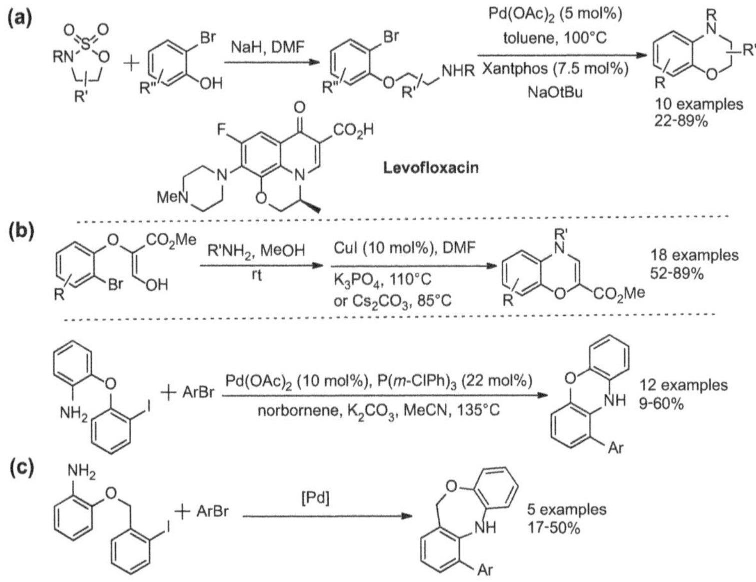

Scheme 3.17 Procedures for the synthesis of 1,4-benzoxazines.

(phenythio)acetates and primary amines, various *N*-substituted 4*H*-1,4-benzoxazine- and 4*H*-1,4-benzothiazine-2-carboxylates were synthesized in moderate to high yields by using a Cu(I)-catalyzed Ullmann-type cyclization as a key step (Scheme 3.17b). The method is simple to operate, tolerates many functional groups and does not require any additives. A number of groups including alkoxy, fluoro, bromo, cyclopropyl, and alkoxycarbonyl are well-tolerated under the reaction conditions. Lautens and co-workers developed a novel procedure for the synthesis of alkyl- and arylbenzomorpholines using palladium-catalyzed *ortho*-functionalization–aromatic amination.[52] A variety of arylphenoxazines and a novel class of dihydrodibenzoxazepines can also be synthesized in moderate yields *via* this approach (Scheme 3.17c).

A palladium-catalyzed arylation of TBDPS-protected 2-bromophenols and 2-bromoanilines *via* C–H activation was reported in 2009.[53] Moderate to good yields of the desired six-membered silicon-containing heterocyclic compounds were produced (Scheme 3.18). After deprotection with F^-, the corresponding 2-phenyl phenols or 2-phenyl anilines were formed.

Whiting and Amos reported a palladium-catalyzed intramolecular arylation in 1987.[54] By treatment of the aryl iodides with palladium acetate, the corresponding tetracyclic stilbenes were formed *via* a stereochemically disallowed Heck reaction in moderate yields. Notably, the method was exploited for the total synthesis of (±)-munduserone (Figure 3.1).

Larock and co-workers described a palladium-catalyzed annulation of 1,4-dienes with *ortho*-functionalized aryl halides in 1993.[55] The reactions go through a (π-allyl)palladium intermediate; the corresponding six-membered

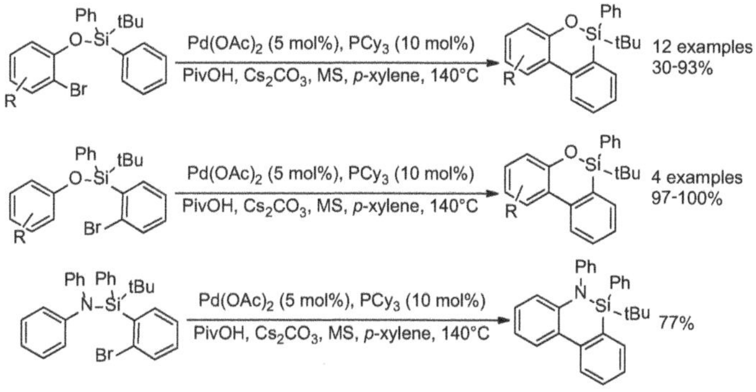

Scheme 3.18 Palladium-catalyzed synthesis of 2-phenyl phenols.

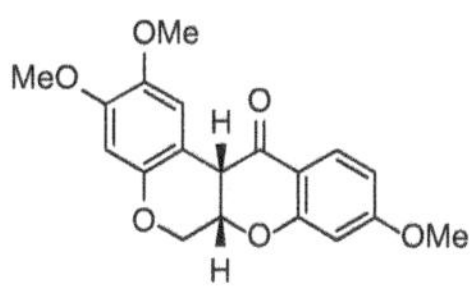

Figure 3.1 (±)-Munduserone.

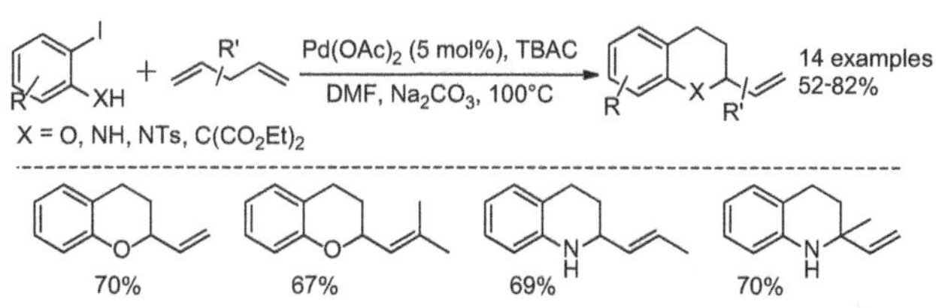

Scheme 3.19 Palladium-catalyzed cyclization of 1,4-dienes.

heterocycles were formed in good yields (Scheme 3.19). The intramolecular cyclization of 2,5-cyclohexadienyl-substituted aryl or vinylic iodides was developed by the same group.[56]

In 2004, Larock's group developed a novel palladium alkyl to aryl migration methodology for the synthesis of complex fused polycycles, which employs sequential Pd-catalyzed C–H activation processes (Scheme 3.20a).[57] The chemistry developed here works best with electron-rich aromatics, which is in agreement with the idea that these palladium-catalyzed C–H activation reactions parallel electrophilic aromatic substitution. In 2006, Lautens and co-workers reported a procedure for alkylidene benzoxacyclic production (Scheme 3.20b).[58] This method utilizes palladium-catalyzed *ortho*-alkylation with an intramolecular Heck reaction as the ring-forming step. In this sequence, two carbon–carbon bonds are formed in one pot, one from an unactivated aryl C–H bond. For six- and seven-membered rings, the intramolecular Heck reaction was slow enough for the reaction to proceed smoothly in the desired sequence. The modularity of this reaction was demonstrated by the variety of chroman-4-ylidenes and 1-benzoxepin-5-ylidenes which

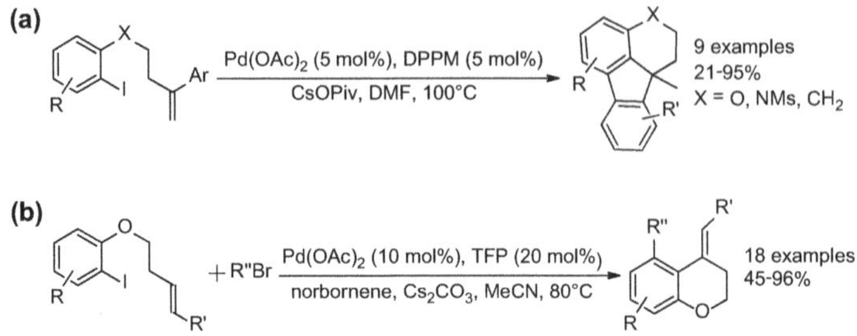

Scheme 3.20 Palladium-catalyzed synthesis of heterocycles *via* C–H activation.

Reaction Procedure (Scheme 3.20a): A mixture of aryl halide (0.25 mmol), $Pd(OAc)_2$ (2.8 mg, 0.0125 mmol), bis(diphenylphosphino)methane (4.8 mg, 0.0125 mmol), caesium pivalate (0.1170 g, 0.5 mmol), and the indicated amounts of alkene (if employed) in dry DMF (4 mL) was quickly flushed with argon and stirred at the indicated reaction temperature. The reaction was monitored by TLC and gas chromatographic analysis. After the reaction was complete, it was diluted with ethyl ether, washed with satd aq. $NaHCO_3$, dried over Na_2SO_4, concentrated, and purified by flash chromatography on silica gel.

(Scheme 3.20b): In a round-bottomed flask with a fused, water cooled condenser were combined $Pd(OAc)_2$ (7.80 mg, 0.0350 mmol, 10 mol%) and $P(2\text{-furyl})_3$ (16.2 mg, 0.0700 mmol, 20 mol%), then the flask was capped and flushed with $N_2(g)$. Dry CH_3CN (0.5 mL) was added, and the mixture was stirred for 10–15 minutes at room temperature, until a homogeneous, orange-red solution was obtained. Next, Cs_2CO_3 (228 mg, 0.700 mmol, 2 equiv.) was added, and the solution turned a dark brown color. Norbornene (132 mg, 1.40 mmol, 4 equiv.) was added, followed by a mixture of aryl iodide (0.350 mmol) and alkyl bromide (1.75 mmol, 5 equiv.) in CH_3CN (0.5 mL). The system was capped, flushed with $N_2(g)$, and then stirred at 80 °C for 18 h. The mixture was quenched with 10 mL of saturated $NH_4Cl(aq)$ and extracted with 3×10 mL of Et_2O. The combined organic phases were dried over $MgSO_4$, filtered, and concentrated *in vacuo*. The crude mixture was then purified by flash chromatography on silica gel to afford the product.

could be synthesized through variation of the alkyl halide and Heck acceptor moieties. Under the optimized conditions [$Pd(OAc)_2$ (10 mol%), $P(2\text{-furyl})_3$ (20 mol%), norbornene (4 equiv.), Cs_2CO_3 (2 equiv.), CH_3CN, 80 °C], aryl iodides with oxygen-tethered Heck acceptors were coupled with alkyl bromides (5 equiv.) to generate the desired benzoxacyclic products in good yields.

A procedure for the highly enantioselective synthesis of chiral chromenes and tetrahydroquinolines was achieved in 2012 by Feringa and co-workers.[59] By combining asymmetric copper-catalyzed allylic substitution with Grignard reagents and an efficient palladium-catalyzed intramolecular Heck reaction, the corresponding products were formed in excellent yields and enantioselectivity (Scheme 3.21a). Additionally, the exocyclic double bond formed in the cyclisation was subjected to RCM (ring-closing metathesis), hydroboration and hydrogenation illuminating the synthetic versatility of these heterocycles. Alami's group succeeded in developing an efficient and flexible synthetic route to *E*- and *Z*-substituted methylene-3,4-dihydro-2*H*-1-benzopyran compounds starting from readily accessible acyclic derivatives.[60] The palladium-catalyzed intramolecular cyclic carbopalladation of iodoalkynes followed by a carbonylation or a hydride ion capture process was the key step for this procedure. Lautens and co-workers developed a diastereoselective approach to isochromans and chromans *via* Pd(0)-catalyzed carboiodination in 2012.[61] The transformations using this methodology display excellent yields and diastereoselectivities as well as broad functional group compatibility (Scheme 3.21b). The selectivity observed in these cyclizations, forming isochroman or chroman targets, is postulated to originate from the minimization of $A^{1,2}$ strain and axial–axial interactions, respectively. This method has also been used to highlight the concept of reversible oxidative addition to carbon–iodine bonds in polyiodinated substrates. A practical synthetic method for the annulation of benzo-rings by the intramolecular coupling of an aryl iodide and a methylene C–H bond was described in 2010 by Chen and co-workers.[62] In this procedure, the palladium-catalyzed C–H functionalization is directed by an aminoquinoline carboxamide group, which can be easily installed and removed. High yields and broad substrate scope were achieved (Scheme 3.21c). An additive of *ortho*-phenyl benzoic acid, identified from a systematic screening, functions as a critical ligand for the catalytic process under mild conditions, even at

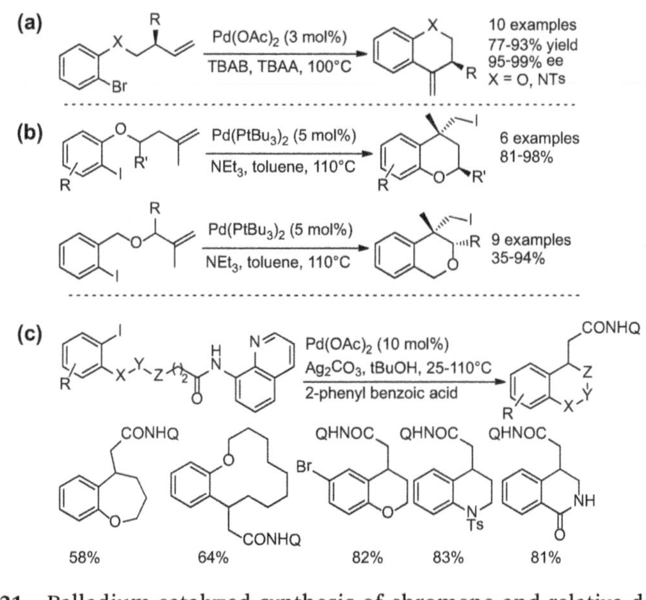

Scheme 3.21 Palladium-catalyzed synthesis of chromene and relative derivatives.

near room temperature. Regarding the reaction mechanism, a Pd^{II}–Pd^{IV} cycle was proposed.

A procedure for the synthesis of substituted trioxa[7]helicenes was developed in 2012[63] by a one-pot palladium catalyzed C–H arylation reaction starting from readily prepared dibenzofuran fragments. On the other hand, an inter-followed by an intramolecular double Heck reaction of bridged *o*,*o*′-dibromobiaryls with ethyl acrylate was described by Prashad's group.[64] 9-(Ethoxycarbonylmethylene)-9*H*-xanthene and 9 and/or 10-substituted anthracene derivatives were prepared in moderate yields. The nature of the bridging atom/group determined the outcome of the reaction.

Ackermann's group reported copper-catalyzed step- and atom-economical sequential transformations involving direct C–H bond arylations in 2012.[65] In detail, CuI enabled the synthesis of fully substituted 1,2,3-triazoles through cascade reactions consisting of copper(I)-catalyzed [3 + 2]-azide–alkyne cycloadditions (CuAAC) and intramolecular C–H bond arylations. Notably, the optimized copper catalyst accelerated two mechanistically distinct transformations, which set the stage for the formation of up to one C–C and three C–N bonds in a chemo- and regioselective fashion, and also allowed for twofold C–H/N–H bond arylations on various azoles (Scheme 3.22a). Meanwhile, a similar sequential copper-catalyzed, one-pot, click reaction–intramolecular direct arylation was developed by Swamy and co-worker.[66] The reaction involves two mechanistically distinct reactions (atom-economical Click reaction and direct arylation of 1,2,3-triazole) and provides [6,6]-, [6,7]-, [6,8]-, and [6,9]ring-fused triazoles in good yields (Scheme 3.22b). Furthermore, a unique divergence of reactivity between the fused triazoles prepared from 2-bromobenzyl azide and 2-bromophenyl azide that leads to a fused pentacyclic heterocycle for the former and a C–C-coupled, biphenyl-fused, tricyclic product for the latter is observed under Pd catalysis. All of the key products have been characterized by single-crystal X-ray crystallography. Lei's group developed a cobalt-catalyzed direct arylation of arenes with aryl halides in 2011 ($Co(acac)_3$ (15 mol%), LiHMDS, benzene, 80 °C).[67] By this procedure, 1-(benzyloxy)-2-bromobenzene was cyclized and gave the corresponding 6*H*-benzo[*c*]chromene in 86% yield.

Li and co-workers developed a procedure for the synthesis of 6*H*-benzo[*c*]-chromenes by palladium-catalyzed annulations of arynes with 2-(2-iodophenoxy)-1-substituted ethanones.[68] This mild route allows the formation of two

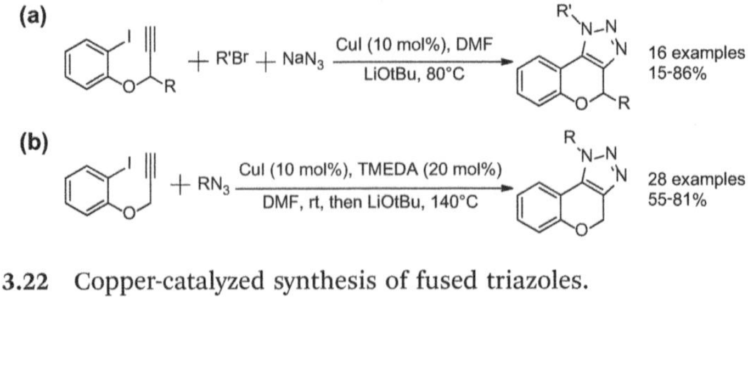

Scheme 3.22 Copper-catalyzed synthesis of fused triazoles.

new carbon–carbon bonds *via* an α-arylation–annulation process; moderate to good yields of the corresponding products were isolated (Scheme 3.23).

Tietze and co-workers developed a methodology for the synthesis of tetrasubstituted alkenes through a palladium-catalyzed domino carbopalladation/C–H activation reaction.[69] Helical tetrasubstituted alkenes were obtained in good to excellent yields through the reaction of propargylic alcohols (Scheme 3.24a). Electron-withdrawing and electron-donating substituents can be introduced onto the upper and lower aromatic rings. The substrates for the domino process were synthesized by addition of the lithiated alkyne to various aldehydes; moreover, the substrates were accessible enantioselectively (in 95 % ee) by reduction of the corresponding ketone using the Noyori procedure. Catellani's group achieved an efficient

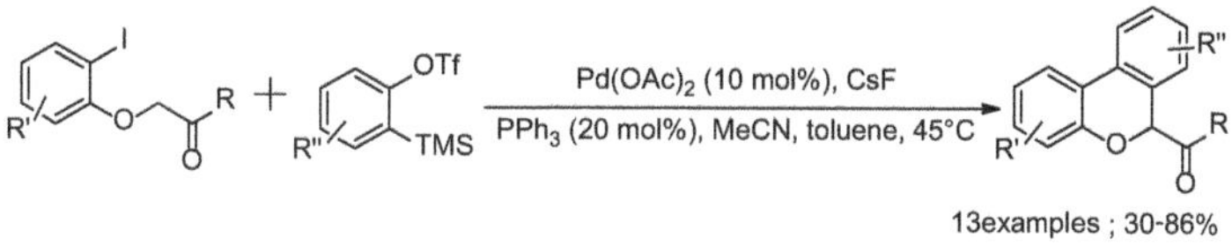

Scheme 3.23 Pd-catalyzed synthesis of chromenes from arynes.

Reaction Procedure (Scheme 3.23): To a Schlenk tube were added 2-(2-iodophenoxy)-1-substituted ethanones (0.2 mmol), 2-(trimethylsilyl) aryltriflate (0.3 mmol), $Pd(OAc)_2$ (2.4 mg, 10 mol%), triphenylphosphine (10.5 mg, 20 mol%), CsF (121 mg, 0.8 mmol), dry MeCN (1 mL) and dry toluene (1 mL). Then the tube was charged with N_2, and was stirred at 45 °C (oil bath temperature) for the indicated time (about 30 h) until complete consumption of starting material as monitored by TLC and GC-MS analysis. After the reaction was finished, the reaction mixture was cooled to room temperature, diluted in diethyl ether, and filtered with a short column on silica gel. The solution was concentrated *in vacuo*, and the resulting residue was purified by silica gel column chromatography (hexane–ethyl acetate) to afford the desired product.

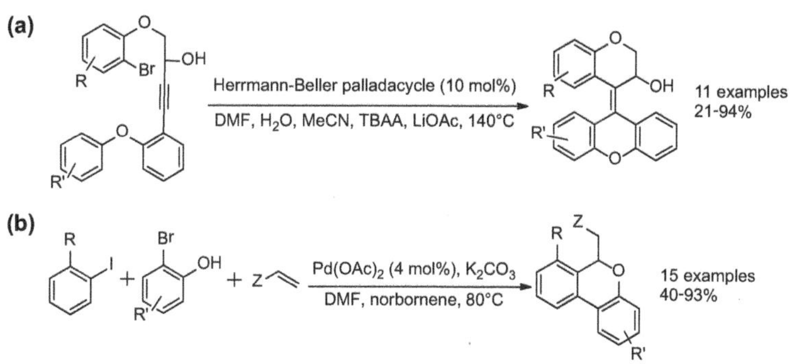

Scheme 3.24 Pd-catalyzed synthesis of pyrans *via* C–H activation.

catalytic process for the synthesis of 6*H*-dibenzopyran derivatives in 2006.[70,71] The reaction started from easily available 2-bromophenols, aryl iodides and alkenes. The reaction proceeds along a rather complex pathway but is very simple from an experimental point of view and allows the formation of two new C–C bonds and one C–O bond in a single operation. 6*H*-Dibenzopyrans were produced in good to excellent yields (Scheme 3.24b). In the catalytic cycle, a palladium (IV) complex was involved.

3.3 2-Haloaniline Derivatives

Quinolines, 4(1*H*)-quinolones and their perhydro derivatives are constituents of several biologically active natural and synthetic compounds. For example, cinchona alkaloids, quinine and cinchonine are well known antimalarial agents. *Galipea longiflora* was found to be a rich source of a number of 2-substituted quinolines, including 2-phenylquinoline, which was found to be as effective as the standard antileishmanial drug glucantime. 2-Substituted quinolines have been described as 5-lipoxygenase inhibitors, leucotriene antagonists and LTD_4 receptor antagonists. The latter may prove useful for controlling asthma. Larock and co-workers reported a palladium-catalyzed intramolecular cyclization of *ortho*-iodo *N*-allylic anilines to the corresponding quinolines in 1987.[72] Later on, they studied the cyclization of 2-iodoaniline with allylic alcohols to quinolines.[73] The reaction followed an intermolecular Heck coupling–isomerization–condensation–dehydrogenation; sequence and the desired products were formed in moderate yields (Scheme 3.25a). Interestingly, different parameters were tested,

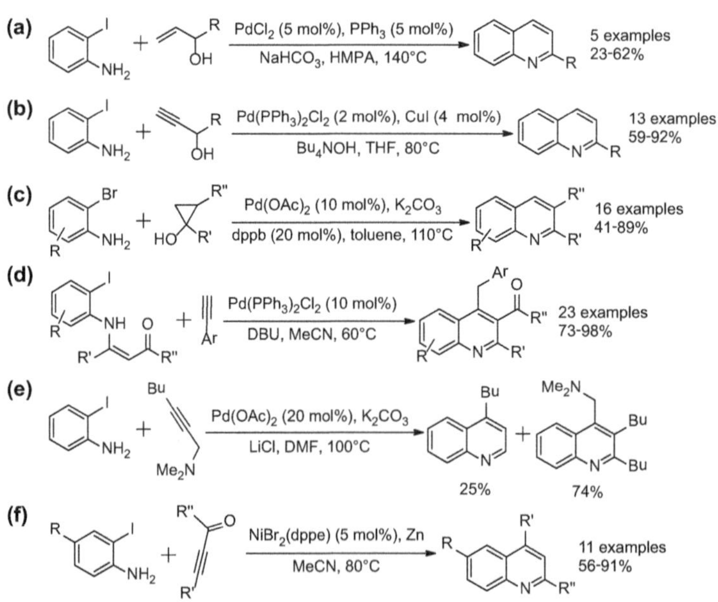

Scheme 3.25 Pd and Ni-catalyzed synthesis of quinolines from alkene and alkynes.

such as palladium precursors, ligands, solvents, bases, and additives, but the parameters have little effect on the yields. The use of 2-bromoanilines as substrates was developed with a palladium catalyst as well; in this case, a dehydrogenation reagent (DIAD; diisopropyl azodicarboxylate) was required.[74] Kundu and co-workers reported a palladium-catalyzed procedure for the synthesis of quinolines from *o*-iodoanilides and terminal acetylenic carbinols.[75] With the assistance of a palladium catalyst, *o*-substituted anilides were produced at the first step and then cyclized to 2-arylquinolines in the presence of NaOEt/EtOH. Also, the anilides, on acid-catalyzed rearrangement, deprotection and cyclisation, led to 2-aryl-2,3-dihydro-4(1*H*)-quinolones. Cho showed that 2-iodoaniline can undergo tandem coupling, isomerization and cyclodehydration with an array of terminal acetylenic carbinols in the presence of a catalytic amount of $PdCl_2(PPh_3)_2$ and CuI along with aqueous tetrabutylammonium hydroxide to give 2-arylquinolines in high yields (Scheme 3.25b).[76] The catalytic pathway seems to proceed *via* a sequence involving initial Sonogashira coupling between 2-iodoaniline and terminal acetylenic carbinols to form coupled acetylenic carbinols, isomerization of coupled acetylenic carbinols to α,β-unsaturated ketones, and cyclodehydration. By introducing microwave heating, the reaction can proceed with high efficiency.[77,78] More recently, Orellana and co-workers reported a palladium-catalyzed one-pot synthesis of quinolines *via* the cross-coupling of 2-bromoanilines with cyclopropanols (Scheme 3.25c).[79] The corresponding quinolines were produced in good yields in a single operation *via* an intramolecular condensation and palladium-catalyzed oxidation sequence. The reaction tolerates a variety of cyclopropanols and substituted bromoanilines. Deuterium-labeling experiments provide direct evidence of a second equivalent of bromoaniline serving as the terminal oxidant. A new and high-yielding method to prepare 3,4-disubstituted 2-trifluoromethylquinolines by a palladium catalyzed tandem Sonogashira–alkyne carbocyclization of β-trifluoromethyl β-enaminoketones with arynes was described in 2010.[80] Moderate to excellent yields of the corresponding products have been achieved under mild conditions (Scheme 3.25d). This reaction can also be expanded to non-fluorine containing substrates. The reaction mechanism was discussed as well. The first step is the oxidative addition of Pd(0) with aryl halide. The presence of an *ortho*-β-acyl-enamine group as a ligand makes this step easier. Coordination of the alkyne to the ArPdOR complex then follows. Upon coordination, the C–H bond of the alkyne is weakened, and HOR is removed from Pd(II) in the presence of a base to form an arylalkynylpalladium species, which undergoes reductive elimination to afford the intermediate and regenerates the catalyst. The intermediate then undergoes a base-catalyzed alkyne carbocyclization process, which then isomerizes to the target compound. The coupling of 2-iodoaniline with alk-2-yn-1-amines can lead to quinolines as well with the presence of $Pd(OAc)_2$ (Scheme 3.25e).[81] Alkoxy-functionalized butadienylboronic esters were synthesized starting from α,β-unsaturated acetals and cross-coupled with both *N*-protected and *N*-unprotected 2-bromo and

2-iodoaniline, and with 2-iodophenol. In particular, *N*-tosyl-protected dienylanilines can be transformed under mild conditions into quinolines and quinolinones, in the presence of a Pd^{II} catalyst.[82] Moreover, the cross-coupling reaction between butadienylboronic esters and iodophenol directly affords chromenes that can be successively transformed into chromenones. A nickel catalyst was explored as well. In 2006, Cheng and co-workers developed an efficient and convenient nickel-catalyzed cyclization of 2-iodoanilines with alkynyl aryl ketones to give 2,4-disubstituted quinolines.[83] The reaction can be employed for the synthesis of naturally occurring quinoline derivatives in good yields (Scheme 3.25f). On the basis of the regiochemistry of the products, a possible pathway for the reaction *via* the formation of *o*-aminochalcone was proposed. The catalytic cycle is likely initiated by the reduction of Ni(II) to Ni(0) by zinc metal powder. The oxidative addition of 2-iodoaniline to Ni(0) species affords an *o*-metalated aniline nickel complex. Coordination of alkyne and insertion of this alkyne into the Ni–carbon bond generates a vinyl nickel intermediate. Subsequent protonation with water yields amino chalcone and a Ni(II) species. Further reduction of the Ni(II) species by zinc regenerates the catalyst Ni(0) along with a zinc(II) derivative. The keto group and the *o*-amino group in the amino chalcone obtained are expected to be *trans* to each other after protonation of the vinyl nickel intermediate. Thus, *trans* to *cis* isomerization to yield *cis*-amino chalcone should occur. Further condensation between the keto and the amino groups yields the final product. The water thus generated is utilized in the protonation process of the vinyl nickel intermediate.

In 1994, a procedure for the synthesis of benzo[*c*]-2,7-naphthyridines from pyridine methylstannanes and *ortho*-bromoacetanilides was reported based on a palladium catalyst.[84] By using CuO as a promoter, good yields of the products were formed (Scheme 3.26a). In 2004, Banwell's group reported the synthesis of quinolines, 2-quinolones, phenanthridines, and 6(5*H*)-phenanthridinones *via* palladium-mediated Ullmann cross-coupling of 1-bromo-2-nitroarenes with β-halo-enals, -enones, or -esters and then Pd on carbon catalyzed nitro reduction and cyclization.[85,86] Hu's group developed a procedure for the one-pot synthesis of a 2,3-dihydro-1*H*-pyrrolo[2,3-*c*]quinoline core from substituted 2-iodoanilines and 2,3-dihydro-1*H*-pyrrole.[87] The reaction was achieved using 10 mol% $Pd(PPh_3)_4$, and K_2CO_3 in 1,4-dioxane at 170 °C for 1 h in a microwave oven. The authors proposed that the reaction mechanism involves a Heck-coupling reaction followed by intramolecular Schiff base formation and double bond migration (Scheme 3.26b). Doucet's group reported a methodology for the one-pot synthesis of furo- or thienoquinolines in 2012.[88] This method involves sequential imination and intramolecular palladium-catalyzed direct arylation. Under appropriate reaction conditions, the palladium-catalyzed intramolecular direct arylation of *in situ* prepared *N*-(2-haloaryl)thiophene imine or *N*-(2-haloaryl)furan imine derivatives proceeded regioselectively, starting from a variety of 2-haloaniline derivatives and thiophene-3-carbaldehyde or furan-3-carbaldehyde (Scheme 3.26c). More recently, a palladium-catalyzed cascade reaction of

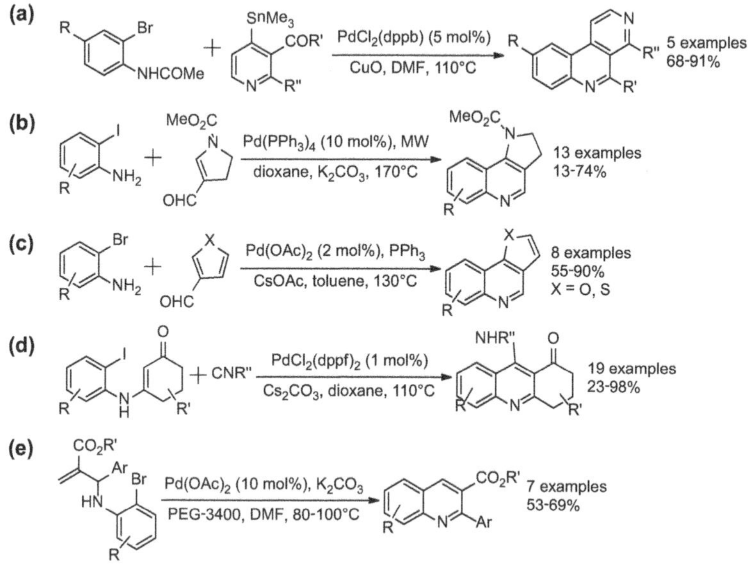

Scheme 3.26 Pd-catalyzed synthesis of quinolines from 2-haloanilines.

isocyanides with enaminones for the [5 + 1] cyclization reaction of enaminones with isocyanides was reported.[89] This protocol provides a new and straightforward approach to 4-aminoquinoline derivatives under mild conditions (Scheme 3.26d). Kim's group developed the synthesis of 2-arylquinolines and tetrahydropyridines *via* palladium-mediated Heck type reactions starting from Baylis–Hillman adducts.[90] 2-Arylquinolines were prepared in moderate yields *via* the Heck type cyclization followed by concomitant aerobic oxidation (Scheme 3.26e).

Quinazolinones are an important class of fused heterocycles that have been reported with remarkable activities in biology and pharmacology such as anticancer, antiinflammatory, anticonvulsant, antibacterial, antidiabetic, hypolipidemic, and protein tyrosine kinase inhibitors. Alper and Zheng reported a palladium-catalyzed cyclocarbonylation of *o*-iodoanilines with imidoyl chlorides to produce quinazolin-4(3*H*)-ones in 2008.[91] A wide variety of substituted quinazolin-4(3*H*)-ones were prepared in 63–91% yields (Scheme 3.27a). The reaction is believed to proceed *via in situ* formation of an amidine, followed by oxidative addition, CO insertion, and intramolecular cyclization to give the substituted quinazolin-4(3*H*)-ones. Later on, a procedure was established based on generating the amidine *in situ* by a copper-catalyzed reaction of terminal alkynes, sulfonyl azide and *o*-iodoanilines.[92] The desired quinazolinones can be produced by carbonylation with $Pd(OAc)_2$–DPPB–NEt_3–THF as the reaction system. In the same year, Alper's group developed a procedure for 2,3-dihydro-4(1*H*)-quinazolinone preparation.[93] The reaction started with the reaction of 2-iodoanilines and *N*-toluenesulfonyl aldimines followed by palladium-catalyzed intramolecular

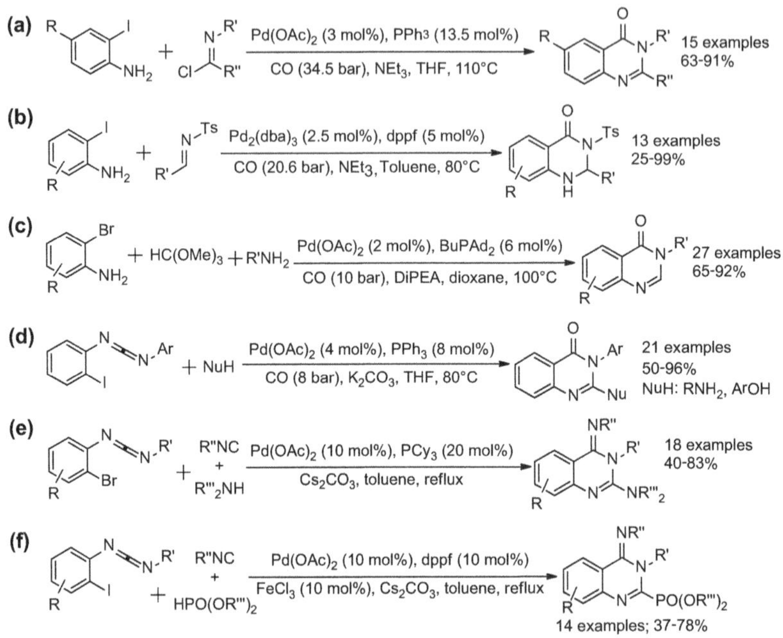

Scheme 3.27 Pd-catalyzed synthesis of quinazolinones and related derivatives.

cyclocarbonylation to give the final product (Scheme 3.27b). More recently, Wu, Beller and their co-workers reported a palladium-catalyzed four-component carbonylative coupling system for the synthesis of diverse 4(3*H*)-quinazolinones in a concise and convergent fashion (Scheme 3.27c).[94] Starting from 2-bromoanilines, trimethyl orthoformate, and amines, under 10 bar of CO, the desired products were isolated in good yields in the presence of $Pd(OAc)_2$ (2 mol%), $BuPAd_2$ (6 mol%) in 1,4-dioxane at 100 °C, using *N*,*N*′-diisopropylethylamine (DiPEA) as the base. Alternatively, Alper and co-workers found that quinazolinones can be produced from carbonylative cyclization of *N*-(2-iodophenyl)-*N*′-phenylcarbodiimide with amines as well (Scheme 3.27d).[95,96] A large number of 2-aminoquinazolin-4(3*H*)-ones and 2-phenoxyquinazolin-4(3*H*)-ones were obtained in good to excellent yields. Later on, the palladium catalyst and carbon monoxide applied here were replaced by $Mo(CO)_6$ using microwave irradiation.[97] As an equivalent of CO, isocyanide was explored in this type of cyclization to produce quinazolinimines as well. In 2012, Wu and co-workers described a novel and efficient route for the preparation of quinazolin-4(3*H*)-imines *via* a palladium-catalyzed three-component reaction of carbodiimide, isocyanide, and amine or alcohol (Scheme 3.27e).[98] In the reaction process, an isocyanide insertion was believed to be the key step for the successful transformation. Later on, the group extended their methodology to 4-imino-3,4-dihydroquinazolin-2-ylphosphonate production by combining 2-iodoarylcarbodiimide, isocyanide, and phosphite.[99] Moderate to good yields of the desired products were isolated (Scheme 3.27f).

4*H*-3,1-Benzoxazin-4-ones are a class of fused heterocycles with various biological activities, such as chymotrypsin inactivators, inhibitors of human leukocyte elastase, serine protease, and 2-aryl derivatives have the ability to lower the concentration of plasma cholesterol and triglyceride. In 1999, Alper and co-workers reported a palladium-catalyzed cyclocarbonylation procedure for the synthesis of benzoxazinones from *o*-iodoanilines and acid chlorides.[100] The reaction was believed to proceed *via in situ* amide formation from an *o*-iodoaniline and an acid chloride, followed by oxidative addition to Pd(0), CO insertion, and intramolecular cyclization to form the 2-substituted-4*H*-3,1-benzoxazin-4-one derivatives in good to excellent yields (Scheme 3.28a). This transformation was reported with a Pd–NHC complex[101] and Pd/C as well.[102] In the case of using Pd/C as the catalyst, the catalyst can be reused twice without losing activity. Beller and co-workers reported a domino synthesis of 2-arylbenzoxazinones in 2011 (Scheme 3.28b).[103] Starting from commercially available 2-bromoanilines and (hetero)aryl bromides, 19 different benzoxazinones were produced in good yields (65–91%). The key step of this transformation is the chemoselective carbonylation of the aryl bromide. Moreover, a one-pot synthesis of 2,3-diarylquinazolinones was demonstrated exemplarily. This type of reaction using 2-iodoaniline and vinyl triflates as substrates was reported before.[104] More recently, Wu, Beller and their co-workers reported procedures to synthesize benzoxazinones by using 2-bromoanilines and acid anhydride or 2-bromoanilines and isocyanates as substrates.[105]

In 2008, Ma and Yuan reported an intramolecular condensation process to assemble pyrrolo[1,2-*a*]quinoxalines from pyrrole-2-carboxylate esters.[106] In the presence of CuI/L-proline with K_2CO_3 as a base in DMSO at 80–90 °C followed by *in situ* hydrolysis at 60 °C, a wide range of these fused heterocycles bearing different functional groups, such as ketone, ester, methoxy, bromo, and chloro, could be elaborated from suitable substrates, and provided the desired quinoxalines in good to excellent yields (Scheme 3.29a). Indole-2-carboxylate esters were compatible with this process, giving the corresponding fused tetracyclic compounds, while imidazole-2-carboxylate esters gave low yields of the desired coupling products. Tanimori and co-workers found that substituted dihydroquinozalin-2-ones can be synthesized

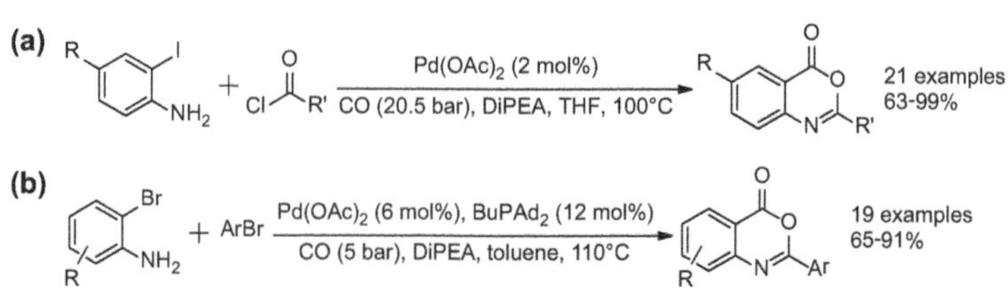

Scheme 3.28 Pd-catalyzed carbonylative synthesis of benzoxazinones.

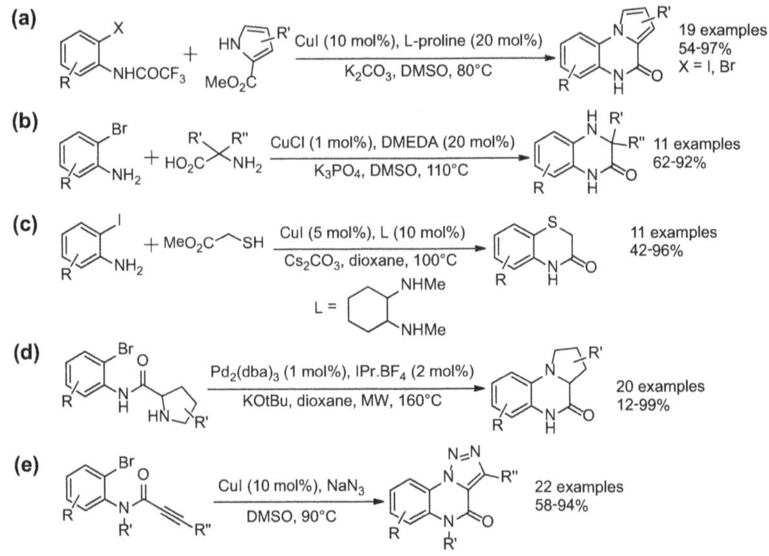

Scheme 3.29 Procedures for the synthesis of quinoxalinones.

by copper-catalyzed coupling of 2-haloanilines with a variety of α-amino acids.[107,108] In the presence of copper(I) chloride, the corresponding 3-substituted dihydroquinozalin-2-ones were produced in up to 86% yield (Scheme 3.29b). Their cytotoxic activity was tested as well and moderate activity was observed. As well as secondary amino acids, primary amino acids can also be applied as substrates with DMEDA as the ligand. This concept was extended to the synthesis of 1,4-benzothiazin-3-ones by Cu-catalyzed coupling of 2-iodoanilines and 2-mercaptoacetate (Scheme 3.29c).[109] Benzothiazepinone can be produced in 51% yield by using 2-iodobenzylamine as the substrate under the same conditions. Palladium catalysts were applied in quinoxalinone synthesis too. In 2010, Luo and co-workers reported a palladium-catalyzed intramolecular *N*-arylation of *ortho*-bromoanilides for the synthesis of quinoxalinones.[110] With the assistance of microwave irradiation, the desired products were formed in excellent yields (Scheme 3.29d). Alternatively, a copper-catalyzed procedure for the synthesis of [1,2,3]-triazolo[1,5-*a*]quinoxalin-4(5*H*)-ones was developed (Scheme 3.29e).[111] The method was based on the copper-promoted reaction of 1-(2-haloaryl)-propamides with sodium azide through a tandem azide–alkyne cycloaddition–Ullmann C–N coupling process, which is applicable to a variety of 1-(2-haloaryl) propiolamides. In detail, the N–Cu species was formed through an azide–alkyne [3 + 2] cycloaddition in the presence of CuI, which was then quickly inserted into the aryl C–halo bond and eventually led to the Ullmann C–N coupling product.

In 2011, Cai's group reported a copper-catalyzed tandem reaction of isocyanides with *N*-(2-haloaryl)propiolamides for the synthesis of pyrrolo[3,2-*c*]quinolin-4-ones.[112] Highly reactive cyclic organocopper intermediates were proposed to be generated in the copper-catalyzed formal [3 + 2] cycloaddition reaction of isocyanides with triple bonds. Intramolecular trapping of the organocopper intermediates can lead to aryl C–C bond formation, which offers an efficient method for constructing fused pyrrole structures. Moderate to good yields of the desired heterocycles were isolated (Scheme 3.30a). Later on, the group developed another tandem reaction for the synthesis of fused quinolinones.[113] By utilizing two famous copper-catalyzed reactions, CuAAC and Ullmann coupling, the corresponding products were produced in good yields (Scheme 3.30b). The process took place efficiently when a variety of *N*-(2-iodoaryl)-propiolamides or 2-iodo-*N*-(prop-2-ynyl)benzenamines were used, and it displayed a wide range of functional group compatibility.

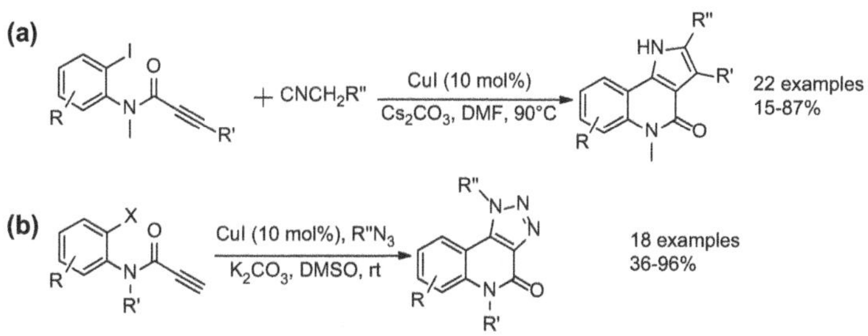

Scheme 3.30 Copper-catalyzed synthesis of fused quinolinones.

Reaction Procedure (Scheme 3.30a): Isocyanides (0.55 mmol) was added to a mixture of caesium carbonate (325 mg, 1.0 mmol, 2.0 equiv.), copper iodide (10 mg, 0.05 mmol, 10% equiv.), *N*-(2-haloaryl)propiolamides (0.5 mmol, 1.0 equiv.) in DMF (1 mL) at 90 °C. The mixture was stirred under air for 10 min. Monitoring by TLC showed that the reaction was complete. Water (5 mL) was added, and the aqueous phase was extracted with ethyl acetate (5 mL×3). The combined organic phase was washed with brine, dried over Na_2SO_4, and concentrated *in vacuo*. The residue was loaded on a silica column and purified by column chromatography (SiO_2, petroleum ether–ethyl acetate, 8 : 1) to afford the final product.

(Scheme 3.30b): A reaction mixture of substrate (0.5 mmol, 1.0 equiv.), RN_3 (0.55 mmol, 1.1 equiv.), CuI (0.05 mmol, 0.1 equiv.) and K_2CO_3 (1.0 mmol, 2.0 equiv.) in DMSO (1.0 mL) was stirred under argon at room temperature and monitored by TLC. After the reaction was complete, $NH_3 \cdot H_2O$ (5.0 mL) and dichloromethane (5.0 mL) were added into the mixture. The organic phase was separated and the aqueous phase was extracted with dichloromethane (5 mL×3). The combined organic phase was washed with H_2O and brine, and dried over Na_2SO_4. The solvent was removed under reduced pressure. The residue was loaded on silica column and purified by flash chromatography (dichloromethane–ethyl acetate, 30 : 1) to afford the product.

In 1998, Rigby's group studied the addition of alkyl and aryl isocyanides to benzynes generated *in situ*.[114] In the case of using 2-bromophenylisocyanide as a substrate, the products formed can be further transformed into the corresponding phenanthridinones in moderate yields (Scheme 3.31a).[115] This transformation was restudied by Cacchi and co-workers with *N*-benzyl-*N*-benzoyl-*o*-iodoanilides as the substrates.[116] In 2014, a novel and efficient procedure for the synthesis of *N*-substituted phenanthridinones *via* palladium-catalyzed annulation of benzynes with *N*-substituted-*N*-(2-halophenyl)formamides was developed by Liang and co-workers.[117] This methodology constructs two new C–C bonds *via* an arylation–annulation process, and provides the desired products in good yields (Scheme 3.31b). Grigg and co-workers reported a palladium-catalyzed carbonylative cyclization of 2-iodoaniline, 2-iodophenol and 2-iodobenzyl alcohol with norbornene.[118] The corresponding six- and seven-membered heterocyclic compounds were produced in moderate yields (Scheme 3.31c). Larock's group reported a palladium-catalyzed annulation of internal alkynes with *N*-substituted *o*-iodoanilines in 2004.[119] Under 1 bar of carbon monoxide, 3,4-disubstituted 2-quinolones were formed in good yields (Scheme 3.31d). The nature of the substituent on the nitrogen was crucial to obtaining high yields of 2-quinolones. The best results were obtained using alkoxycarbonyl, *p*-tolylsulfonyl, and trifluoroacetyl substituents. The nitrogen substituent is lost during the course of the reaction resulting in the formation of *N*-unsubstituted 2-quinolones. A variety of internal alkynes, bearing alkyl, aryl, heteroaryl, hydroxyl, and alkoxyl substituents, were effective in this process. Electron-rich and electron-poor *N*-substituted *o*-iodoanilines, as well as heterocyclic analogues, can be employed as annulating agents. Cho and Kim found that quinolones and quinolones can be produced from 2-iodoaniline and

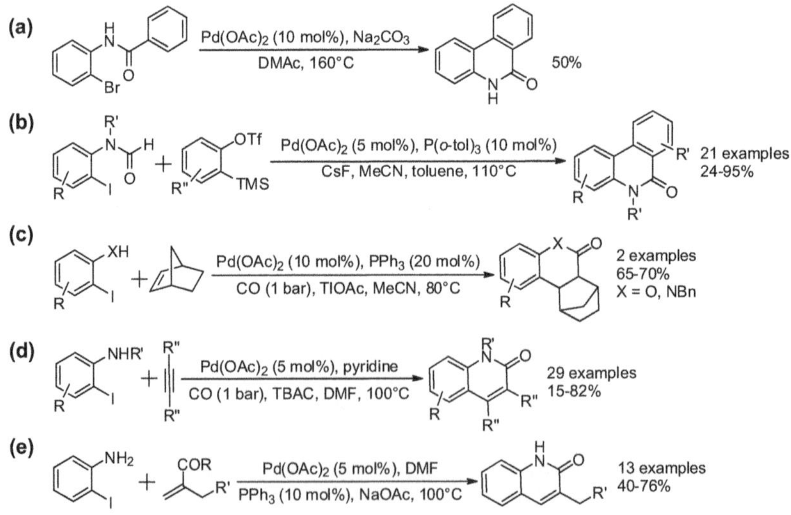

Scheme 3.31 Pd-catalyzed synthesis of phenanthridinones.

α,β-unsaturated carbonyl compounds as well.[120] With the assistance of a palladium catalyst in DMF at 100 °C, the corresponding products were formed in moderate to good yields (Scheme 3.31e). A one-pot two-step heterogeneous Pd/$NiFe_2O_4$-catalyzed reaction was also developed.[121]

In 2010, Jackson and co-workers studied the Negishi cross-coupling of iodoalanine-derived zinc reagents with aryl iodides and aryl bromides.[122] In that work, they showed that a 3,4-dihydro-2(^{1}H)-quinolinone derivative can be produced by using 2-iodoaniline as a substrate. An interesting study on 3,4-dihydro-2(^{1}H)-quinolinone preparation was reported by Tsuritani and co-workers in 2009.[123] By applying *N*-(1′-alkoxy)cyclopropyl-2-haloanilines as substrates *via* palladium-catalyzed cyclopropane ring expansion, the desired 3,4-dihydro-2(^{1}H)-quinolinones were produced in good yields (Scheme 3.32). The reaction tolerates a variety of functional groups such as ester, nitrile, ether, and ketone groups. 5,6-Dihydrobenzimidazo[1,2-*a*]quinoline can be formed as well by using the corresponding 1,1-dianilinocyclopropane as a substrate.

An efficient method to prepare 1′*H*-spiro-[indoline-3,3′-quinoline]-2′,4′-diones and their trifluoromethylated products was developed by Wang and co-workers in 2014.[124] The reaction proceeds *via* a palladium-catalyzed Sonogashira coupling–Wacker-type oxypalladation–cyclization cascade reaction sequence. The amount of water in the reaction system played an important role in the *in situ* trifluoromethylation reaction, and the trifluoromethylation exhibited excellent molecular self-induced stereoselectivity.

In 2011, a palladium-catalyzed ring opening of furans as a source of α,β-unsaturated aldehydes was developed.[125] In the presence of a palladium catalyst, heterocyclic compounds can be produced. In this procedure, furans may be ring-opened *via* pallado-catalyzed reactions leading to α,β-unsaturated aldehydes and ketones tethered to indole and isoquinoline moieties. Fu and co-workers developed a simple and efficient copper-catalyzed one-pot tandem method for the synthesis of benzimidazo[1,2-*b*]isoquinolin-11-one derivatives

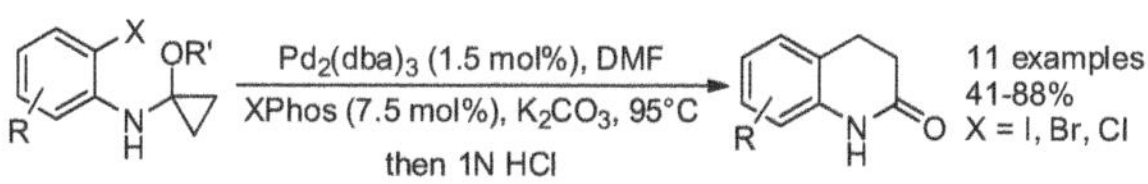

Scheme 3.32 Pd-catalyzed synthesis of 3,4-dihydro-2(^{1}H)-quinolinones.

Reaction Procedure (Scheme 3.32): A mixture of *N*-cyclopropyl haloaniline (1.0 mmol, 0.2 M), K_2CO_3 (1.5 equiv.), $Pd_2(dba)_3$ (1.5 mol%) and XPhos (7.5 mol%) in DMF (5.0 mL) was stirred at 95 °C under a nitrogen atmosphere in a round-bottomed flask for 12 h. The reaction mixture was then poured into aqueous 1 N HCl, and the resulting mixture was extracted with AcOEt three times. The combined organic extracts were washed with brine twice. The solution obtained was dried with Na_2SO_4, and concentrated *in vacuo* with a rotary evaporator. The residue was purified by flash chromatography to give the pure product.

in 2010.[126] By using substituted 2-halo-*N*-(2-halophenyl)benzamides and alkyl 2-cyanoacetates or malononitrile as substrates under mild conditions without the addition of any ligand or additive, the corresponding products were produced in good yields (Scheme 3.33).

In 1990, Torii's group reported the carbonylative cyclization of 3-substituted 3-(2-haloarylamino)prop-2-enoates to synthesize quinolines (Scheme 3.34a).[127] Soon after, they developed another methodology for quinolone synthesis.[128,129] By using 2-haloaniline and terminal alkynes as starting materials, the corresponding products were formed in good yields (Scheme 3.34b). Kundu and co-workers reported the synthesis of quinolones and 2,3-dihydro-4(1*H*)-quinolones from 2-iodoanilides and acetylenic carbinols.[130] Ennis's group reported a convenient procedure for the preparation of 3-carbomethoxymethyl-2,3-dihydro-4-quinolones.[131] With *N*-allyl-2-iodoaniline as a substrate, the desired products were formed in the presence of a palladium catalyst, carbon monoxide and methanol. In 2007, Alper and Ye reported a palladium-catalyzed cyclocarbonylation reaction of *o*-iodoanilines with allenes and CO in 1-butyl-3-methylimidazolium hexafluorophosphate ($BMIM \cdot BF_6$) to produce 3-methylene-2,3-dihydro-1*H*-quinolin-4-ones.[132] The products were formed in moderate to excellent yields under a low pressure (5 bar) of CO (Scheme 3.34c). The ionic liquid, as the solvent and promoter, enhances the efficiency of the cyclocarbonylation reaction. Notably, the reaction system can be reused six times with a slight activity decrease. The group developed a catalyst system comprising palladium acetate–bidentate phosphine (DPPB) for the cyclocarbonylation of *o*-iodoanilines and heterocumulenes as well.[133] The corresponding 4(3*H*)-quinazolinone derivatives were produced in good yields at 70–100 °C for 12–24 h. Utilizing

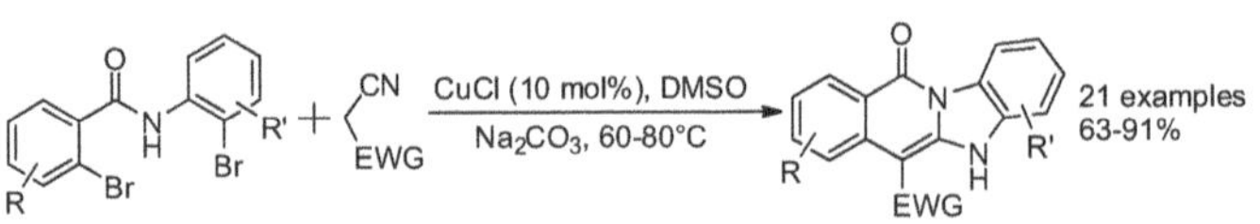

Scheme 3.33 Cu-catalyzed synthesis of isoquinolinones.

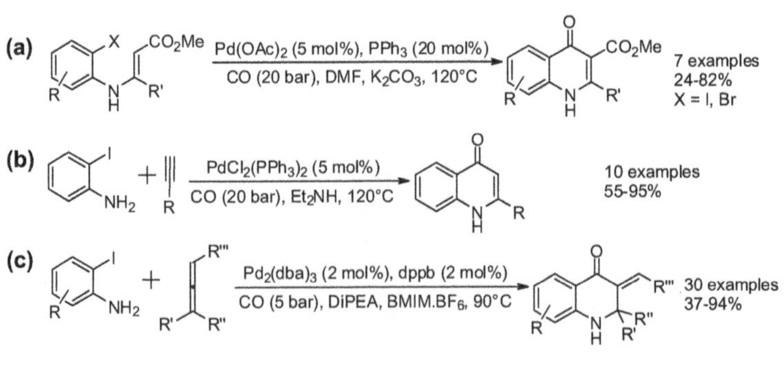

Scheme 3.34 Pd-catalyzed carbonylative synthesis of quinolines.

o-iodoaniline with isocyanates, carbodiimides, and ketenimines for the reaction, 2,4-(1*H*,3*H*)-quinazolinediones, 2-amino-4(3*H*)-quinazolinones and 2-alkyl-4(3*H*)-quinazolinones were obtained, respectively. The nature of the substrates, including the electrophilicity of the carbon center of the carbodiimide and the stability of the ketenimine, influence the product yields of this reaction. Urea-type intermediates were believed to be generated first *in situ* from the reaction of *o*-iodoanilines with heterocumulenes, followed by palladium-catalyzed carbonylation and cyclization to yield the desired products.

In 2012, a palladium-catalyzed ring opening of aminocyclopropyl Ugi adducts was presented.[134] Dihydropyridine and benzoazepinone derivatives were produced in moderate yields (Scheme 3.35a). Catellani and co-workers reported a palladium-catalyzed synthesis of substituted phenanthridine derivatives in 2008.[135] Selectively substituted phenanthridine derivatives were obtained by a facile reaction of *o*-alkylated aryl iodides, *o*-bromoarenesulfonylanilines and activated olefins in the presence of palladium and norbornene as catalysts. The reaction takes place under mild conditions to give the products in satisfactory yields using readily available starting materials (Scheme 3.35b).

Substituted phenazine cores are important biologically active motifs, and are usually found in natural products, dyestuffs, pesticides, and antibiotics. In 2010, a simple one-step copper-catalyzed synthesis of pyrrolo[1,2-*a*]quinoxalines from *o*-aminoiodoarenes and 2-formylpyrroles was developed (Scheme 3.36a).[136] The method is amenable to structural variation of both reaction partners and consequently serves as a useful alternative to existing routes to pyrrolo[1,2-*a*]quinoxalines. Indolo[1,2-*a*]quinoxalines can be prepared under the same reaction conditions by starting with 2-formylindoles. In addition, the reaction can be extended to 2-formylimidazole, 2-formylbenzimidazole, and a 3-formylpyrazole to generate a range of unique tricyclic heterocycles. Later on, a procedure based on L-proline as a ligand

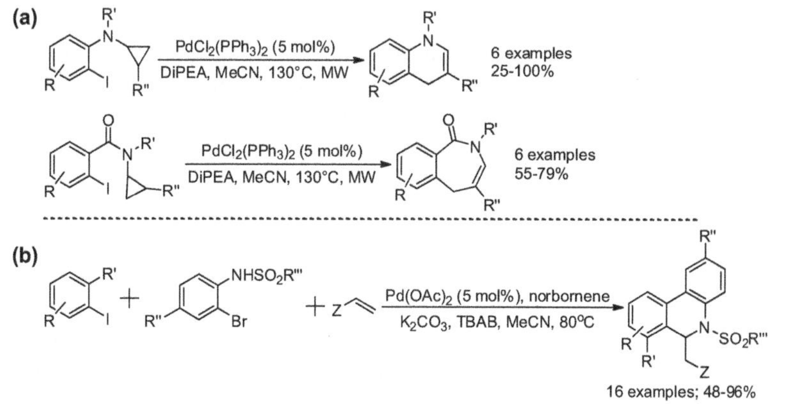

Scheme 3.35 Pd-catalyzed synthesis of dihydropyridines and related derivatives.

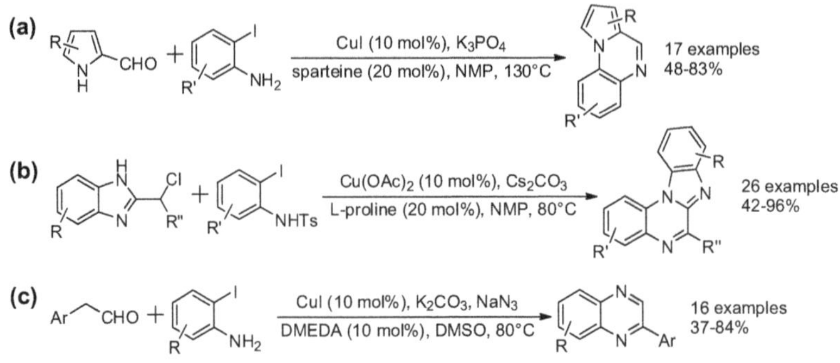

Scheme 3.36 Cu-catalyzed synthesis of quinoxalines.

was reported.[137] Ma and co-workers developed a convenient, efficient Cu-catalyzed domino process for the construction of benzo[4,5]imidazo[1,2-*a*]quinoxalines from *N*-tosyl-2-haloanilines and 2-(chloromethyl)-1*H*-benzo[*d*]imidazoles under mild conditions.[138] A variety of benzo[4,5]imidazo-[1,2-*a*]quinoxaline derivatives were obtained in good to excellent yields (Scheme 3.36b). Notably, *N*-tosyl-2-chloroanilines were compatible with this process, giving the corresponding products in good yields. Zhou and co-workers reported the synthesis of phenazines by Cu-catalyzed homo-coupling of 2-haloanilines in water.[139] Chen's group developed a novel pathway for the synthesis of quinoxalines.[140] The reaction involves condensation and C–N bond formation in a copper-catalyzed, one-pot, three-component reaction. The reaction was optimized when 2-iodoanilines (1.0 equiv.), arylacetaldehydes (2.0 equiv.), sodium azide (1.2 equiv.), CuI (10 mol%), DMEDA (10 mol%), and K_2CO_3 (1.0 equiv.) were reacted in DMSO at 80 °C for 20 hours. A variety of quinoxalines were produced in moderate to good yields (Scheme 3.36c).

Ghorai and co-workers developed a simple and practical protocol for the synthesis of racemic and nonracemic substituted and fused tetrahydroquinoxalines in 2011.[141] The reaction proceeds through a solvent and catalyst free S_N2-type ring-opening of *N*-activated aziridines with 2-bromo anilines followed by Pd-catalyzed intramolecular C–N bond formation. This method allows the use of a wide range of aziridines and 2-bromoanilines to construct tetrahydroquinoxalines in excellent yields and enantioselectivities (Scheme 3.37a). Chen and Liu developed a general, more sustainable methodology for the copper-catalyzed intramolecular *O*-arylation to synthesize 2,3-dihydro-1,4-benzoxazines (Scheme 3.37b).[142] The process tolerates variation of both aryl iodide and amino alcohol portions of the substrate. Huang's group developed a simple and efficient copper-catalyzed method for the synthesis of benzo[*e*][1,2,4]triazine derivatives in 2012 (Scheme 3.37c).[143] The protocol uses cheap and less toxic Cu_2O as the catalyst, readily available 2-haloanilines and hydrazides as the starting materials, and the reaction conditions are mild. Many substrates are suitable for this procedure, affording various benzo[*e*][1,2,4]triazine derivatives in moderate yields.

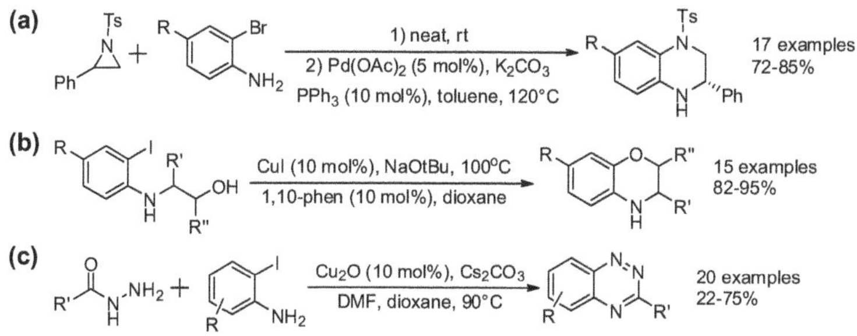

Scheme 3.37 Heterocycle synthesis from 2-haloanilines.

3.4 1-Carbon-2-haloarenes

In 1996, Buchwald and co-workers reported the first palladium-catalyzed intramolecular C–O coupling reaction to generate cyclic ethers.[144] The reaction showed good functional group tolerance; five-, six-, and seven-membered heterocycles were produced in moderate to good yields. DPPF, Tol-Binap, DavePhos, and JohnPhos are all suitable phosphine ligands for this transformation. By applying enantio pure substrates, chromans can be prepared in an enantioselective manner.[145] In 2012, a synthetic route for the preparation of 2-aryl-substituted chromans from commercially available starting materials and utilizing either a palladium- or copper-catalyzed intramolecular cyclization of aryl bromides was described.[146] Chromans with stereocontrol at C-2 can thus be obtained *via* a palladium-catalyzed asymmetric allylic etherification procedure utilizing a chiral indolephosphine oxazoline (IndPHOX) ligand. In the case of copper catalysts, 2-aminopyridine or 2,2′-bipyridine can be applied as a ligand.[147] In 2004, Whitby's group reported a palladium-catalyzed synthesis of benzo-fused cyclic amidines and imidates.[148] Isonitriles were inserted into the corresponding 2-bromobenzyl amine and 2-bromobenzyl alcohol derivatives.

Benzopyrans have been reported with various applications in various areas. In 2006, Li and Fang reported a copper-catalyzed intramolecular *O*-arylation of ketones.[149] With CuI (10 mol%) as the catalyst, *N*,*N*′-dimethylethylenediamine (DMEDA) as the ligand, and Cs_2CO_3 as the base, the reactions of α-(2-bromobenzyl)-β-keto esters in THF at refluxing temperature afforded the corresponding substituted 4*H*-1-benzopyrans in high yields *via* *O*-arylation (Scheme 3.38a). On the other hand, the reactions of δ-(2-bromophenyl)-β-keto esters in refluxing dioxane led to the formation of 3,4-dihydronaphthalen-2(1*H*)-one derivatives *via* *C*-arylation. In 2011, Beifuss and co-workers reported that a Cu(I)-catalyzed domino reaction between bromobenzyl bromides and β-keto esters exclusively yields either 4*H*-chromenes or naphthalenes (Scheme 3.38b).[150,151] Good yields of chromenes were formed, and naphthalenes can be produced by increasing the ratio of 1,3-diketone and adding 2-picolinic acid as a ligand. More recently, the group succeeded in exploring the use of 2-halobenzyl tosylates as substrates

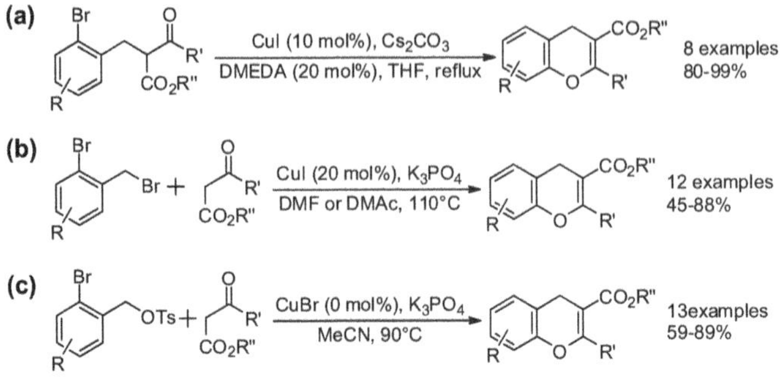

Scheme 3.38 Copper-catalyzed synthesis of chromenes.

Reaction Procedure (Scheme 3.38a): *N,N′*-Dimethylethylenediamine (0.04 mmol) in THF (2 mL) was added to a mixture of CuI (3.8 mg, 0.02 mmol), aryl bromide (0.2 mmol), and Cs_2CO_3 (0.4 mmol) in a round flask under nitrogen. The mixture was stirred at a refluxing temperature for 1 h. TLC monitoring indicated that all the starting material was consumed. The resulting mixture was then cooled to room temperature, and ethyl acetate (20 mL) was added. The mixture was filtered, and the filtrate was concentrated under reduced pressure. The crude product was then purified by column chromatography on silica gel with hexane–ethyl acetate (15 : 1, v/v) as the eluent to give the cyclized product.

(Scheme 3.38b): An oven-dried 10-mL vial was charged with CuI (98%) (39 mg, 0.2 mmol), potassium phosphate (97%) (875 mg, 4.0 mmol), 2-bromobenzyl bromide (249 mg, 1.0 mmol) and 1–2 mmol freshly distilled β-keto ester under air. The vial was sealed, evacuated and backfilled with argon (six times), then 3 mL freshly distilled dry DMF were added. The reaction mixture was stirred at 110 °C until the 2-bromobenzyl bromide was consumed (TLC). After cooling to room temperature, the reaction mixture was partitioned between 50 mL EtOAc and 20 mL saturated brine. The aqueous phase was extracted with EtOAc (2×40 mL). The combined organic layers were dried over $MgSO_4$ and concentrated *in vacuo*. The residue thus obtained was purified by flash column chromatography over silica gel (cyclohexane–EtOAc, 70 : 2, unless otherwise indicated).

in copper-catalyzed domino intermolecular substitution–intramolecular arylation processes for the efficient and selective preparation of heterocycles.[152] The reaction of 2-halobenzyl tosylates with β-keto esters delivers 4*H*-chromenes with yields ranging between 59 and 89% (Scheme 3.38c). Imidazobenzothiazines are formed with yields of up to 82% upon reaction of 2-halobenzyl tosylates with 2-mercaptoimidazoles. When 2-halobenzyl tosylates are reacted with benzamidines the corresponding quinazolines are obtained as well. It was reported that chromenes can be produced from the condensation of 1,3-diketones and 2-hydroxybenzaldehydes too.[153]

In 2011, Wu and co-workers developed a palladium-catalyzed reaction of 2-alkynylhalobenzenes with 2-alkynylphenols in the presence of PCy_3 for the synthesis of indeno[1,2-*c*]chromenes (Scheme 3.39).[154] This reaction works efficiently, leading to the corresponding products in good to excellent yields. Additionally, the group reported later that the scaffold of indeno[1,2-*c*]-chromene could be constructed *via* a palladium-catalyzed reaction of 2-alkynylbromobenzene with water, in which four bonds are formed with high efficiency.

The intramolecular cyclization of allenyl(2-iodobenzyl)ethers with nucleophiles provides a promising procedure for heterocycle synthesis. In the presence of palladium catalysts, organoboron reagents, tin reagent *etc*, the desired products can be formed in a highly selective manner.[155–159] Jia and co-workers reported an efficient palladium-catalyzed domino reaction involving a C–H activation process.[160] The palladacycle intermediate was successfully trapped by cyanation, Heck reaction, secondary C–H activation and Suzuki coupling. The regioselectivity was controlled by manipulating the reaction conditions to give either an aryl functionalized product with yields in the 47–95% range or an alkyl-functionalized product with yields in the 60–95% range. Diverse products can be prepared starting from the same substrate in good yields (Scheme 3.40a). The reaction between

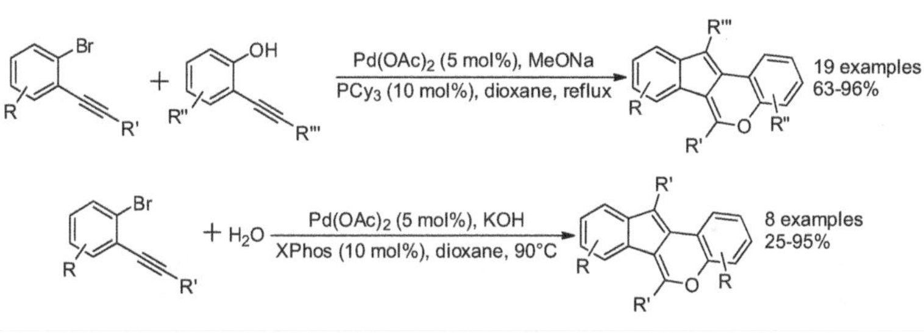

Scheme 3.39 Palladium-catalyzed synthesis of chromenes.

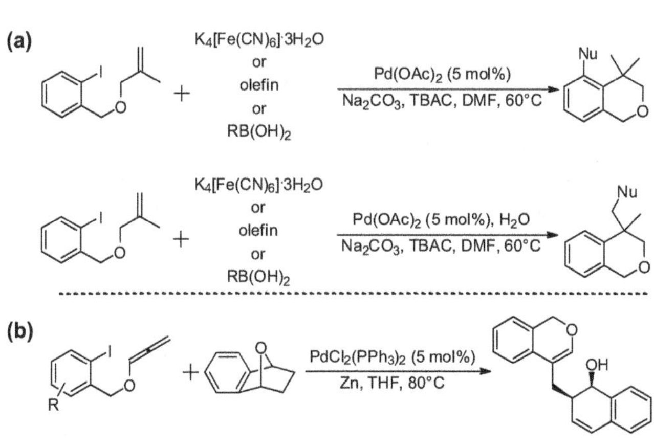

Scheme 3.40 Palladium-catalyzed cyclization of 2-halophenoxyallenes.

2-iodophenoxyallenes and ring opening of bicyclic alkenes was explored as well.[161] Cheng and co-workers developed a new palladium-catalyzed multistep reaction involving ring closure of 2-iodophenoxy-, 2-iodobenzyloxy-, and 2-iodobenzylaminoallenes and ring opening of bicyclic alkenes in 2006 (Scheme 3.40b). This method provides the construction of two different and new C–C bonds in one pot and allows an efficient synthesis of various benzo[*b*]furan, 1*H*-isochromenyl, and 1,2-dihydroisoquinoline-substituted 1,2-dihydro-1-naphthalenols in good yields. In this procedure, the role of zinc metal is crucial in the present catalytic reaction. First, it is used to reduce Pd(II) to Pd(0) to initiate the reaction and as a reducing agent to regenerate the Pd(0) catalyst. Second, the zinc halide produced from the oxidation of zinc metal during the reaction likely acts as a mild Lewis acid to remove a halide from the Pd(II) center and to assist the coordination of the bicyclic alkenes. Evidence to support this function came from the observation that the addition of triethylamine, which is expected to form a Lewis acid–base pair with zinc halide, to the catalytic reaction solution suppressed entirely the formation of product.

In 1997, a palladium-catalyzed cyclization of 3-(2-halobenzyloxy)phenols was reported by Rawal and co-workers.[162] In the presence of Herrmann–Beller palladacycle catalyst with Cs_2CO_3 as a base in DMAc at 95–115 °C, the desired 6*H*-benzo[*c*]chromenes were produced in good yields. Later on, Daugulis's group found that by using KO*t*Bu as the base at 140 °C in 1,4-dioxane, this reaction can proceed in the absence of a transition-metal catalyst.[163] The authors proposed that the reaction proceeds by an initial formation of a benzyne intermediate followed by an aromatic sp^2 C–H functionalization. The benzyne intermediate generated in the presence of a strong base undergoes cyclization *via ortho*(and/or *para*)-arylation of the phenolate. Subsequent protonation by *tert*-butyl alcohol and aromatization leads to the final products. Catalysis by trace amounts of palladium is unlikely since aryl chloride affords the cyclized products in a good yield as well. In 2004, the palladium-catalyzed intramolecular cyclization of (2-halobenzyloxy)arenes was developed by Fagnou and co-workers.[164] By using $Pd(OAc)_2$–DavePhos–K_2CO_3–DMAc–130 °C as the catalyst system, the desired products were formed in good yields. Since then, several modified systems were developed by them and other groups as well, such as the $Pd(OH)_2$/C system, which was found to be leaching palladium as the real catalyst, NHC–Pd complex (IPrPd$(OAc)_2 \cdot H_2O$) and ((SIPr)Pd(cinnamyl)Cl),[165] and so on.[166] Ultrasonic irradiation and a continuous flow capillary microreactor were found to promote this transformation at 90 °C.[167] By using DCPP as the ligand, tosylate and mesylate analogues of the substrates mentioned can also be transferred to the same kinds of products.[168] Additionally, *ortho*-tosylate/mesylate biphenyl ethers can be applied as substrates, and give the corresponding carbazoles in good yields. Procedures based on the generation of substrates *in situ* from 2-bromobenzyl bromides and phenols were developed as well.[169–171] Recently, Li and co-workers reported an iron-catalyzed, microwave-promoted cascade benzylation–cyclization of phenols and a variety of benzylating reagents such as benzyl acetates, benzyl bromides, and benzyl carbonates.[172] The reactions proceeded to afford both 9-aryl and 9-alkyl xanthene derivatives in good to high yields using $FeCl_3$ as the catalyst with the assistance of MW irradiation. Catellani and co-workers reported a new one-pot method for preparing 6*H*-dibenzo[*b*,*d*]pyran with good to excellent yields (Scheme 3.41a).[173] The reaction proceeded through

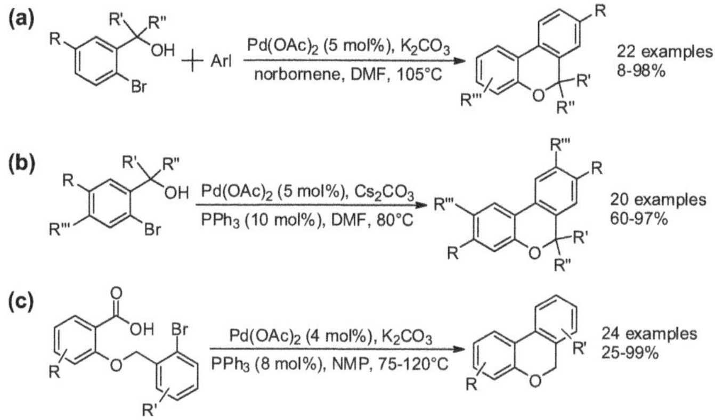

Scheme 3.41 Palladium-catalyzed synthesis of dibenzopyran and related derivatives.

Reaction Procedure (Scheme 3.41a): To a Schlenk-type flask containing $Pd(OAc)_2$ (4 mg, 0.018 mmol) and K_2CO_3 (125 mg, 0.90 mmol), was added a DMF solution (8 mL) of the *o*-substituted aryl iodide (0.36 mmol), 2-bromobenzyl alcohol (63 mg, 0.36 mmol) and norbornene (34 mg, 0.36 mmol). The reaction mixture was stirred under nitrogen at 105 °C for 24 h. After cooling to room temperature, the organic layer was diluted with EtOAc (30 mL), washed twice with water (25 mL) and dried over Na_2SO_4. The solvent was removed under reduced pressure and the resulting residue was purified by flash chromatography on silica gel using mixtures of hexane–EtOAc as eluent.

(Scheme 3.41b): In an oven dried round-bottomed flask under a nitrogen atmosphere were added $Pd(OAc)_2$ (5 mol%), PPh_3 (10 mol%), Cs_2CO_3 (0.40 to 0.94 mmol), and tertiary alcohol (100 mg, 0.20 to 0.47 mmol) followed by the addition of DMF (2 mL). The resulting reaction mixture was stirred at 80 °C for 12 to 27 h. The progress of the reaction was monitored by TLC until the reaction was complete. The reaction mixture was quenched by the addition of aq. NH_4Cl solution and extracted with CH_2Cl_2 (3×10 mL). The organic layer was dried over Na_2SO_4 and concentrated under reduced pressure. Purification of the residue on silica gel column chromatography using petroleum ether–ethyl acetate as eluent furnished the cyclic ethers.

(Scheme 3.41c): Under a nitrogen atmosphere, a mixture of $Pd(OAc)_2$ (0.9 mg, 4 mol%), PPh_3 (2.1 mg, 8 mol%), substrate (0.1 mmol), and K_2CO_3 (0.12 mmol) was dissolved in NMP (0.5 mL). The reaction system was stirred at 75 °C and monitored by TLC. Upon completion, the mixture was poured into aqueous HCl (1 N, 5 mL), and extracted with ethyl acetate (3×15 mL). The combined organic layers were dried over anhydrous Na_2SO_4, filtered, and the volatiles were removed *in vacuo*. The residue was purified by column chromatography (ethyl acetate–petroleum ether).

oxapalladacycles formed in the final stage of the reaction sequence that lead to biarylpalladium intermediates; Pd(IV) intermediates were involved. In 2012, an efficient Pd-catalyzed domino reaction of α,α-dialkyl-(2-bromoaryl)methanols to 6,6-dialkyl-6*H*-benzo[*c*]chromenes was presented (Scheme 3.41b).[174] Their formation can be explained *via* a five membered Pd(II)-cycle that efficiently involves a domino homocoupling with the second molecule, β-carbon cleavage, and finally intramolecular Buchwald–Hartwig cyclization. This domino process effectively involves the breaking of five σ-bonds (2C–Br, 2O–H, and a C–C) and formation of two new σ-bonds (C–C and C–O). In the same year, a palladium-catalyzed intramolecular decarboxylative coupling of arene carboxylic acids/esters with aryl bromides was developed by Shen and co-workers.[175] Various cyclized products were isolated in good yields (Scheme 3.41c).

A highly practical and sustainable methodology for the synthesis of a xanthone framework through a copper-catalyzed intramolecular *O*-arylation process was presented in 2009.[176] The target tricyclic skeletons were formed in good to excellent yields exclusively in an aqueous medium (Scheme 3.42a). The recyclability, scalability and the easy reaction setup combined with the low cost of the starting materials make this simple method transferable to pharmaceutical purposes. Recently, a palladium-catalyzed intramolecular cyclization of 1,3-diketones was developed.[177,178] By using readily available *C*-acylated cycloalkanones as substrates, 7-substituted tetrahydroxanthones were formed in a single operation in good yields (Scheme 3.42b). One of the synthesized derivatives was shown to selectively kill pancreatic cancer (PANC-1) cells under conditions of nutrient deprivation indicating that the tetrahydroxanthone is responsible, in part, for the "antiausterity" effects of the naturally occurring kigamicins. The cyclization of the methoxy substituted derivative was realized as well.[179] A copper-catalyzed *ortho*-acylation of phenols with aryl aldehydes and its application in a one-step preparation of xanthones was reported in 2012.[180] By using copper(II) as the catalyst in the presence of triphenylphosphine, the corresponding xanthones were produced in one step in high yields (Scheme 3.42c).

Isocoumarins are an important class of heterocyclic scaffolds that exhibit a wide range of biological activities, such as antifungal, antibacterial, and antidiabetic effects. In 2012, several copper-catalyzed procedures for the synthesis of isocoumarins were developed. Xi and co-workers developed an

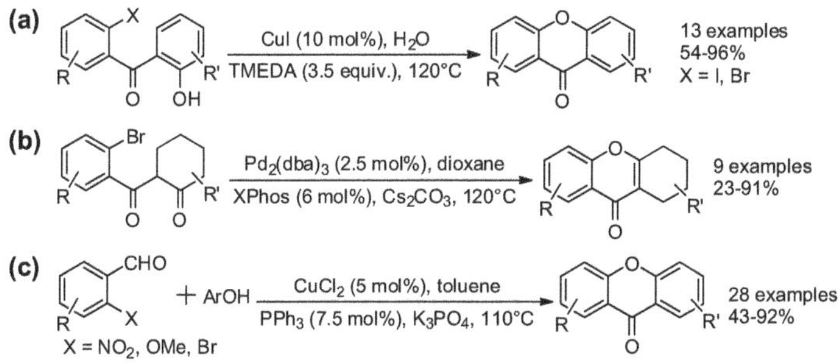

Scheme 3.42 Procedures for the synthesis of xanthones.

Reaction Procedure (Scheme 3.42c): Phenols (1.3 equiv.), 2-nitrobenzaldehydes (1 equiv.), K_3PO_4 (2.2 equiv.), $CuCl_2$ (5 mol%) and PPh_3 (7.5 mol%) were added together in 3 mL toluene, and then the reaction solution was stirred at 110 °C for 24 h. The mixture was extracted with DCM, washed with water, brine, and then the combined organic layer was dried by anhydrous Na_2SO_4. After evaporation of solvents, the crude product was purified by flash chromatography to afford the products.

efficient strategy for the synthesis of a variety of 3-substituted isocoumarins from *o*-halobenzoic acids and 1,3-diketones.[181] The reaction proceeded *via* a copper(I)-catalyzed domino reaction in DMF under the action of K_3PO_4 at 90–120 °C without a ligand to afford the corresponding isocoumarin derivatives in good to excellent yields. The *o*-halobenzoic acids could be *o*-iodobenzoic acid, *o*-bromobenzoic acid, and *o*-chlorobenzoic acid derivatives. The 1,3-diketones could be alkyl- and aryl-substituted 1,3-diketones. Later on, the group found that 2-bromobenzamides can be applied as substrates as well.[182] Meanwhile, Yao's group reported the synthesis of isocoumarin derivatives from 2-iodo-*N*-phenyl benzamides and 1,3-diketones with copper as the catalyst.[183] Interestingly, in the same period, Liang and co-workers found that these reactions can be performed in transition-metal-free conditions by using Cs_2CO_3 as the promoter in MeCN at 100 °C.[184] At the same time, a highly efficient strategy for the synthesis of 3-substituted isocoumarins through a copper(I)-catalyzed reaction of 1-(2-halophenyl)-1,3-diones was developed.[185] The procedure is based on a cascade copper-catalyzed intramolecular Ullmann-type *C*-arylation and rearrangement process and tolerates a wide range of substrates. By using readily available 1-(2-halophenyl)-1,3-diones (obtained from 2-haloarylcarboxylic acid chloride with ketones) as starting materials, the desired products were formed in good to excellent yields (Scheme 3.43).

A palladium-catalyzed intramolecular cyclization of aryl 2-bromobenzoates to the corresponding dibenzo[*b*,*d*]pyran-6-ones *via* C–H activation was reported in 1998.[186] By using $PdCl_2(PPh_3)_2$ as the catalyst and NaOAc as a

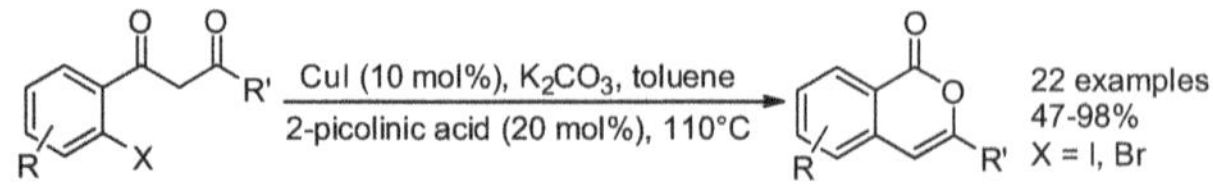

Scheme 3.43 Copper-catalyzed synthesis of isocoumarins from 1,3-diketones.

Reaction Procedure (Scheme 3.43): A sealed tube was charged with a magnetic stir bar, 1,3-dione, (0.5 mmol), CuI (10 mg, 0.05 mmol), 2-picolinic acid (12 mg, 0.1 mmol), K_2CO_3 (138 mg, 1.0 mmol), and anhydrous toluene (2 mL). The tube was purged with nitrogen gas, and its contents were stirred at 110 °C for the indicated time. After reaction completion, the mixture was filtered through a short plug of Celite and washed with EtOAc (2×3 mL). The combined filtrates were concentrated on a rotary evaporator and purified on a silica gel column.

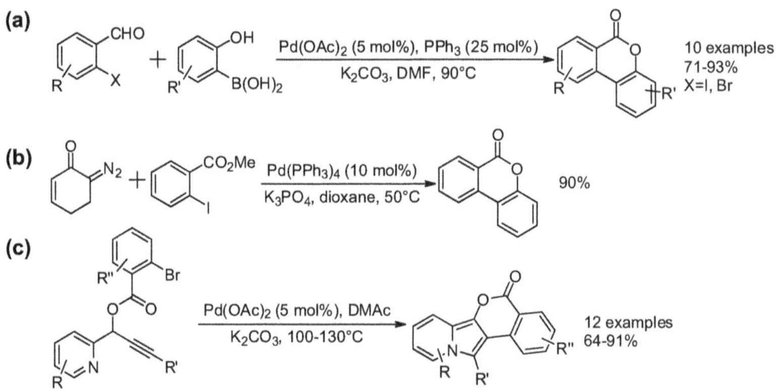

Scheme 3.44 Pd-catalyzed synthesis of isocoumarins from aryl halides.

base in DMAc, low yields of the products were formed. This method was applied to the synthesis of arnottin I and WS-5995A by different groups.[187,188] The Hurtley reaction was applied as well by using 3-hydroxyphenol and 2-bromobenzoic acids as substrates.[189–191] Alternatively, Ray and co-workers developed a one-pot synthesis of benzo[*c*]chromen-6-ones and its analogues *via* tandem Suzuki–Miyaura cross-coupling followed by oxidative lactonization in good to excellent yields.[192]

Starting from 2-hydroxyphenylboronic acid and *o*-bromobenzaldehyde or *o*-bromonaphthalene carboxaldehyde derivatives as substrates, a number of 6*H*-benzo[*c*]chromene-6-ones, 5*H*-naphtho[1,2-*c*]chrome-5-ones, and 6*H*-naphtho[2,1-*c*]chromene-6-one were synthesized in good yields (Scheme 3.44a). The overall transformation consists of three reactions: Suzuki–Miyaura cross-coupling, hemi-acetal formation, and oxidation. By combining 6-diazo-2-cyclohexenone and 2-iodo methylbenzoate as substrates, benzo[*c*]chromen-6-ones were produced in good yields as well

(Scheme 3.44b).[193] Another efficient palladium-catalyzed intramolecular carbopalladation–cyclization cascade toward tetra- and pentacyclic N-fused heterocycles was developed in 2010.[194] This transformation proceeds *via* the palladium-catalyzed coupling of aryl halides with internal propargylic esters or ethers followed by a 5-*endo-dig* cyclization leading to polycyclic pyrroloheterocycles in moderate to excellent yields (Scheme 3.44c).

Cheng and co-workers reported a nickel-catalyzed cyclization of oxa-bicyclic alkenes with β-iodo-(*Z*)-propenoates and *o*-iodobenzoate to the corresponding coumarins.[195] In the presence of Ni(dppe)Br_2 and Zn powder in acetonitrile at 80 °C, the desired benzocoumarin derivatives were obtained in moderate to good yields (Scheme 3.45).

In 2005, Larock and co-workers reported a palladium-catalyzed carbonylative synthesis of coumestans and coumestrols.[196] The reaction involves iodocyclization of acetoxy-containing 2-(1-alkynyl)anisoles and subsequent direct palladium-catalyzed carbonylation–lactonization. The desired coumestans and coumestrols and their related analogues were produced in low to excellent yields (Scheme 3.46a). In 2009, Willis's group demonstrated that

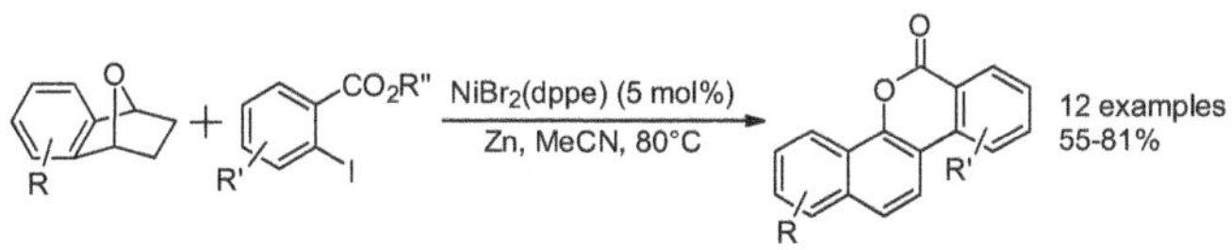

Scheme 3.45 Ni-catalyzed synthesis of coumarins.

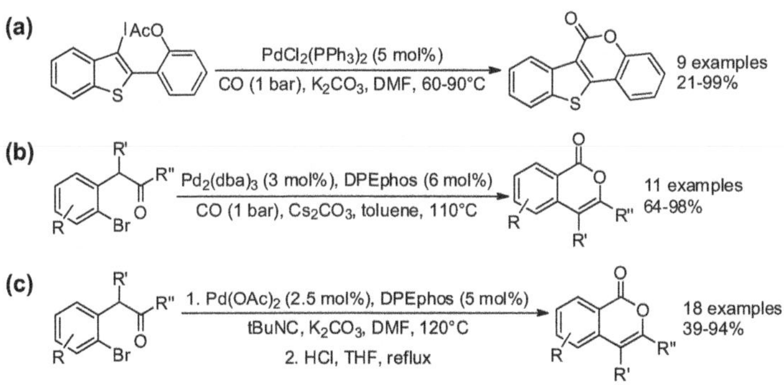

Scheme 3.46 Pd-catalyzed carbonylative synthesis of coumarin derivatives.

Reaction Procedure (Scheme 3.46a): DMF (1.0 mL), $PdCl_2(PPh_3)_2$ (0.0125 mmol), K_2CO_3 (0.5 mmol), and the aryl iodide (0.25 mmol) were stirred under an argon atmosphere at room temperature for 5 min. The mixture was flushed with CO, and the flask was fitted with a balloon of CO. The reaction mixture was heated at the specified temperature with vigorous stirring for 6 h. The reaction mixture was then cooled to room temperature, diluted with diethyl ether (35 mL), and washed with brine (30 mL). The aqueous layer was extracted with diethyl ether (15 mL). The organic layers were combined, dried over anhydrous Na_2SO_4, and filtered, and the solvent was removed under reduced pressure. The residue was purified by column chromatography on a silica gel column.

(Scheme 3.46b): Caesium carbonate (1.42 mmol) was added to an oven dried Schlenk tube (30 mL) charged with $Pd_2(dba)_3$ (13.0 mg, 0.014 mmol), DPEphos (15.3 mg, 0.028 mmol) and 2-(2-bromophenyl)-cyclohexanone (120 mg, 0.474 mmol) under argon. To this anhydrous toluene (1.0 mL) was added. A balloon fitted with a glass tap attachment was filled with argon and evacuated three times. The balloon was then filled with carbon monoxide from a lecture bottle and attached to the top of the Schlenk tube. The inert atmosphere was then exchanged for carbon monoxide by briefly exposing the reaction vessel to a vacuum through the side arm of the Schlenk tube and filling the vessel with carbon monoxide through the balloon on the top. The evacuation of the atmosphere and filling with carbon monoxide was performed three times. The reaction was then left open to the balloon and stirred vigorously, heating at 110 °C for 16 hours. After this time the Schlenk tube was allowed to cool and the carbon monoxide balloon removed. The reaction mixture was then diluted with ethyl acetate (*ca.* 10 mL), filtered through a Celite pad, washed with ethyl acetate (*ca.* 40 mL), and concentrated under reduced pressure. The crude material was purified *via* flash column chromatography (hexane–ethyl acetate, 4 : 1) to yield the isocoumarin.

(Scheme 3.46c): A sealed tube was charged with a magnetic stir bar, substrate (0.5 mmol), *tert*-butyl isocyanide (85 μL, 0.75 mmol), $Pd(OAc)_2$ (6 mg, 2.5 mol%), DPEPhos (27 mg, 5 mmol%), K_2CO_3 (138 mg, 1.0 mmol) and anhydrous DMF (3 mL). The tube was purged with nitrogen gas and stirred at 120 °C for 2 h. After reaction completion, the mixture was filtered through a short plug of Celite, and DMF was removed by a vacuum. The combined filtrates were refluxed in THF (15 mL) and hydrochloric acid (1 M, 3 mL) for 2 h. Then, the mixture was extracted with EtOAc, dried (Na_2SO_4) and evaporated. The residue was purified on a silica gel column using petroleum ether–EtOAc as the eluent to give the pure target product.

O-enolates can be employed as intramolecular nucleophiles in Pd-catalyzed aryl-carbonylation reactions. Under a balloon pressure of CO, α-(*o*-haloaryl)ketones were transformed into the corresponding isocoumarins in high yields (Scheme 3.46b). Both cyclic and acyclic ketones can be applied as substrates.[197] A concise synthesis of natural product thunberginol A was performed as well. Recently, a novel and highly efficient strategy for the synthesis of isocoumarins and phthalides through a palladium(0)-catalyzed reaction incorporating *tert*-butyl isocyanide was developed as well.[198] This process involves two steps including a cyclization reaction and simple acid hydrolysis. The methodology tolerates a wide range of substrates and is applicable to library synthesis. By using $Pd(OAc)_2$/DPEPhos as the catalyst system and THF/hydrochloric acid as the hydrolysis condition, good to excellent yields of the corresponding products were produced (Scheme 3.46c). Soon after, the group reported a nickel-catalyzed version of this reaction. In the presence of $NiCl_2$ (2.5 mol%)/DPPE (5 mol%) as the catalyst system with K_2CO_3 as a base in DMF at 135 °C, the desired products were formed in good yields.[199]

As early as 1993, Miura and co-workers reported a copper-catalyzed reaction of aryl and vinyl iodides with terminal alkynes.[200] One example of using 2-iodobenzoic acid and phenylacetylene as substrates to produce the corresponding isocoumarins was described. In the presence of CuI/PPh_3, 19% of the desired product was isolated. In 2009, a practical and general catalytic system [Cu(I) salts in DMF] for the efficient tandem coupling–heterocyclization reaction of (*Z*)-3-iodopropenoic acid derivatives with terminal alkynes was developed.[201] The reaction led to the easy formation of an important range of (*Z*)-5-alkylidene- or arylidene-5*H*-furan-2-ones and can also be extended to the preparation of 3-substituted isocoumarins. In the case of reactions between 2-iodobenzoic acid with terminal alkynes, aliphatic alkynes led to 3-substituted isocoumarins selectively while a 1:1 mixture of two isomers resulted with aromatic alkynes. Interestingly, this problem was solved in 2013 by Lee and co-workers.[202] They developed a simple copper catalytic system for the regioselective synthesis of isocoumarins and phthalides from the coupling reaction of 2-iodobenzoic acids and alkynes (Scheme 3.47). The isocoumarins were obtained at 100 °C, and

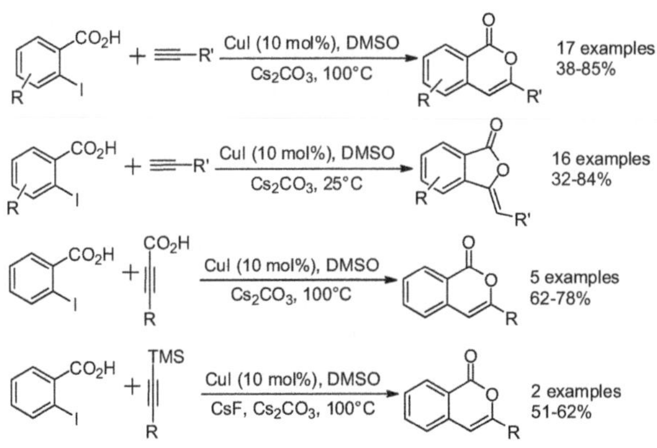

Scheme 3.47 Cu-catalyzed synthesis of isocoumarin derivatives from 2-iodobenzoic acids.

Reaction Procedure (Scheme 3.47): General procedure for the synthesis of isocoumarin: 2-iodobenzoic acid derivative (2.0 mmol) and Cs_2CO_3 (1.30 g, 4.0 mmol) were added to a vial containing anhydrous DMSO

(8 mL), followed by the addition of the appropriate acetylene derivative (2.2 mmol) and CuI (38 mg, 0.2 mmol). The suspension was stirred for 12 h at 100 °C. After cooling, the mixture was poured into EtOAc (50 mL) and washed with water (2×25 mL), brine (2×25 mL), then dried over Na_2SO_4. Evaporation of the solvent under reduced pressure provided the crude product, which was purified by column chromatography (hexane–EtOAc, 20 : 1) or re-crystallized from hexane to afford the final product.

General procedure for synthesis of phthalides: 2-iodobenzoic acid derivative (2.0 mmol) and Cs_2CO_3 (1.30 g, 4.0 mmol) were added to a vial containing anhydrous DMSO (8 mL), followed by the addition of the appropriate acetylene derivative (2.2 mmol) and CuI (38 mg, 0.2 mmol). The suspension was stirred for 12 h at 25 °C. After cooling, the mixture was poured into EtOAc (50 mL) and washed with water (2×25 mL), brine (2×25 mL), then dried over Na_2SO_4. Evaporation of the solvent under reduced pressure provided the crude product, which was purified by column chromatography (hexane–EtOAc, 20 : 1) or re-crystallized from hexane to afford the final product.

the phthalides were obtained at 25 °C in the presence of Cs_2CO_3 and DMSO. The authors found that: (i) the electronic properties of different substituents on the aryl rings did not affect the formation of isocoumarins at high temperature very much, but they did affect the formation of phthalide at low temperature; (ii) arylalkynyl carboxylic acid, and trimethylsilylacetylene were also good substrates for the formation of isocoumarins.

Palladium catalysts were explored as well. In 1993, Kundu and Pal studied the application of palladium catalyst in the coupling of 2-iodobenzoic acids with terminal alkynes.[203] In the presence of $PdCl_2(PPh_3)_2$–CuI–NEt_3–DMF–60 °C, the corresponding phthalides were formed as the major product and isocoumarins as the minor product. Cheng's group found that the addition of $ZnCl_2$ instead of CuI can drive the reaction to isocoumarins selectively.[204] Pd/C as a heterogeneous catalyst was also applied, but the recyclability was not demonstrated.[205] Larock and co-workers reported a palladium-catalyzed cyclization of 2-iodobenzoates with internal alkynes in 1999 (Scheme 3.48).[206] A number of 3,4-disubstituted isocoumarins and polysubstituted α-pyrones were prepared in good yields by treating halogen- or triflate-containing aromatic and α,β-unsaturated esters, respectively, with internal alkynes in the presence of a palladium catalyst. Mechanistically, the reaction was believed to proceed through a seven-membered palladacyclic complex in which the regiochemistry of the reaction was controlled by steric factors. The group found that allenes can be applied instead of alkynes as well.[207] In 2006, Swamy's group studied the cyclization of 2-iodobenzoic acids with allenylphosphonates.[208] Good yields of the corresponding 3-phosphorus substituted isocoumarins were produced. Alternatively, a route to these compounds from (*Z*)-iodovinylic acids or 2-iodobenzoic acids and allenyltributyltin reagents was described in 2005.[209] Treatment of

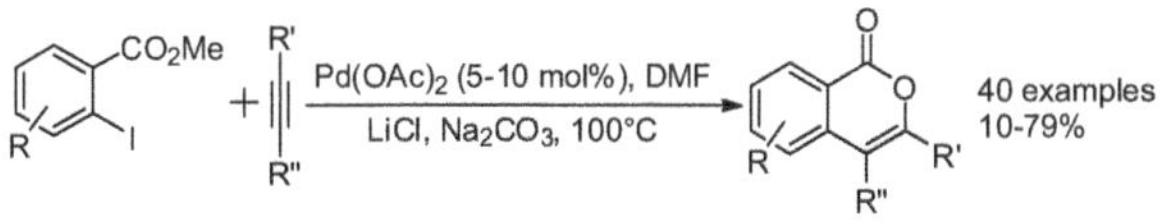

Scheme 3.48 Pd-catalyzed synthesis of isocoumarins from internal alkynes.

Reaction Procedure (Scheme 3.48): $Pd(OAc)_2$ (3 mg, 0.013 mmol), Na_2CO_3 (26.5 mg, 0.25 mmol), DMF (5 mL), LiCl (10.6 mg, 0.25 mmol), the alkyne (0.5 mmol), and the ester (0.25 mmol) were placed in a 2-dram vial (the isocoumarins and bicyclic pyrones were prepared on twice this scale). The vial was heated in an oil bath at 100 °C for the necessary period of time. The reaction was monitored by TLC to establish completion. The reaction mixture was cooled, diluted with ether, washed with saturated NH_4Cl, dried over anhydrous Na_2SO_4, and filtered. The solvent was evaporated under reduced pressure, and the product was isolated by chromatography (EtOAc–hexanes) on a silica gel column.

(*Z*)-β-iodovinylic acids or 2-iodobenzoic acids with various allenyltributyltin reagents in the presence of palladium acetate, triphenylphosphine, and tetrabutylammonium bromide in dimethylformamide provided good yields of the corresponding α-pyrones or 3-substituted isocoumarins *via* tandem Stille reaction and 6-*endo-dig* oxacyclization. Notably, Larock and co-workers found that isocoumarins can be produced from *ortho*-thallated benzoic acids and olefins too.[210]

Tetrahydroquinolines and related derivatives are important heterocyclic compounds with various known biological activities. In 1983, a copper-catalyzed intramolecular C–N bond coupling to synthesize fused β-lactams from the corresponding azetidinones was reported.[211] In the presence of copper powder in DMF at 100 °C, the corresponding products were formed in low to moderate yields. In 2011, a facile copper(I)-catalyzed tandem reaction for the synthesis of 4,5-dihydropyrazolo[1,5-*a*]quinolines and pyrazolo[1,5-*a*]indoles was reported.[212] High efficiency and good yields are displayed in this transformation under mild reaction conditions (Scheme 3.49a). Recently, a new strategy to synthesize tetrahydroquinolines *via* the sequential functionalization of two different types of remote C–H bonds was developed (Scheme 3.49b).[213] Picolinamide (PA) was used as the only directing/protecting group to effect Pd-catalyzed γ-C(sp^3)–H arylation, metal-free ε-C(sp^2)–H iodination, and Cu-catalyzed intramolecular C–N cross-coupling. The PA group of cyclized products could be easily removed. Meanwhile, a copper-catalyzed domino S_N2'–coupling procedure to prepare cyclic compounds from Baylis–Hillman acetates and amines was developed.[214] A variety of substituted quinoline/pyridine, thiochromene and naphthalene derivatives, which might be of biological and medicinal value, were synthesized by copper-catalyzed domino S_N2'–coupling,

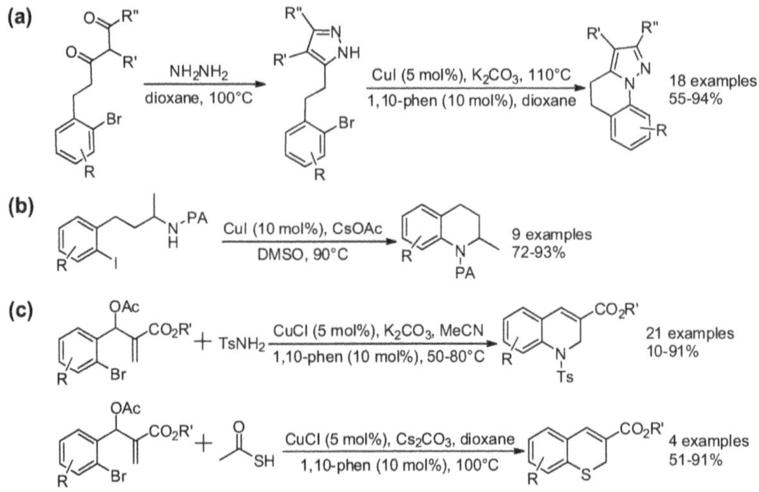

Scheme 3.49 Cu-catalyzed synthesis of tetrahydroquinolines.

S_N2'-deacylation–coupling and S_N2'-coupling–elimination reactions from the corresponding Baylis–Hillman (B–H) acetates and N-/S-/C-nucleophiles (Scheme 3.49c).

Palladium catalysts were also explored in tetrahydroquinoline synthesis. In 2005, a palladium-catalyzed intramolecular cross-coupling reaction between aryl iodides and allyl moieties was successfully demonstrated.[215] This synthetic approach gave a variety of *trans*-2,4-disubstituted 1,2,3,4-tetrahydroquinolines with excellent diastereoselectivities as well as various five- to seven-membered carbo- and heterocycles *via* conventional thermal and microwave-assisted conditions. Solé's group studied the Pd(0)-catalyzed intramolecular coupling reactions of amino-tethered aryl halides and ketones.[216] Two different and competitive reaction pathways starting from ω-(2-haloanilino)alkanones, enolate arylation and addition to the carbonyl group, have been observed, while ω-(2-halobenzylamino)alkanones exclusively underwent the enolate arylation process. The dichotomy between ketone α-arylation and carbonyl-addition in the reactions of ω-(2-haloanilino)alkanones has been rationalized by the intermediacy of unprecedented four-membered azapalladacycles, from which X-ray data and chemical behavior were reported. Li's group developed a new and general method for the synthesis of spiro[4,5]trienones by the intramolecular *ipso*-halocyclization of 4-(*p*-unsubstituted-aryl)-1-alkynes.[217] In the presence of halide electrophiles, a variety of 4-(*p*-unsubstituted-aryl)-1-alkynes underwent intramolecular *ipso*-halocyclization with water smoothly, affording the corresponding halo-substituted spiro[4,5]trienones in moderate to good yields. The obtained spiro[4,5]trienones can be applied in constructing the azaquaternary tricyclic skeleton *via* Pd-catalyzed Heck reaction. The mechanism of the intramolecular *ipso*-halocyclization reaction was also discussed according to ^{18}O-labeling experiments and DFT calculations. Palladium-catalyzed cyclization of *N*-bromobenzylnaphthylamines to naphthobenzazepines by intramolecular C–H activation was reported in 2003.[218] Utilizing the intramolecular coordination of an amine to a palladium center, the reactions proceeded selectively at the 8 position. In some cases, the product formed by coupling at the 2 position can be detected as well. Here, the *N*-bromobenzylnaphthylamines applied are produced from the reaction between 2-bromobenzyl bromides and naphthylamines. A similar procedure was used for the preparation of indolo[3,2,1-*d*,*e*]phenanthridines and isochromeno[3,4-*a*]carbazoles *via* palladium catalyzed intramolecular C–H coupling of the products from the reaction between 2,3,4,9-tetrahydro-1*H*-carbazol-1-ones and 2-bromobenzyl bromide.[219] The coupling reactions proceeded smoothly and in high yields under ligand-free conditions with the catalytic system $Pd(OAc)_2$–Cs_2CO_3–TBAB–110 °C. Interestingly, by applying KO*t*Bu as the promoter in mesitylene at 110 °C, the reactions can proceed in a transition-metal-free manner.[220]

In 2003, Grigg's group reported a one-pot Petasis reaction–palladium-catalyzed process involving 2-iodo/bromo benzylamine, ethyl glyoxalate and aryl/heteroaryl boronic acids (Scheme 3.50a).[221] The carbinolamine/imine generated *in situ* undergoes the Petasis reaction to afford the corresponding 2-halobenzylamine derivatives, which react with carbon monoxide or allene (1 bar) in the presence of Pd(0) to generate acyl palladium or allyl palladium species, which are intercepted intramolecularly by the proximal secondary amine to afford isoindolone/4-methylene-3,4-dihydroisoquinoline α-amino acid derivatives in good yields. These types of products can be produced *via* palladium-catalyzed Heck–aza-Michael reaction as well.[222] Starting from 2-(2-bromoaryl)ethanamines and electron-deficient terminal alkenes, the

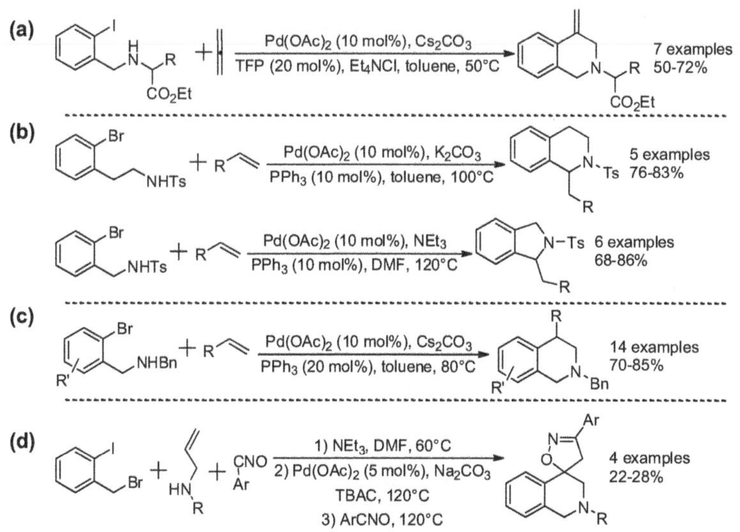

Scheme 3.50 Pd-catalyzed synthesis of tetrahydroquinolines from alkenes.

desired products were isolated in good yields (Scheme 3.50b). In the case of using 2-bromobenzylamines as substrates, the corresponding isoindolines were formed. Interestingly, tetrahydroisoquinolines can be produced from the same substrates by slightly changing the reaction conditions (Scheme 3.50c).[223] Remarkably, Broggini's group reported a simple and efficient procedure for tetrahydroisoquinoline synthesis in 2006.[224] Starting from secondary *N*-allylamines and involving sequential 2-iodobenzylation–intramolecular Heck reaction-1,3-dipolar cycloaddition, a variety of 4-spiroannulated tetrahydroisoquinolines were produced in good yields by a one-pot sequential procedure (Scheme 3.50d). Isolation and characterization of δ-alkylpalladium Heck intermediates were achieved as well.

Perumal and co-workers developed a procedure for the synthesis of functionalized tetrahydroisoquinolines in 2011.[225] The reaction goes through palladium-catalyzed 6-*exo-dig* carbocyclization of 2-bromo-*N*-propargylbenzylamines to give the products in good yields (Scheme 3.51). The substrates were produced by a CuI-catalyzed three-component coupling reaction of terminal alkyne, aldehyde and amine. More recently, the group reported a palladium-catalyzed highly regio- and stereoselective 6-*exo-dig* and 7-*endo-dig* cyclization of functionalized propargylic compounds for the synthesis of (*E*)-4-(isobenzofuran-1(3*H*)-ylidene)-1,2,3,4-tetrahydroisoquinolines and aze/oxepinoindoles too.[226] Alternatively, a copper-catalyzed synthesis of highly fused perimidine derivatives was achieved *via*

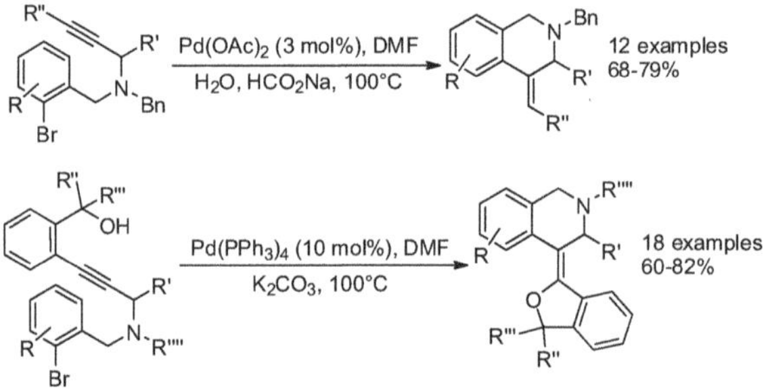

Scheme 3.51 Pd/Cu-catalyzed synthesis of tetrahydroquinolines from alkynes.

Reaction Procedure (Scheme 3.51): $Pd(PPh_3)_4$ (3 mol%) and HCOONa (1.5 equiv.) were added into a two-necked round-bottomed flask. The flask was evacuated and flushed with nitrogen. Propargylamine (0.4 mmol) dissolved in DMF (4.5 ml) was added, followed by distilled water (1.5 ml). The flask was heated to 100 °C in an oil bath for 3 h under a N_2 atmosphere. Upon completion of the reaction, the mixture was diluted with CH_2Cl_2. The organic phase was washed several times with brine, dried (anhydrous Na_2SO_4) and concentrated under reduced pressure. The crude product was chromatographed (petroleum ether–ethyl acetate as eluent).

two steps from 2-alkynylbenzaldehydes as reported in 2011.[227] By copper-catalyzed annulation of 2-[(2-bromophenyl)ethynyl]benzaldehydes with 1,8-diaminonaphthalene, the desired dihydroisoquinolino[2,1-*a*]perimidines bearing a 2-bromophenyl group were isolated in good to excellent yields. Subsequent palladium-catalyzed C–H arylation provided dibenzo[1,2 : 7,8]quinolizino[3,4,5,6-*kla*]perimidine derivatives in moderate to good yields.

In 2010, a methodology for the synthesis of 1,2,3-triazole-fused isoindolines and dihydroisoquinolines was reported.[228,229] Starting from easily available terminal alkynes and (2-haloaryl)alkylazides, the desired products were formed in good to excellent yields. This method is based on a cycloaddition reaction, *via* Click chemistry, followed by a transition-metal-catalyzed functionalization of a C–H bond. Later on, a one-pot procedure was reported as well.[230]

Xi and co-workers reported a copper-catalyzed domino N–H/C–H bond activation for the synthesis of nitrogen-bridgehead heterocycles by using azoles and 1,4-dihalo-1,3-dienes as substrates.[231] The reaction exhibits good regioselectivity when unsymmetrical 1,4-dihalo-1,3-diene was employed. A variety of azolopyridine derivatives were produced in good yields (Scheme 3.52).

Larock and Dai developed the synthesis of isoquinolines *via* palladium-catalyzed coupling of organic halides with 2-(1-alkynyl)benzaldimines.[232] Starting from *N-tert*-butyl-2-(1-alkynyl)benzaldimines and aryl, allylic, benzylic, alkynyl halides, as well as a vinylic halide, the corresponding 3,4-disubstituted isoquinolines with aryl, allylic, benzylic, 1-alkynyl, and vinylic substituents, respectively, in the 4-position were produced in good yields. The reaction appears to require an aryl group on the end of the acetylene furthest from the imine functionality.

In 2010, a palladium-catalyzed procedure for the synthesis of substituted isoquinoline derivatives from *o*-bromoarylaldehydes, terminal alkynes, and aqueous ammonia was reported.[233] In the presence of a palladium catalyst, the desired isoquinolines were synthesized in moderate yields with the assistance of microwave heating (Scheme 3.53a). The strategy tolerates a selection of substituents on both alkynyl and aldehyde partners. The use of ammonium acetate was shown to be possible as well.[234] Wu's group developed another methodology for the efficient synthesis of isoquinolines in 2011.[235] The reaction proceeds *via* a tandem reaction of imination of *o*-halobenzaldehydes with *tert*-butyl amine and subsequent palladacycle-catalyzed iminoannulation of internal alkynes; the desired isoquinoline derivatives were isolated in moderate to good yields with high

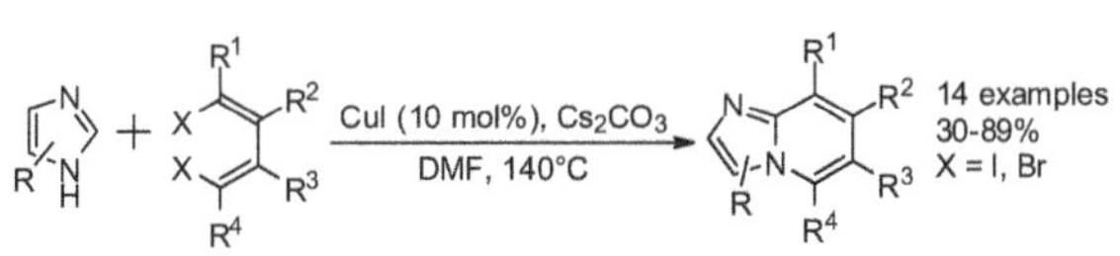

Scheme 3.52 Cu-catalyzed synthesis of tetrahydroquinolines from alkynes.

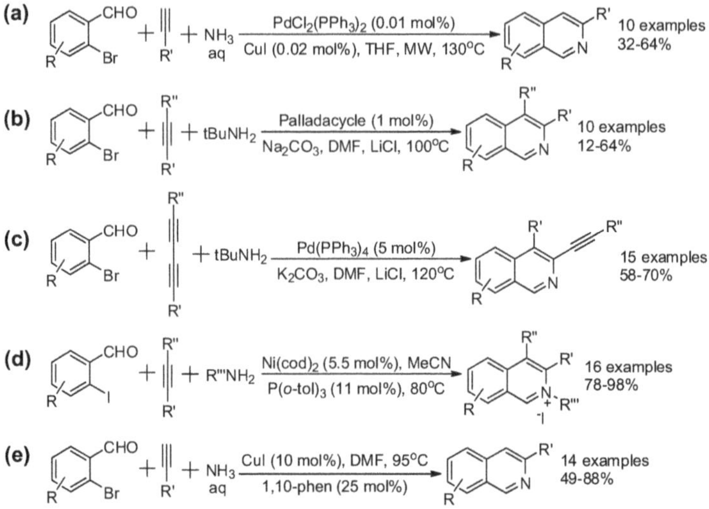

Scheme 3.53 Transition-metal-catalyzed three-component synthesis of isoquinolines.

regioselectivity (Scheme 3.53b). Here, the cyclopalladated ferrocenylimines were used as the stable palladium precursor. A one-pot one-step procedure with microwave irradiation was also developed.[236] Later on, Kundu's group found that 1,3-diynes can be applied as substrates as well.[237] A series of benzo[*a*]phenanthridines were constructed *via* multicomponent tandem reaction–carbocyclization in good yields (Scheme 3.53c). The reaction proceeds initially *via* the formation of a 4-aryl-3-arylethynyl-isoquinoline from 2-bromobenzaldehyde/*tert*-butylamine/1,3-diyne in a three component format followed by a second ring closure either *via* gold/silver-catalyzed intramolecular hydroarylation or *via* iodo-catalyzed regioselective 6-*endo-dig* electrophilic cyclization. This type of reaction can be performed in an aqueous medium with a palladacycle as the catalyst.[238] A cobalt catalyst [$CoCl_2$(DPPE)] was explored in this topic by Cheng and co-workers. Instead of isoquinolines, indenamines were produced.[239] Notably, isoquinolinium salts were formed when nickel complexes were applied as catalysts (Scheme 3.53d).[240] Copper as an inexpensive catalyst was tested in this transformation by Ray and co-workers.[241] By the combination of substituted *o*-bromoaldehyde, terminal alkyne, and aq. NH_3 in a CuI/1,10-phenanathroline catalytic system, 3-substituted isoquinolines were formed regioselectively in good to excellent yields (Scheme 3.53e).

In 2008, Ma and co-workers developed a CuI-catalyzed process for the synthesis of substituted isoquinolines from *o*-halobenzylamines and β-keto esters.[242] A number of functional groups in both the benzylamine and the β-keto ester moieties were tolerated by the reaction conditions. The reactions were performed in *i*PrOH with K_2CO_3 as the base; 1,2-dihydroisoquinolines were formed as the coupling–condensative cyclization products, which underwent smooth dehydrogenation in air to afford the targeted substituted isoquinolines (Scheme 3.54).

In 2009, Verma, Larock and their co-workers developed a route for the direct synthesis of diversely-substituted indolo- and pyrrolo[2,1-*a*]isoquinolines.[243–245] In the presence of a copper catalyst, the desired products were formed in good yields with excellent regioselectivity (Scheme 3.55a).

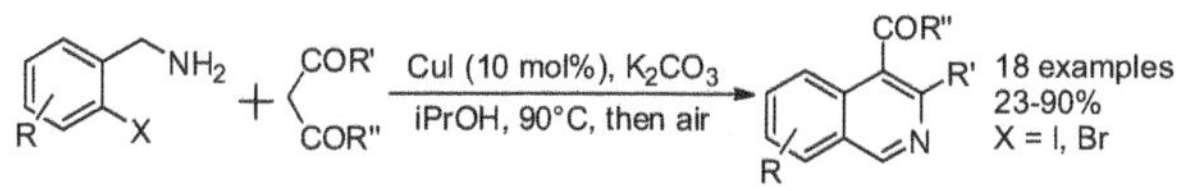

Scheme 3.54 Cu-catalyzed two-step synthesis of isoquinolines.

Reaction Procedure (Scheme 3.54): An oven-dried Schlenk tube was charged with CuI (10 mg, 0.05 mmol), and potassium carbonate (1.5 mmol), and 2-bromobenzylamine (0.5 mmol). The tube was evacuated and backfilled with argon (3 times), and then ethyl acetoacetate (1 mmol) and *i*-PrOH (1.5 mL) were added. The reaction mixture was stirred at 90 °C for 24–28 h. The cooled mixture was stirred at room temperature in air for 12–24 h before it was partitioned between ethyl acetate (40 mL) and saturated brine (10 mL). The organic layer was dried over Na_2SO_4, and concentrated *in vacuo*. The residue was purified by silica gel chromatography to give the desired product.

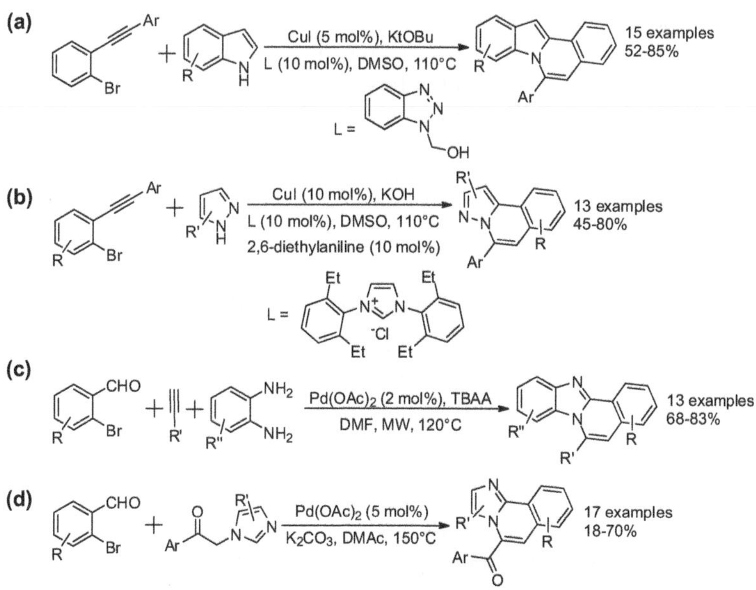

Scheme 3.55 Transition metal-catalyzed synthesis of fused isoquinolines.

This methodology involves the preferential nucleophilic addition of indoles and pyrroles onto the *ortho*-haloarylalkynes over *N*-arylation of the aryl halide. Recently, pyrazole was shown to be a suitable partner as well. In the presence of a copper catalyst, 2-alkynylbromobenzene and pyrazole, pyrazolo[5,1-*a*]isoquinolines were formed in good yields (Scheme 3.55b).[246] Here, copper(I)-catalyzed hydroamination and C–H activation were involved. Wu and co-workers reported a tandem four-component reaction of 2-bromobenzaldehyde, alkyne, amine, and diethyl phosphite transformation in 2009.[247] By a combination of palladium and copper salts, 1,2-dihydroisoquinolin-1-ylphosphonates were produced in moderate to good yields. A cascade Sonogashira–intramolecular cyclization–nucleophilic addition may be involved in the reaction process. Meanwhile, Yanada's group developed the copper-catalyzed synthesis of benzimidazo[2,1-*a*]isoquinolines from 2-bromoarylaldehydes, terminal alkynes, and 1,2-phenylenediamines (Scheme 3.55c).[248] In this procedure, Sonogashira coupling, 5-*endo* cyclization, oxidative aromatization, and 6-*endo* cyclization were performed in a single synthetic operation. More recently, Kumar and co-workers developed a method for the synthesis of aroyl-substituted imidazo-/benzimidazo-fused isoquinolines.[249] The cascade reaction proceeds *via* a cross-aldol condensation of 2-(1*H*-imidazol-1-yl/benzimidazolyl-1-yl)-1-arylethanones and 2-bromobenzaldehyde followed by palladium-catalyzed intramolecular C–H functionalization. The desired imidazo/benzimidazo[2,1-*a*]isoquinolines were isolated in moderate to good yields (Scheme 3.55d). In 2013, Batra and co-workers reported a cascade imination–decarboxylative coupling of potassium 5-amino-1-phenyl-1*H*-pyrazole-4-carboxylates and 3-aminothiophene-2-carboxylate with 2-haloarylaldehydes.[250] 3*H*-Pyrazolo[3,4-*c*]isoquinolines and thieno[3,2-*c*]isoquinolines were produced in good yields. The reactions of pyrazole-based substrates require a Pd/Cu bimetallic system for superior yields whereas the thienyl-based substrates afford the products in excellent yields with a palladium catalyst only.

Kirsch and Hesse studied the palladium-catalyzed amination of β-chloroacroleins with substituted anilines in 2005.[251] By using 2′-aminoacetophenone as the coupling partner, a quinoline derivative was produced (Scheme 3.56a). Another method for the synthesis of phenanthridines from 2-bromobenzylamines and aryl iodides was developed in 2010.[252] The domino sequence ends *via* an intramolecular amination and an oxidative dehydrogenation. No protecting group or prefunctionalization of the amine was required, and the process uses dioxygen as the terminal oxidant. Good to excellent yields of the desired products were isolated (Scheme 3.56b). Ding, Cai and their co-workers reported a method for the synthesis of benzimidazo[1,2-*a*]quinolines and other aza-fused polycyclic quinolones in 2010 (Scheme 3.56c).[253] Copper salt was used as the catalyst and 2-iodo, 2-bromo, and 2-chloro aryl aldehydes were applied as substrates. The reaction displayed great functional group compatibility. Recently, Wu's group reported a novel and efficient route for the synthesis of 5*H*-cyclopenta[*c*]quinoline derivatives *via* a palladium-catalyzed domino reaction of *o*-alkynylhalobenzene

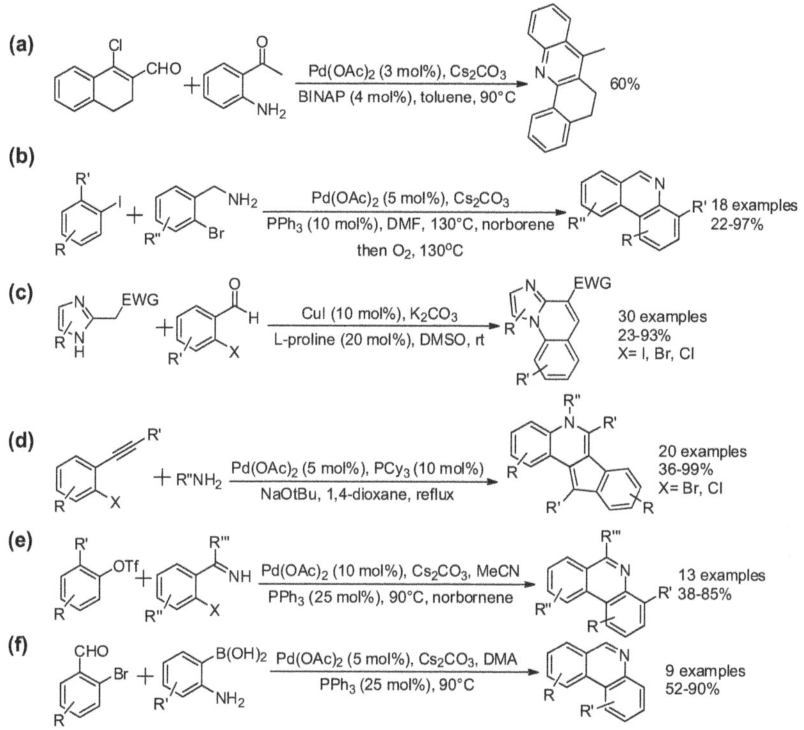

Scheme 3.56 Transition-metal-catalyzed synthesis of quinolines.

with amines (Scheme 3.56d).[254] The desired products were isolated in good yields with good functional group tolerance. The generation of *o*-alkynylbenzeneamine from *o*-alkynylhalobenzene and amines was believed to be the first step. Based on this hypothesis, the group developed the synthesis of 11*H*-indeno[1,2-*c*]quinolin-11-ols from various 2-alkynylhalobenzenes and 2-alkynylanilines.[255] By adding *N*-fluorobenzenesulfonimide to the reaction system, 11-fluoro-11*H*-indeno[1,2-*c*]quinolones can be selectively produced.[256] Additionally, the group found that NBS, NCS and allylic bromide are suitable electrophiles as well. The corresponding 11-Br/Cl/allylic-11*H*-indeno[1,2-*c*]quinolones were formed in good yields.[257] Lautens and co-workers developed a palladium-catalyzed cyclization of aryl triflates and imines.[258] The desired quinolones were formed in moderate to good yields. The application of this synthetic procedure in the concise synthesis of Nitidine and NK109 was also demonstrated (Scheme 3.56e). Jiang and co-workers reported the cyclization of 2-ethynylanilines with isocyanides in 2013.[259] 4-Halo-2-aminoquinolines were produced in good yields. More recently, a direct pathway for the synthesis of phenanthridines and their analogous benzo[*k*] and benzo[*i*] derivatives *via* Suzuki coupling and condensation between suitably substituted aromatic *ortho*-bromoaldehydes and *ortho*-aminobenzeneboronic acids in good to excellent yields was also developed (Scheme 3.56f).[260]

Quinazoline derivatives are known as powerful inhibitors of the epidermal growth factor (EGF) receptors of tyrosine kinase, ligands for benzodiazepine and GABA receptors in the CNS system or as DNA binders, and some quinazoline derivatives have shown remarkable activity as anticancer, antiviral, and antitubercular agents. Fu and co-workers developed an efficient copper-catalyzed method for the synthesis of 4-aminoquinazoline and 2,4-diaminoquinazoline derivatives in 2010.[261] The method proceeded *via* the reactions of substituted 2-bromobenzonitriles with amidines or guanidine (Scheme 3.57a). The reaction of 2-iodobenzaldehydes with amidines can lead to the preparation quinazolines in good to excellent yields (Scheme 3.57b).[262] The use of 2-bromobenzaldehydes as starting materials was developed in the same period.[263] In 2012, Hua and co-workers developed the one-pot CuCl-catalyzed synthesis of multi-substituted quinazolines by the three-component reaction of either *o*-bromo aromatic ketones/aldehydes, aromatic aldehydes and ammonia water or *o*-bromo aromatic ketones/aldehydes, primary alcohols and ammonia water.[264] The substituted quinazoline derivatives were formed in moderate yields (Scheme 3.57c). Fu's group reported that quinazolines can be produced from 2-bromobenzylamines and amidines with copper as the catalyst as well (Scheme 3.57d).[265] More recently, an efficient procedure for the synthesis of diversely

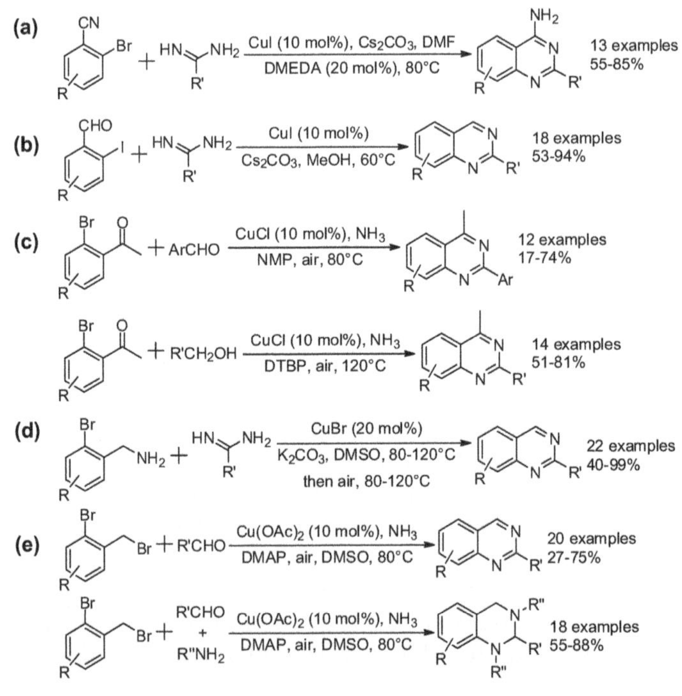

Scheme 3.57 Copper-catalyzed synthesis of quinazolines.

substituted quinazolines and 1,2,3,4-tetrahydroquinazolines was developed.[266] The reaction undergoes copper-catalyzed tandem reactions of the readily available 2-bromobenzyl bromides, aldehydes, and aqueous ammonia or amines. By using ammonia and simple aliphatic amines as the nitrogen source, the desired quinazolines and 1,2,3,4-tetrahydroquinazolines were produced in good yields (Scheme 3.57e).

Cinnolines are known to exhibit anticancer, fungicidal and bactericidal, and antiinflammatory activity as well as luminescent and optical properties. In 2012, Willis and co-workers reported a one-pot two-step route to cinnolines.[267] By using 2-(2-bromoalkenyl) aryl bromide and diethyl-1,2-hydrazine dicarboxylate as the starting materials and copper as the catalyst, the desired dihydrocinnoline derivatives can be produced in moderate to good yields, which can give cinnolines after deprotection (Scheme 3.58).

Recently, Beifuss and co-workers reported the synthesis of 2-aryl-1,2-dihydrophthalazines by the reaction of 2-(bromomethyl)benzaldehydes with arylhydrazines. The reaction employing K_2CO_3 as a base and $FeCl_3$ as a Lewis catalyst in CH_3CN at 100 °C gave the desired 2-aryl-1,2-dihydrophthalazines with yields ranging from 60 to 91%.[268] The transformation is considered to proceed as an intermolecular condensation–intramolecular nucleophilic substitution, and the role of $FeCl_3$ is to promote the condensation step. Zhang and co-workers developed a copper-catalyzed tandem reaction of

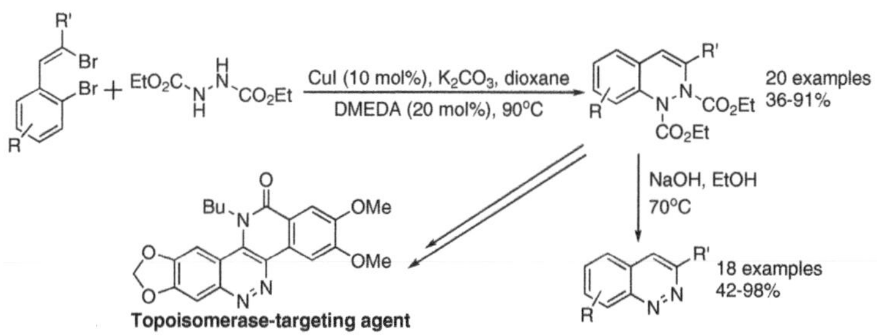

Scheme 3.58 Copper-catalyzed synthesis of cinnolines.

Reaction Procedure (Scheme 3.58): Bromide (0.4 mmol, 1.0 equiv.), diethyl hydrazine-1,2-dicarboxylate (0.8 mmol, 2.0 equiv.), K_2CO_3 (1.0 mmol, 2.5 equiv.) and CuI (0.04 mmol, 0.1 equiv.) were combined in a vial. The mixture was evacuated under vacuum and charged with nitrogen three times before 1,4-dioxane (0.5 mL) and DMEDA (10 µl, 0.08 mmol, 0.2 equiv.) were added. The reaction mixture was stirred in a preheated oil bath at 90 °C for 18 h. After cooling to room temperature, the reaction mixture was diluted with DCM (10 mL) and filtered through a pad of Celite, washing with further DCM (15 mL). The resulting filtrate was concentrated *in vacuo*. Column chromatography (15% acetone in petroleum ether) yielded the pure product.

2-bromobenzaldehydes/ketones with aminopyrazoles toward the synthesis of pyrazolo[1,5-*a*]quinazolines.[269] Good yields of the desired products were formed in the presence of copper and without a ligand (Scheme 3.59a). More recently, a copper(I)-catalyzed domino reaction of *N*-(2-benzimidazolyl)-2-aminobenzamide and 2-halogenated benzaldehyde was studied by Wang and co-workers.[270] The procedure is based on a sequential CuI/L-proline - catalyzed Ullmann reaction (C–N bond formation) and two bond cleavage reactions, and provides an efficient strategy for the synthesis of benzimidazo[1,2-*a*]quinazolines in good yields (Scheme 3.59b).

In 2006, Willis and co-workers reported a tandem palladium-catalyzed regioselective synthesis of 3-alkylated 2,4-quinazolinediones *via* urea arylation–intramolecular ester amidation.[271] *o*-Halo benzoates were reacted with monoalkyl ureas and gave the corresponding quinazolinedione products in good to excellent yields (Scheme 3.60). The reactions are

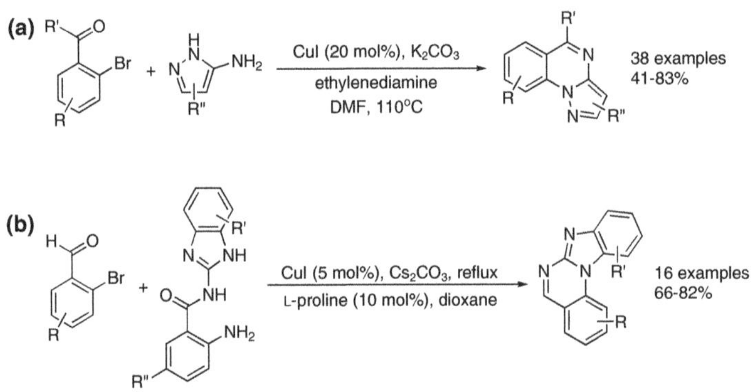

Scheme 3.59 Copper-catalyzed synthesis of fused quinazolines.

Reaction Procedure (Scheme 3.59a): To a solution of 2-bromobenzaldehyde (1 mmol) and 1*H*-pyrazol-5-amine (1.2 mmol) in DMF (5 mL) were added K_2CO_3 (2 mmol), CuI (0.2 mmol) and ethylenediamine (0.2 mmol). The mixture was stirred at 110 °C until a complete conversion as indicated by TLC. It was cooled to room temperature and saturated brine was added, then it was extracted with ethyl acetate. The combined organic phase was concentrated under vacuum. The crude product was purified by column chromatography eluting with petroleum ether–ethyl acetate (10 : 1) to give the desired product.

(Scheme 3.59b): *N*-(2-Benzimidazolyl)-2-aminobenzamide (1.0 mmol), *o*-halogenated benzaldehyde (1.0 mmol), L-proline (12 mg, 0.1 mmol), CuI (10 mg, 0.05 mmol), Cs_2CO_3 (2.0 mmol, 652 mg), and dioxane (10 mL) were added into a 25 mL flask. The reaction mixture was stirred at reflux for 14–24 h before reaching completion, which was monitored by TLC. The product was purified by column chromatography using ethyl acetate and petroleum ether (1 : 1) as eluent.

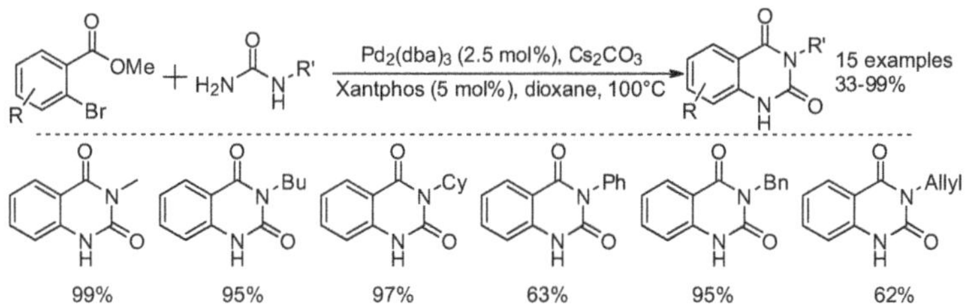

Scheme 3.60 Pd-catalyzed synthesis of quinazolinediones.

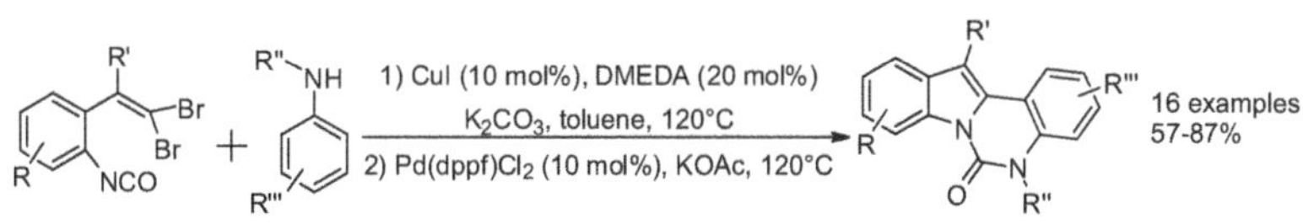

Scheme 3.61 Synthesis of pyrimido[1,6-*a*]indol-1(2*H*)-ones.

regioselective for formation of the 3-*N*-alkyl isomers. Significant variation of both coupling partners was possible, allowing the synthesis of a diverse array of substituted quinazolinediones, exemplified by the preparation of a simple unsymmetric-dialkylated natural product. This procedure was applied for starting material preparation later on.[272]

Bao and co-workers reported a one-pot synthesis of pyrimido[1,6-*a*]indol-1(2*H*)-one derivatives by a nucleophilic addition–Cu-catalyzed *N*-arylation–Pd-catalyzed C–H activation sequential process.[273] By the reaction of easily prepared *ortho*-*gem*-dibromovinyl isocyanates with *N*-alkylanilines, desired indole derivatives were produced in moderate to good yields (Scheme 3.61). More recently, a one-pot two-step process for the synthesis of annulated indoles was reported by Kundu's group.[274] The reaction initially proceeds through the annulation of 2-(2,2-dibromovinyl)aniline, an isocyanate, and a terminal alkyne in a three-component tandem format *via* Cu/Pd-catalyzed cross-coupling to afford N-1 and C-2 functionalized indoles. In the subsequent step, the enyne-urea derivative undergoes chemo- and regioselective 6-*endo* cyclization to afford an O-cyclized product in the presence of Au(I)/$AgNO_3$ and N-cyclized product in the presence of Au(I)/AgOTf in a post-MCR modification step.

In 2011, Xi and co-workers described a copper-catalyzed one-pot synthesis of 2-thioxo-2,3-dihydroquinazolin-4(1*H*)-ones by using *ortho*-bromobenzamides and isothiocyanates as substrates.[275] Various 2-thioxo-2,3-dihydroquinazolin-4(1*H*)-ones were conveniently synthesized in moderate to excellent yields with CuI as a precatalyst, Cs_2CO_3 as a base, *N*,*N*′-dimethylethane-1,2-diamine as a ligand, and toluene as a solvent, with the reaction temperature at 120 °C. Yu's group reported a novel procedure for the synthesis of 4-substituted-1,4-dihydroisoquinolin-3-ones and evaluated their antitumor activities (Scheme 3.62).[276]

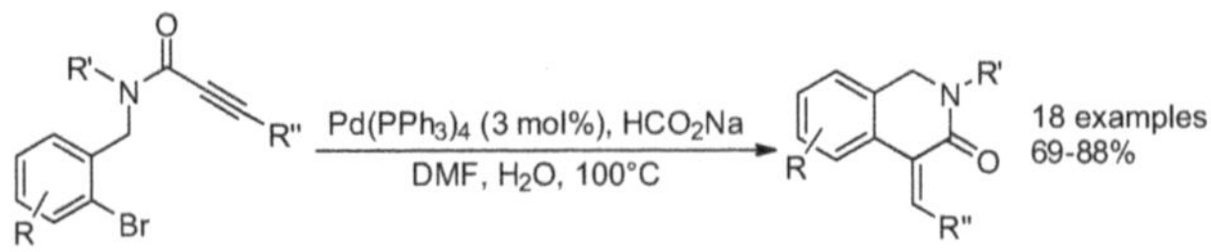

Scheme 3.62 Synthesis of 4-substituted-1,4-dihydroisoquinolin-3-ones.

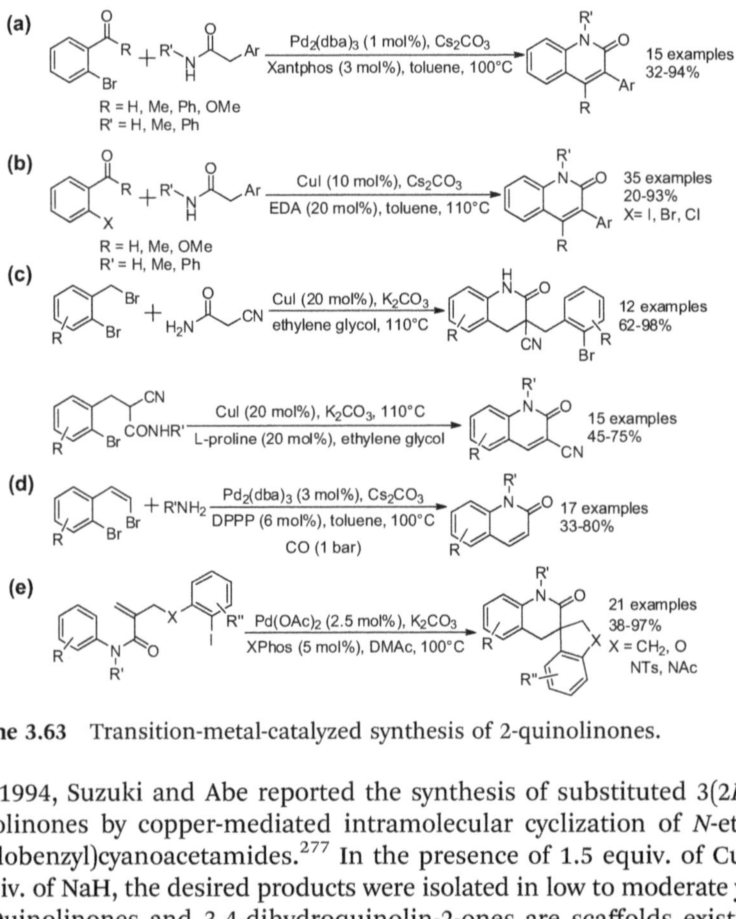

Scheme 3.63 Transition-metal-catalyzed synthesis of 2-quinolinones.

In 1994, Suzuki and Abe reported the synthesis of substituted 3(2*H*)-isoquinolinones by copper-mediated intramolecular cyclization of *N*-ethyl-*N*-(2-iodobenzyl)cyanoacetamides.[277] In the presence of 1.5 equiv. of CuI and 2 equiv. of NaH, the desired products were isolated in low to moderate yields.

2-Quinolinones and 3,4-dihydroquinolin-2-ones are scaffolds existing in numerous natural products, showing potent biological and pharmaceutical properties, and are valuable synthetic intermediates for the construction of sophisticated heterocyclic systems as well. In 2004, Manley's group developed a new strategy for the synthesis of 3-aryl naphthyridinones and quinolinones using a palladium-catalyzed cross-coupling reaction between enolizable primary and secondary amides and *o*-carbonyl substituted aryl halides.[278] A variety of functionalized 2-quinolinones were isolated in low to moderate yields (Scheme 3.63a). In this procedure, non-cyclized products were formed when propionamide and 2-cyclopropylacetamide were coupled with 2-bromobenzaldehyde. The use of CuI/DMEDA as a catalyst at 150 °C was also mentioned. In 2011, the same procedure with CuI as the catalyst was reported by Ding and co-workers (Scheme 3.63b).[279] Recently, copper-catalyzed intramolecular *N*-arylation and *O*-arylation reactions of amides were reported. Through these reactions, a variety of 3,3-disubstituted 3,4-dihydroquinolin-2-ones, 3-substituted quinolin-2-ones, and 12*H*-chromeno[2,3-*b*]quinolin-12-ones were obtained in moderate to excellent yields by using readily available 2-bromobenzyl bromides and cyanoacetamides as starting materials (Scheme 3.63c).[280] Alternatively, Willis and co-workers developed palladium-catalyzed intermolecular aminocarbonylation–intramolecular amidation cascade sequences, which can be used to convert a range of 2-(2-haloalkenyl)aryl halide substrates to the corresponding 2-quinolones in an efficient and selectively manner (Scheme 3.63d).[281] Delaying the introduction of the CO atmosphere allows an amination–carbonylation sequence and the preparation of an isoquinolone. 2-Quinolinones can be produced from the corresponding aryl bromides *via* intramolecular C–N coupling as well.[282] Additionally, a palladium-catalyzed spirocyclization was developed in 2012, which provides direct access to dihydrofuranyls, indolinyls, and indanyls spiro-fused to dihydroquinolin-2-ones at the 3,3′-position from linear substrates.[283] The domino process, involving an intramolecular Heck reaction and a C–H functionalization step, allowed the one-pot creation of two cycles and one quaternary center in a highly efficient manner. All the desired products were isolated in moderate to excellent yields (Scheme 3.63e).

Cacchi and co-workers reported the synthesis of 1,2-disubstituted 4-quinolones *via* copper-catalyzed cyclization of 1-(2-halophenyl)-2-en-3-amin-1-ones in 2009.[284] Here, the 1-(2-bromophenyl)- and 1-(2-chlorophenyl)-2-en-3-amin-1-ones were prepared from alkynones and primary amines. The reaction tolerates a variety of useful functionalities including ester, keto, cyano, and chloro substituents. Good to excellent of the desired products were formed (Scheme 3.64a). One year later, Xu and Zhao developed a palladium-catalyzed procedure that could directly use *o*-haloaryl acetylenic ketones and primary amines (Scheme 3.64b).[285]

Quinazolinone derivatives have been reported with various biological and medicinal properties, such as inhibitors of the epidermal growth factor (EGF) receptors of tyrosine kinase, and as anticancer, antiviral, and antitubercular agents. They are also used as ligands for benzodiazepine and

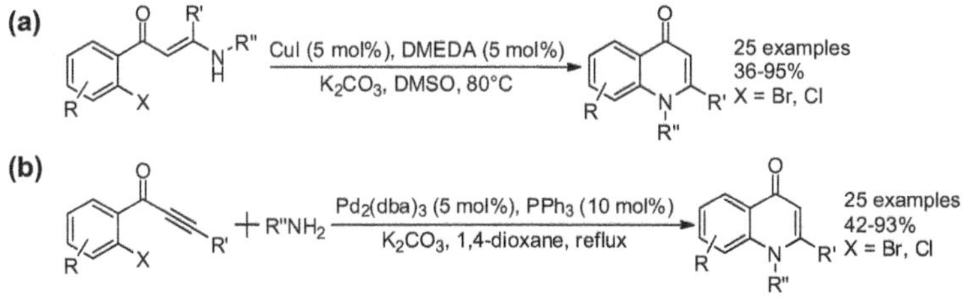

Scheme 3.64 Pd/Cu-catalyzed synthesis of 4-quinolones.

GABA receptors in the CNS system or as DNA binders. Remarkably, quinazolinone was found as a building block for approximately 150 naturally occurring alkaloids. Pellón and co-workers reported a synthetic method based on microwave- and ultrasonic irradiation-promoted Ullmann condensation of 2-aminopyridines with 2-chlorobenzoic acids to yield various substituted 11*H*-pyrido[2,1-*b*]quinazolin-11-ones in dry DMF under catalyst-free conditions.[286,287] In 2008, Fu, Jiang and their co-workers reported a copper-catalyzed synthesis of quinazolinones from methyl 2-halobenzoates and acetamidine hydrochloride.[288] With the assistance of a copper catalyst, the desired products were formed in moderate to good yields (Scheme 3.65a). In the case of using 2-bromobenzaldehydes and 2-bromoacetophenones as

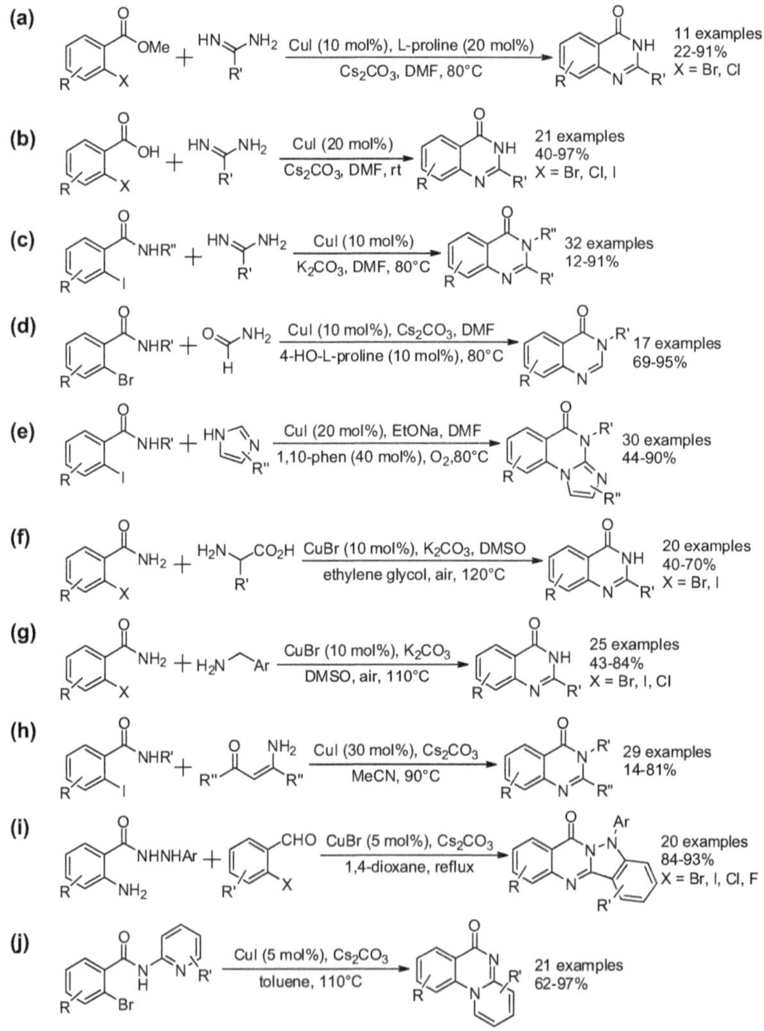

Scheme 3.65 Cu-catalyzed synthesis of quinazolinones.

substrates, the corresponding quinazolines were formed. In 2009, the group reported another procedure for preparing quinazolinone derivatives from 2-bromo- and iodobenzoic acid derivatives with amidines.[289,290] The coupling reactions performed well at room temperature without the addition of a ligand or an additive. The target products were also obtained in higher yields from nonactive substrates, such as 2-chlorobenzoic acid or guanidines, by simply increasing the reaction temperature to 80 °C (Scheme 3.65b). This transformation was studied further with an iron catalyst in water[291] and recyclable heterogeneous copper catalysts (for example Fe_3O_4 nanoparticle-supported copper).[292,293] Applying 2-iodobenzamides and amidines as substrates was also explored.[294] This provided another efficient synthesis of substituted quinazolin-4(3*H*)-ones by a one-pot ligand-free CuI-catalyzed coupling–condensative cyclization under mild conditions (Scheme 3.65c). Ma and co-workers developed a CuI/4-hydroxy-L-proline-catalyzed coupling of *N*-substituted *o*-bromobenzamides with formamide to quinazolinones (Scheme 3.65d).[295] The reaction takes place at 80 °C and tolerates a variety of functional groups. Imidazole derivatives and pyrazole were explored in azoquinazolinone synthesis with 2-halobenzamides as coupling partners.[296] The desired products were isolated in good yields under mild conditions (Scheme 3.65e). Recently, the use of 2-halobenzamides as substrates with various coupling partners based on a copper catalyst was described, such as applying amino acids (Scheme 3.65f),[297] benzylamines (Scheme 3.65g),[298,299] and enaminones (Scheme 3.65h).[300] Wang and co-workers reported the synthesis of fused quinazolinones from 2-amino-*N*′-arylbenzohydrazides and *o*-halogenated benzaldehydes.[301] In the presence of CuBr and Cs_2CO_3, the desired 5-arylindazolo[3,2-*b*]quinazolin-7(5*H*)-ones were isolated in high yields (Scheme 3.65i). A copper-catalyzed intramolecular cyclization of 2-bromo-*N*-(pyridin-2-yl)benzamides to the corresponding pyridoquinazolones was developed in 2011.[302] The corresponding products were formed in good to excellent yields (Scheme 3.65j). The use of 2-bromobenzylbromides and 2-aminopyridine as starting materials based on a multi-step operation was reported as well.[303] Palladium catalysts were explored in quinazolinones as well. In 2009, a palladium-catalyzed intramolecular cyclization of 2-bromo-*N*-(2-iodobenzyl)benzamides underwent sequential cyanation–*N*-addition–*N*-arylation to isoindolo[1,2-*b*]quinazolin-10(12*H*)-ones in a two-stage, one-pot manner.[304] This method also allowed the convenient synthesis of luotonin A and its derivatives. Reactions starting with 2-aminopyridines or aminoquinolines with substituted bromothiophene-2- or -3-carboxylates with the assistance of a palladium catalyst were described as well.[305]

Phthalazinones constitute an interesting family of bioactive N-heterocycles and have been reported with diverse pharmacological activities. Selected derivatives are used in the treatment of asthma, diabetes, hepatitis B, arrhythmia, and vascular hypertension. In addition, certain phthalazinones show antimicrobial activity and represent potent inhibitors of poly(ADP-ribose)polymerase-1 (PARP-1). In 2011, Kollár's group reported the synthesis of tetrahydrophthalazines from 2-iodobenzylbromides and hydrazines.[306] In the presence of a palladium catalyst [$Pd(OAc)_2$, PPh_3] and carbon monoxide (1 bar), the desired products were formed in good yields. Later on, Wu, Beller and their co-workers reported the use of 2-bromobenzaldehydes with hydrazines as substrates.[307] The corresponding phthalazinones were formed in good yields under carbonylative conditions (Scheme 3.66a). Afterwards, the exploration of $Mo(CO)_6$ or $Co_2(CO)_8$ as solid CO sources was reported too.[308,309] Meanwhile, a procedure based on isocyanide as an analogue of the carbonyl source was developed.[310,311] 4-Aminophthalazin-1(2*H*)-ones were produced in good yields (Scheme 3.66b). Regarding the reaction mechanism, the reaction starts with the oxidative addition of bromides to the Pd(0) catalyst followed by isocyanide insertion leading to palladium species. Coordination of hydrazine to Pd(II) and subsequent deprotonation followed by reductive elimination to give the intermediate, which then cyclizes and tautomerizes under the reaction conditions to form the final product. Alternatively, hydrazide formation by hydrazinolysis of the ester functional group may occur as the primary step.

Phenanthridinone derivatives have been reported to be found in a number of natural alkaloids and exhibit a wide range of biological activities. In the case of starting from bi-functionalized arenes with transition metal catalysts, the intramolecular cyclization of 2-bromo-*N*-arylbenzamides *via* C–H activation is the most direct pathway,[312–315] which has been applied in the synthesis of anti-hepatitis C virus agents and materials.[316–318] Remarkably, Yao, Xu and their co-workers developed a one-pot procedure for the synthesis of a pyrrolophenanthridone skeleton *via* an intramolecular Heck reaction and oxidation of *N*-(2-bromobenzyl) substituted indoles.[319] Moderate to good yields of the desired products were isolated in one step (Scheme 3.67). From the point view of academic interest, the

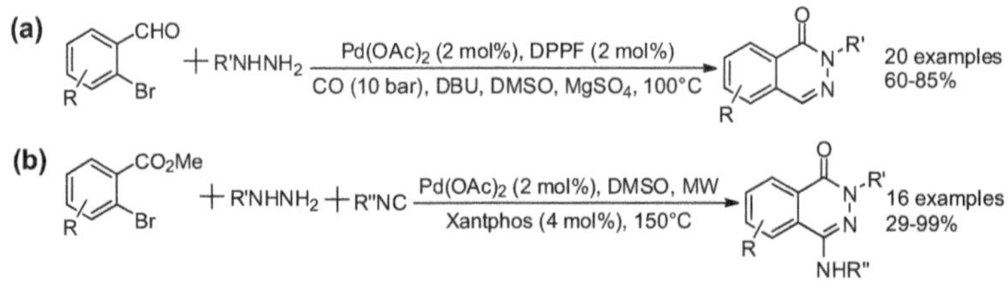

Scheme 3.66 Pd-catalyzed carbonylative synthesis of phthalazinones.

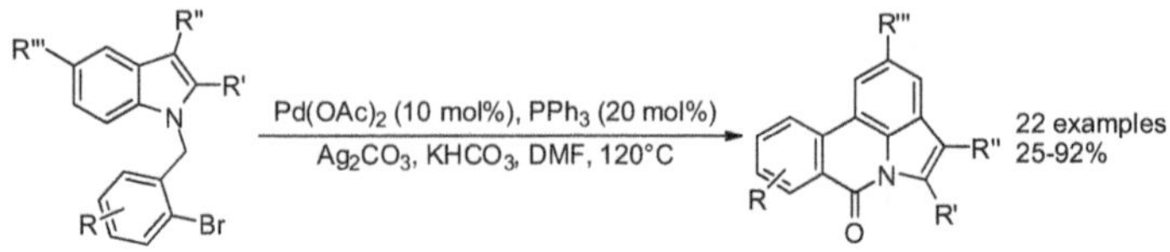

Scheme 3.67 Pd-catalyzed synthesis of phenanthridinones.

palladium-catalyzed cross-coupling reaction and oxidation reaction were combined here in one step. Mechanistically, the cross-coupling occurred in advance of oxidation.

In 2004, a new palladium-catalyzed method for synthesising 6-phenanthridinones and heterocyclic condensed quinolones was reported by Ferraccioli, Catellani and their co-workers.[320] The reaction was based on consecutive palladium-catalyzed regioselective aryl–aryl and *N*-aryl coupling of electron-rich *o*-iodoarenes with *o*-bromobenzamides and/or their heterocyclic analogues. Moderate to good yields of the products can be formed and a palladium(IV) metallacycle was proposed as the reaction intermediate (Scheme 3.68a). Later on, the group found that the same product can be

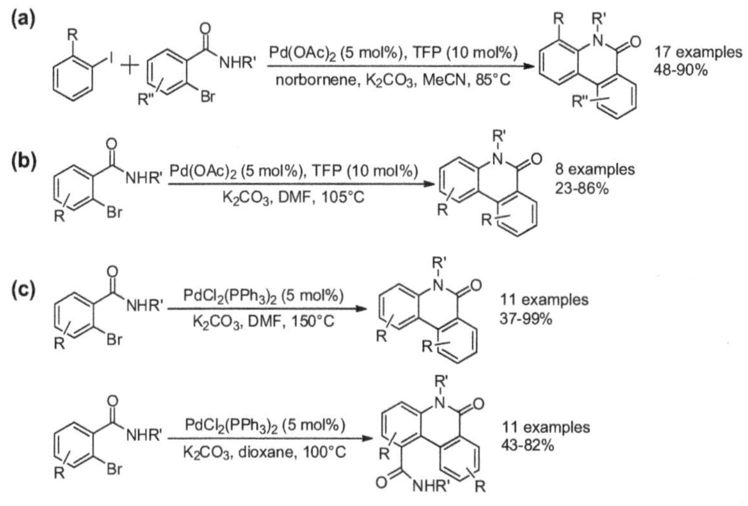

Scheme 3.68 Pd-catalyzed synthesis of phenanthridinones from 2-bromobenzamides.

Reaction Procedure (Scheme 3.68a): A Schlenk-type flask was charged under nitrogen with $Pd(OAc)_2$ (3.0 mg, 0.013 mmol), tri-2-furylphosphine (6.2 mg, 0.027 mmol), K_2CO_3 (72.3 mg, 0.52 mmol), the amide (0.26 mmol), a solution of norbornene (26.9 mg, 0.286 mmol) in anhydrous solvent (5.8 mL), and the *o*-iodoarene (0.26 mmol). The reaction mixture was heated with stirring at 85 or 105 °C, then cooled to r.t. After the addition of saturated NH_4Cl (30 mL) and extraction with AcOEt (3×15 mL), the combined organic extracts were washed with brine (30 mL) and dried over Na_2SO_4. Removal of the solvent under reduced pressure gave a crude product, which was purified by flash chromatography on silica gel (hexane–AcOEt).

(Scheme 3.68b): A Schlenk-type flask was charged under nitrogen with $Pd(OAc)_2$ (5.04 mg, 0.0225 mmol), tri-2-furylphosphine (10.4 mg, 0.045 mmol), K_2CO_3 (125 mg, 0.90 mmol), substrate (0.45 mmol) and anhydrous DMF (10 mL). The reaction mixture was heated with stirring at 105 °C for 20–48 h, then cooled to r.t. The reaction was quenched with saturated aq. NH_4Cl (75 mL) and the aqueous layer was extracted with AcOEt (3×40 mL). The combined organic extracts were washed with brine, separated and dried over Na_2SO_4. Removal of the solvent under reduced pressure gave a solid material which was purified by flash chromatography on silica gel (petroleum ether–AcOEt).

(Scheme 3.68c): K_2CO_3 (3 equiv.) and [$Pd(PPh_3)_2Cl_2$] (0.05 equiv.) were successively added to a solution of the amide (1 equiv.) in anhydrous DMF or 1,4-dioxane (20 mL). The resulting suspension was purged three times with argon then heated at reflux in a preheated oil bath at 155 °C for 3 h (DMF) or at 105 °C for 24 h (1,4-dioxane). The reaction mixture was then cooled to r.t., and the solvent was removed *in vacuo*. H_2O was added to the residue, which was extracted with CH_2Cl_2 (3 times). The combined organic extracts were washed with brine, dried over $MgSO_4$, filtered, and evaporated. The residue was purified by column chromatography on silica gel to afford the corresponding phenanthridinone.

produced in the absence of iodides as well (Scheme 3.68b).[321] Furuta, Kan and their co-workers developed a new ligand for this transformation.[322] More interestingly, Porée and co-workers discovered another reaction system that could tune the products by slightly changing the reaction conditions (Scheme 3.68c).[323] In the process, the outcome was strongly influenced by the nature of the substituents on the aromatic core, and the best yields were reached in the presence of electron-donating groups. Studies on the reaction parameters highlighted the dramatic influence of the carbonate base–solvent alliance in the selective formation of two products. Thus, in a weakly dissociating medium, 1-carboxamide phenanthridinones were preferentially formed, and conversely, formation of phenanthridinones was favored in a dissociating medium. Addition of a soluble carboxylate or the use of a benzamide that bears bulky substituents leads to a reversal of the selectivity observed in a weakly dissociating medium, thus outlining the crucial role of the alkaline agent in the formation of phenanthridinones. One of the most direct pathways is the reaction between methyl 2-bromobenzoate and 2-aminophenylboronic acid using palladium as the catalyst through Suzuki coupling and then transamidation.[324] Alternatively, Xu and co-workers developed the synthesis of benzo[*c*]phenanthridine derivatives by the annulation reaction of methyl *o*-halobenzoates with azabicyclic alkenes. The corresponding products were formed in good to excellent yields under base free conditions.[325]

In 2002, a procedure based on the palladium-catalyzed Heck reaction of 2-iodobenzoate with 2-aminoacrylate was developed.[326] In the presence of $Pd(OAc)_2$/TBAC/DMF/85–90 °C, the corresponding methyl isoquinolin-1-one-3-carboxylate can be isolated in 61% yield. This procedure was applied in the synthesis of the peripheral benzodiazepine receptor ligand, PK11195.[327]

Grigg's group reported the intramolecular cyclization of 2-iodo-*N*-allenylbenzamides to the corresponding isoquinolin-1-ones (Scheme 3.69a).[328] The palladium-catalyzed Stille coupling of 2-iodobenzamides with allenyltributyltin reagents to isoquinolin-1-ones was developed by Abarbri's group in 2005 (Scheme 3.69b).[329] Alternatively, 2-iodo-*N*-allylbenzamides can be applied as substrates for the palladium-catalyzed intramolecular cyclization as well.[330–334] In the absence of a ligand, the reaction occurred under mild reaction conditions (Scheme 3.69c). Lautens and co-workers reported the production of isoquinolin-1-ones from 2-halobenzamides and norbornadiene *via* a palladium-catalyzed coupling reaction followed by a retro-Diels–Alder reaction.[335] Moderate to good yields of the desired products were isolated (Scheme 3.69d). More recently, Cheng and co-workers reported a nickel-catalyzed annulation of substituted 2-halobenzamides with alkynes.[336] The corresponding isoquinolinones were isolated in good yields (Scheme 3.69e). In 2014, a copper-catalyzed cross-coupling of aryl iodides with terminal alkynes was applied in isoquinolinone preparation.[337] By using 2-iodobenzamides as substrates, the targeted products were produced in good to excellent yields (Scheme 3.69f). Notably, Pal and co-workers showed that even 2-iodobenzoyl azides can be applied as substrates.[338]

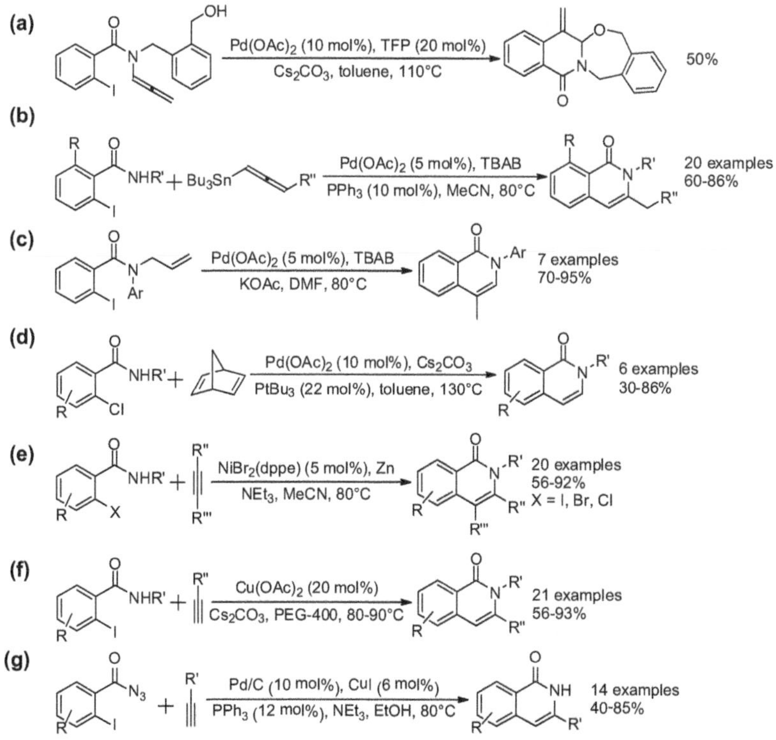

Scheme 3.69 Synthesis of isoquinolinones from unsaturated substrates.

In the presence of Pd/C and a copper catalyst, good yields of isoquinolinones can be formed *via* a Schmidt reaction (Scheme 3.69g).

Fu, Jiang and their co-workers reported a simple and efficient copper-catalyzed method for the synthesis of 3,4-disubstituted isoquinolin-1(2*H*)-one derivatives.[339] The reaction undergoes cascade reactions of substituted 2-halobenzamides with β-keto esters at 80 °C without the addition of any ligand or additive. The corresponding products were formed in moderate to good yields (Scheme 3.70a). Later on, Zhu's group reported the same analogues of compounds from 2-halobenzamides and ketones (Scheme 3.70b).[340] Recently, a copper-catalyzed procedure for the synthesis of fused isoquinolinones was developed by Huang, Yao and their co-workers.[341] Using these protocols, a wide variety of imidazolone-fused isoquinolinone derivatives, pyrimidinone-fused isoquinolinone derivatives, a diazapinone-fused isoquinolinone derivative and iminoisoindolinone derivatives were accessed in excellent to moderate yields (Scheme 3.70c).

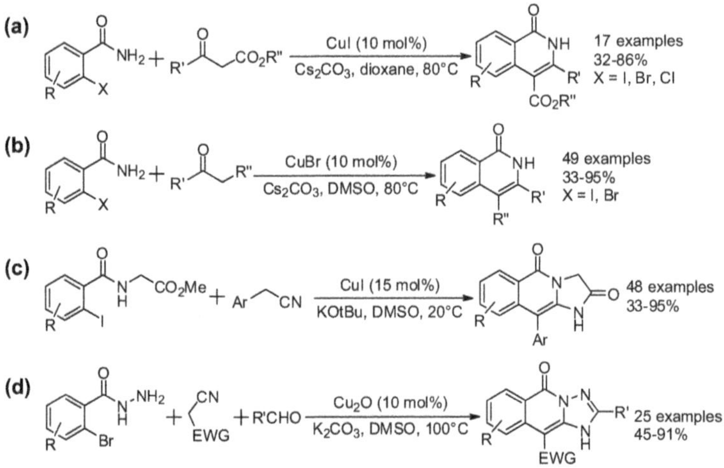

Scheme 3.70 Copper-catalyzed synthesis of isoquinolinones from 2-halobenzamides.

Reaction Procedure (Scheme 3.70): A reaction tube was charged with *N*-(2-bromobenzyl)-2,3,5-trisubstituted-1*H*-indole substrates (0.5 mmol, 1.0 equiv), $Pd(OAc)_2$ (12 mg, 0.05 mmol, 10 mol%), PPh_3 (27 mg, 0.1 mmol, 20 mol%), Ag_2CO_3 (276 mg, 1 mmol, 2.0 equiv), $KHCO_3$ (101 mg, 1 mmol, 2.0 equiv), and DMF (2 mL) under argon atmosphere. The mixture was vigorously stirred at 120 °C (oil temperature). After stirring for 15 h, the mixture was cooled to room temperature, diluted with ethyl acetate, and filtered. The filtrate was concentrated *in vacuo* to give dark residue, which was purified by flash chromatography on silica gel to afford the corresponding products.

In these reaction sequences it was found that benzyl cyanide acts as a nucleophile in the presence of strong bases while it acts as a cyanide donor in the presence of weak bases. The application of this method in the synthesis of inhibitors of PDE4 was reported as well.[342] More recently, Wu's group established a highly efficient three-component domino protocol for the synthesis of [1,2,4]triazolo[1,5-*b*]isoquinolin-5(1*H*)-ones using readily available *o*-halogenated benzohydrazides, aldehydes and nitriles as basic building blocks (Scheme 3.70d).[343] This domino process involves sequential selective condensation, copper-catalyzed intermolecular C-arylation, intramolecular addition of NH with CN, nucleophilic attack of amino to imine and final oxidative dehydrogenation. It is notable that this reaction performs well with varying functional group tolerance in the absence of a ligand in air.

As isoquinolinones contain a carbonyl group, the application of carbonylation in this topic will be interesting and straightforward. As early as 1998, Ryu, Komatsu and their co-workers described the intramolecular addition of acyl radicals onto the C–N double bond to provide nitrogen-containing heterocycles. One example of a six-membered heterocycle was synthesized (Scheme 3.71a).[344] In 2008, Alper's group reported another palladium-catalyzed carbonylation procedure for the synthesis of isoquinolinones.[345,346] By a combination of aminocarbonylation and aldol condensation reactions, the corresponding fused isoquinolinones were isolated in good yields (Scheme 3.71b). A one-pot two-step procedure based on *o*-halo-(2,2-dihalovinyl)-benzene was established as well (Scheme 3.71c).[347]

Benzoxazine derivatives have been shown to possess pharmacological properties and 1,3-oxazine derivatives are widely known for their potential as antibiotics, antitumor compounds, analgesics, and anticonvulsants. In 2012, Bao and co-workers developed a novel and efficient synthesis of benzoxazine and 1,3-oxazine derivatives *via* a ligand-free copper(I)-catalyzed one-pot cascade addition–cyclization reaction (Scheme 3.72).[348] A variety of carbodiimides coupled with *o*-halophenylmethanols and/or substituted (*Z*)-3-iodoprop-2-en-1-ols to give the corresponding products in moderate to excellent yields under mild conditions.

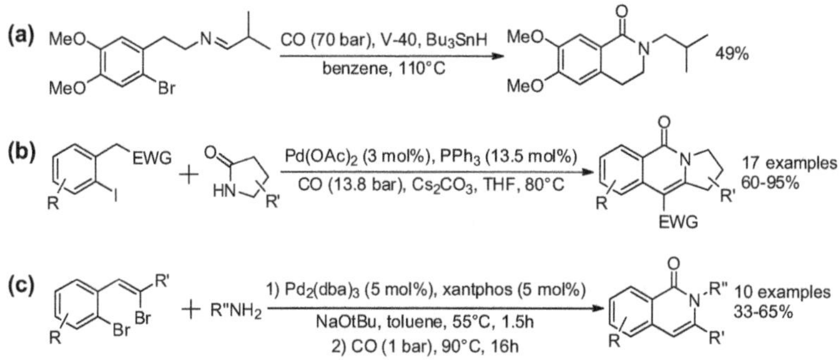

Scheme 3.71 Pd-catalyzed carbonylative synthesis of isoquinolinones.

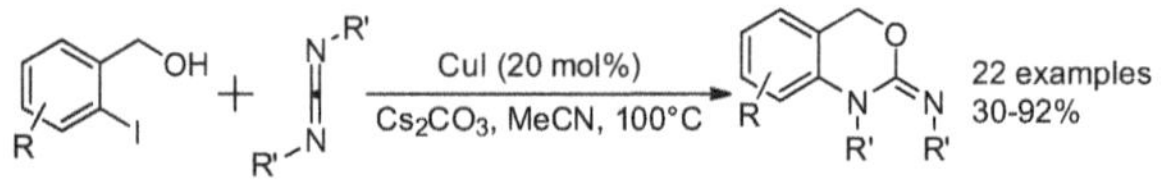

Scheme 3.72 Cu-catalyzed synthesis of benzoxazines.

Reaction Procedure (Scheme 3.72): An oven-dried Schlenk tube equipped with a Teflon valve was charged with a magnetic stir bar, CuI (0.1 mmol, 20 mol%), Cs_2CO_3 (0.5 mmol, 100 mol%), and *o*-halophenylmethanol and/or (*Z*)-3-iodo-3-phenylprop-2-en-1-ol (0.5 mmol). The tube was evacuated and backfilled with N_2 (this procedure was repeated three times). Carbodiimide (0.75 mmol) was added *via* syringe under a counter flow of N_2. Then under a counter flow of N_2, MeCN (2.0 mL) was added by syringe. The reaction mixture was sealed and stirred for 24 h at 100 °C. Then another part of carbodiimide (0.25 mmol) was added and the mixture was stirred for 12 h at 100 °C. The reaction was monitored by TLC. After the starting material was consumed completely, the resulting suspension was cooled to room temperature and filtered through a pad of filter paper with the help of 30 mL of ethyl acetate. After evaporating the solvent under reduced pressure, the residue was purified by column chromatography on silica gel to give the pure product (AcOEt–petroleum ether, v/v = 1 : 10).

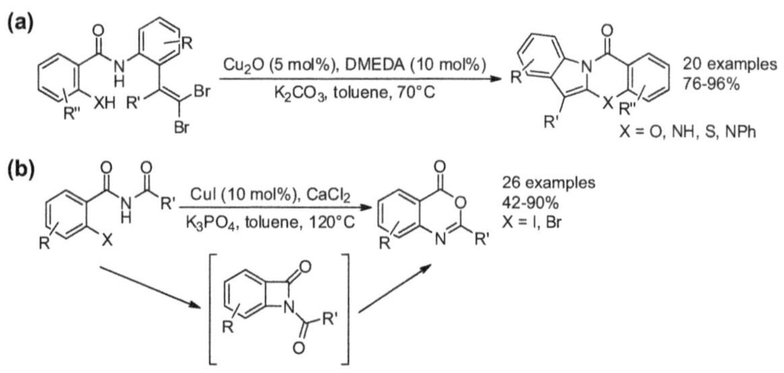

Scheme 3.73 Cu-catalyzed synthesis of benzoxazinones.

4*H*-3,1-Benzoxazin-4-ones are a class of fused heterocycles that have been reported with a varieties of biological activities, such as chymotrypsin inactivators, HSV-1 protease inhibitors, and inhibitors of human leukocyte elastase, cathepsin G, C1r serine protease, and human chymase, and so on. In 2012, a copper-catalyzed domino intramolecular cyclization was established by Wang, Lv and their co-workers.[349] By a combination of Cu_2O and DMEDA, the desired products were formed in good yields (Scheme 3.73a). Recently, Zhu's group developed an efficient and rapid method for the synthesis of 4*H*-3,1-benzoxazin-4-one derivatives. Starting from easily available *N*-acyl-2-halobenzamides, good to excellent yields of the products were isolated (Scheme 3.73b).[350] The key step of this transformation is the C–N bond formation–rearrangement sequence.

In 1999, Harmata's group reported the synthesis of benzothiazines by palladium-catalyzed aryl amination of *ortho*-bromobenzaldehyde with sulfoximine followed by an intramolecular condensation reaction.[351] The target molecules were isolated in good yields in an enantiomerically pure manner (Scheme 3.74a). Bolm and co-workers developed the palladium-catalyzed intramolecular α-arylation of sulfoximines.[352] Through this palladium-catalyzed cyclization of *N*-(2-bromobenzyl)- and *N*-(2-bromobenzoyl)-sulfoximines, the corresponding six-membered heterocycles were formed in moderate to good yields (Scheme 3.74b). In both cases, the α-arylations of the sulfoximine methyl groups were catalyzed by a combination of $Pd(OAc)_2$ and *rac*-BINAP in the presence of a base such as Cs_2CO_3 or K_2CO_3. Through *N*-arylation was developed by the same group as well (Scheme 3.74c).[353] In 2008, a palladium-catalyzed intramolecular cyclization of the adduct from 2-halobenzyl amines with thioisocyanate was developed.[354] Substituted benzothiazines were selectively produced in good yields (Scheme 3.74d). The use of *N*-substituted 2-bromobenzylamines as substrates was developed as well, and benzothiazineimines can be formed.[355] In 2012, Bao and co-workers developed the synthesis of aza[2,1,-*b*][1,3]-benzothiazinones from

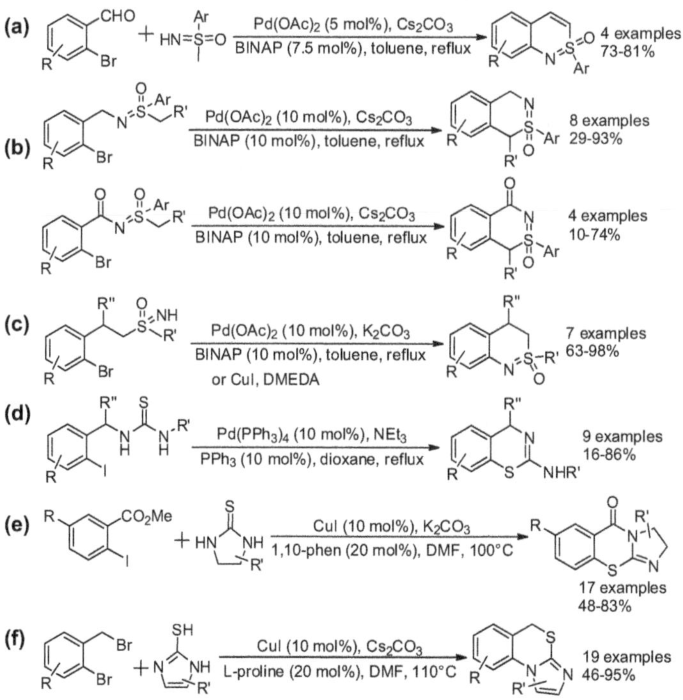

Scheme 3.74 Synthesis of S-containing heterocycles.

cyclic thiourea and methyl 2-iodobenzoate *via* the Cu(I)-catalyzed C–S coupling–amidation process.[356] A number of aza[2,1,-*b*][1,3]-benzothiazinones containing five-, six- and seven-membered rings and different substituents were obtained in good yields (Scheme 3.74e). Later on, the reaction starting from 2-iodobenzoic acid and 2-mercaptobenzimidazole was reported too.[357] In the same year, Bao's group also reported the preparation of various imidazobenzothiazine and primidobenzothiazine derivatives from readily available 2-mercaptoimidazoles (thiouracils) and bromobenzyl bromides *via* a copper(I)-catalyzed one-pot cascade process.[358] The reaction involves a S_N2 process of the benzyl bromides and an intramolecular C–N cross-coupling cyclization of aryl bromides (Scheme 3.74f). More recently, a convenient one-pot method for the synthesis of thiochromen-4-ones was developed.[359] By the condensation of 2′-haloacetophenone and dithioesters at room temperature in the presence of NaH in DMF, the desired thiochromen-4-ones were formed in moderate to good yields. The method involves unusual intramolecular *S*-arylation.

Denmark's group reported the preparation of geometrical isomers of 6-ethylidenedioxadisilacyclohexane by intramolecular hydrosilylation of an unsymmetrical disiloxane with the assistance of Pt (*syn*) and Ru (*anti*) catalysts. This new class of organosilicon reagent underwent cross-coupling reactions with a range of aryl iodides to afford (*E*)- and (*Z*)-trisubstituted allylic alcohols in a highly stereospecific fashion. Interestingly, in the presence of $Pd(dba)_2$ and methyl 2-iodobenzoate, the corresponding benzo-fused six-membered lactone can be formed in good yields.[360] Phosphorus analogues of α-methylene lactones can be produced from the corresponding *o*-bromobenzyl phenylvinylphosphinate as well. $Pd(OAc)_2$ and PPh_3 was used as the catalyst and NEt_3 as the base in MeCN at 100–120 °C.[361] Numbers of benzothiolane and isothiochromane scaffolds with a stereodefined tetrasubstituted exocyclic double bond were produced recently *via* cyclocarbopalladation and cross-coupling reactions.[362] In detail, the process works efficiently *via* either 5-*exo* or 6-*exo-dig* cyclization and is compatible with Stille and Suzuki–Miyaura cross-coupling reactions. By this method, benzene-fused 5- or 6-membered sulfur heterocycles and a stereodefined tetrasubstituted exocyclic double bond were selectively formed.

3.5 Miscellaneous

In 2009, a simple and practical copper-catalyzed method for the synthesis of 1,2,4-benzothiadiazine 1,1-dioxide derivatives was developed.[363,364] The reaction proceeded *via* cascade reactions of substituted 2-halobenzenesulfonamides with amidines, and the corresponding products were formed in good yields (Scheme 3.75). The reaction was proposed to start with C–N coupling then intramolecular transamination to release ammonia and provide the final product. Later on, a procedure for the synthesis of the same products *via* intramolecular cyclization of *o*-bromoarylsulfonylated amidines was developed as well.[365]

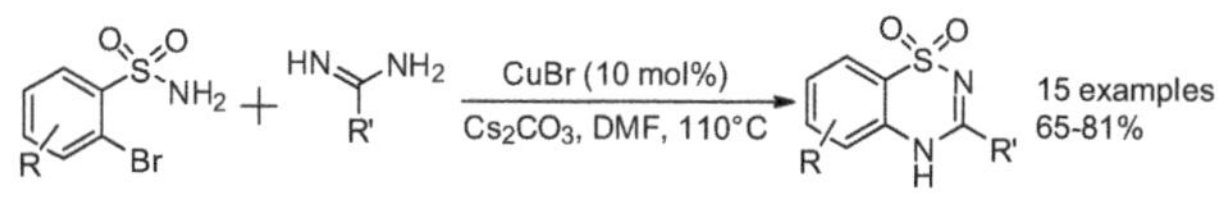

Scheme 3.75 Cu-catalyzed synthesis of benzothiadiazine 1,1-dioxides.

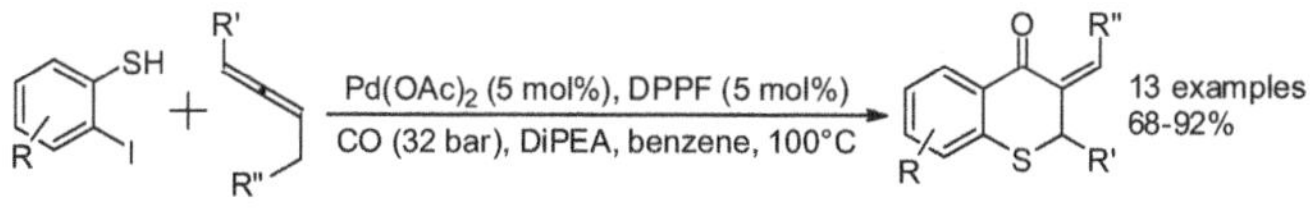

Scheme 3.76 Pd-catalyzed synthesis of thiochroman-4-ones.

Reaction Procedure (Scheme 3.76): A mixture of the allene (2–5 mmol), *o*-iodothiophenol (1 mmol), $Pd(OAc)_2$ (0.05 mmol), dppf (0.05 mmol), *i*-Pr_2NEt (1.5 mmol), and anhydrous benzene (5.0 mL) was heated to 100 °C with stirring, under an atmosphere of carbon monoxide (32 bar), in a stainless steel autoclave for 24–60 h. After the reaction, the autoclave was cooled to room temperature, excess carbon monoxide was released, and the crude reaction mixture was passed through a plug of Florisil, using hexane–ethyl acetate (1 : 1) as the eluent. The light yellow solution was evaporated to dryness and purified by column chromatography on silica gel using *n*-hexanes–ethyl acetate (95 : 5) as an eluent.

In 1999, Alper and Xiao developed a novel access to thiochroman-4-one derivatives by palladium-catalyzed carbonylative heteroannulation of *o*-iodothiophenols with allenes.[366] The reaction afforded the thiochroman-4-ones in good to excellent isolated yields with high regioselectivity (Scheme 3.76). The catalytic heteroannulation may involve regioselective addition of the sulfur moiety of the reactants on the more positive end of the allene, arylpalladium formation, CO insertion, subsequent intramolecular cyclization, and then reductive elimination.

Benzothiazine derivatives represent one of the most important classes of organic molecules, which have profound biological activities including antiinflammatory, antihypertensive, anticancer, antifungal, antitumor, immunostimulating, anti-aldoso-reductase, anti-rheumatic, anti-allergic, vasorelaxant, anti-arrhythmic, neuroprotective, cytotoxic, K_{ATP}channel openers, and anti-HIV activities. In 2009, Panda and Parai reported the synthesis 3,4-dihydro-2*H*-benzo[*b*][1,4]thiazine derivatives *via* a copper-catalyzed intramolecular *N*-aryl amination reaction.[367] Substituted 2-(2-bromophenylthio)-ethanamines, which were synthesized by the nucleophilic substitution reaction of 2-bromobenzenethiol with Boc-protected amino alcohol derivatives,[368] were applied as the starting materials and the desired products were formed in good yields (Scheme 3.77a). Sekar's group developed a method for the synthesis of benzothiazines from aziridines and *o*-halothiophenols.[369] With copper as the catalyst, the products were formed in

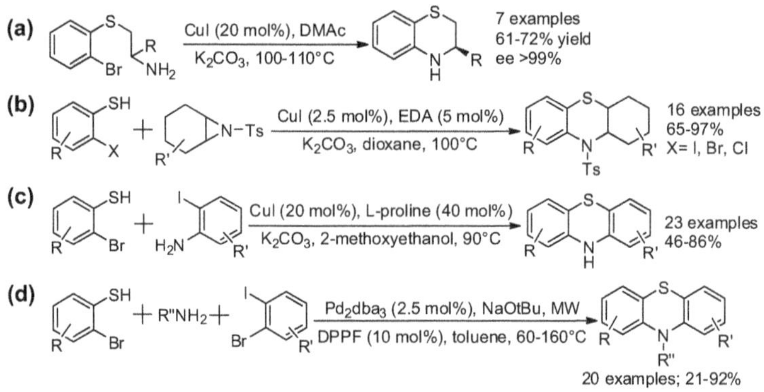

Scheme 3.77 Synthesis of phenothiazines and related derivatives.

good to excellent yields (Scheme 3.77b). Less reactive *o*-chlorothiophenols can also be successfully utilized for this transformation. Ma's group developed a new approach to construct functionalized phenothiazines.[370] Starting from substituted 2-iodoanilines and 2-bromobenzenethiols, and based on CuI/L-proline as the catalyst, the corresponding products were formed in good yields (Scheme 3.77c). Alternatively, a procedure relying on a palladium catalyst with 2-bromoiodobenzenes, 2-bromothiophenols and primary amines as the substrates, was also developed (Scheme 3.77d).[371]

In summary, the main achievements in the transition-metal-catalyzed synthesis of heterocycles from double-functionalized arenes have been summarized and discussed.

References

1. X. Jing, Y. Shi, Y. Liu, Y. Han, C. Yan and L. Wang, *Synth. Commun.*, 2004, **34**, 1723–1727.
2. B. Jiang and Z. G. Huang, *Tetrahedron Lett.*, 2007, **48**, 1703–1706.
3. S. Wang, K. Xie, Z. Tan, X. An, X. Zhou, C. C. Guo and Z. Peng, *Chem. Commun.*, 2009, 6469–6471.
4. T. Beresneva and E. Abele, *Chem. Heterocycl. Compd.*, 2012, **48**, 1420–1422.
5. C. Dai, X. Sun, X. Tu, L. Wu, D. Zhan and Q. Zeng, *Chem. Commun.*, 2012, **48**, 5367–5369.
6. A. Kamal, V. Srinivasulu, J. N. S. R. C. Murty, N. Shankaraiah, N. Nagesh, T. S. Reddy and A. V. S. Rao, *Adv. Synth. Catal.*, 2013, **355**, 2297–2307.
7. D. Kundu, T. Chatterjee and B. C. Ranu, *Adv. Synth. Catal.*, 2013, **355**, 2285–2296.
8. J. K. Laha, K. S. S. Tummalapalli and A. Gupta, *Eur. J. Org. Chem.*, 2013, 8330–8335.

9. (a) J. Barluenga, N. Quiñones, M. Cabal, F. Aznar and C. Valdés, *Angew. Chem., Int. Ed.*, 2011, **50**, 2350–2353; (b) L. Florentino, F. Aznar and C. Valdés, *Chem. - Eur. J.*, 2013, **19**, 10506–10510.
10. Z. Huang, Y. Yang, Q. Xiao, Y. Zhang and J. Wang, *Eur. J. Org. Chem.*, 2012, 6586–6593.
11. D. Takeda, K. Hirano, T. Satoh and M. Miura, *Heterocycles*, 2012, **86**, 487–496.
12. L. Yan, D. Zhao, J. Lan, Y. Cheng, Q. Guo, X. Li, N. Wu and J. You, *Org. Biomol. Chem.*, 2013, **11**, 7966–7977.
13. J. Liu, N. Zhang, Y. Yue, G. Liu, R. Liu, Y. Zhang and K. Zhuo, *Eur. J. Org. Chem.*, 2013, 7686–7687.
14. C. Chen, G. Shang, J. Zhou, Y. Yu, B. Li and J. Peng, *Org. Lett.*, 2014, **16**, 1872–1875.
15. I. Cerna, R. Pohl, B. Klepetárová and M. Hocek, *J. Org. Chem.*, 2010, **75**, 2302–2308.
16. K. Pericherla, A. Jha, B. Khungar and A. Kumar, *Org. Lett.*, 2013, **15**, 4304–4307.
17. V. N. Kalinin, M. V. Shostakovsky and A. B. Ponomaryov, *Tetrahedron Lett.*, 1990, **31**, 4073–4076.
18. B. Liang, M. Huang, Z. You, Z. Xiong, K. Lu, R. Fathi, J. Chen and Z. Yang, *J. Org. Chem.*, 2005, **70**, 6097–6100.
19. Q. Yang and H. Alper, *J. Org. Chem.*, 2010, **75**, 948–950.
20. E. Awuah and A. Capretta, *Org. Lett.*, 2009, **11**, 3210–3213.
21. H. Miao and Z. Yang, *Org. Lett.*, 2000, **2**, 1765–1768.
22. J. Liu, M. Liu, Y. Yue, N. Zhang, Y. Zhang and K. Zhuo, *Tetrahedron Lett.*, 2013, **54**, 1802–1807.
23. K. Okuro and H. Alper, *J. Org. Chem.*, 1997, **62**, 1566–1567.
24. R. Grigg, A. Liu, D. Shaw, S. Suganthan, D. E. Woodall and G. Yoganathan, *Tetrahedron Lett.*, 2000, **41**, 7125–7128.
25. R. Grigg, A. Liu, D. Shaw, S. Suganthan, M. L. Washington, D. E. Woodall and G. Yoganathan, *Tetrahedron Lett.*, 2000, **41**, 7129–7133.
26. (a) D. V. Kadnikov and R. C. Larock, *Org. Lett.*, 2000, **2**, 3643–3646; (b) D. V. Kadnikov and R. C. Larock, *J. Org. Chem.*, 2003, **68**, 9423–9432; (c) D. V. Kadnikov and R. C. Larock, *J. Organomet. Chem.*, 2003, **687**, 425–435.
27. H. Cao and W. J. Xiao, *Can. J. Chem.*, 2005, **83**, 826–831.
28. J. E. Rixson, B. W. Skelton, G. A. Koutsantonis, K. M. Gericke and S. G. Stewart, *Org. Lett.*, 2013, **15**, 4834–4837.
29. K. H. Park, I. G. Jung and Y. K. Chung, *Synlett*, 2004, 2541–2544.
30. Z. An, M. Catellani and G. P. Chiusoli, *J. Organomet. Chem.*, 1989, **371**, C51–C52.
31. M. Catellani and A. Del Rio, *Russ. Chem. Bull.*, 1998, **47**, 928–931.
32. C. Moinet and J. C. Fiaud, *Synlett*, 1997, 97–99.
33. M. Catellani, G. P. Chiusoli, M. C. Fagnola and G. Solari, *Tetrahedron Lett.*, 1994, **35**, 5919–5922.

34. M. Catellani, G. P. Chiusoli, G. Marzolini and E. Rossi, *J. Organomet. Chem.*, 1996, **525**, 65–69.
35. T. de A. Fernandes, R. de C. C. Carvalho, T. M. D. Goncalves, A. J. M. da Silva and P. R. R. Costa, *Tetrahedron Lett.*, 2008, **49**, 3322–3325.
36. T. de A. Fernandes, B. G. Vaz, M. N. Eberlin, A. J. M. da Silva and P. R. R. Costa, *J. Org. Chem.*, 2010, **75**, 7085–7091.
37. E. Feng, H. Huang, Y. Zhou, D. Ye, H. Jiang and H. Liu, *J. Org. Chem.*, 2009, **74**, 2846–2849.
38. D. Chen, G. Shen and W. Bao, *Org. Biomol. Chem.*, 2009, **7**, 4067–4073.
39. G. Feng, S. Wang, W. Li, F. Chen and C. Qi, *Synthesis*, 2013, 2711–2718.
40. R. Grigg, I. Köppen, M. Rasparini and V. Sridharan, *Chem. Commun.*, 2001, 964–965.
41. V. A. Vaillard, R. A. Rossi and S. E. Martin, *Org. Biomol. Chem.*, 2011, **9**, 4927–4935.
42. C. Larksarp and H. Alper, *J. Org. Chem.*, 1999, **64**, 9194–9200.
43. W. Bao, Y. Liu, X. Lv and W. Qian, *Org. Lett.*, 2008, **10**, 3899–3902.
44. Y. Liu and W. Bao, *Org. Biomol. Chem.*, 2010, **8**, 2700–2703.
45. A. R. Das, G. Pal, P. Bhattacharyya, A. K. Ghosh, D. Mukherjee and D. Bandyopadhyay, *Tetrahedron Lett.*, 2012, **53**, 7060–7066.
46. M. N. Reddy and K. C. K. Swamy, *Org. Biomol. Chem.*, 2013, **11**, 7350–7360.
47. (a) K. E. Torraca, S. I. Kuwabe and S. L. Buchwald, *J. Am. Chem. Soc.*, 2000, **122**, 12907–12908; (b) S. Kuwabe, K. E. Torraca and S. L. Buchwald, *J. Am. Chem. Soc.*, 2001, **123**, 12202–12206.
48. K. A. Tony, S. K. Chittimalla, G. A. Rajkumar and A. Chakrabarti, *Tetrahedron Lett.*, 2010, **51**, 4350–4353.
49. A. B. Naidu, D. Ganapathy and G. Sekar, *Synthesis*, 2010, 3509–3519.
50. J. F. Bower, P. Szeto and T. Gallagher, *Org. Lett.*, 2007, **9**, 3283–3286.
51. F. Melkonyan, A. Topolyan, A. Karchava and M. Yurovskaya, *Tetrahedron*, 2011, **67**, 6826–6832.
52. P. Thansandote, E. Chong, K. O. Feldmann and M. Lautens, *J. Org. Chem.*, 2010, **75**, 3495–3498.
53. C. Huang and V. Gevorgyan, *J. Am. Chem. Soc.*, 2009, **131**, 10844–10845.
54. P. C. Amos and D. A. Whiting, *J. Chem. Soc., Chem. Commun.*, 1987, 510–511.
55. R. C. Larock, N. G. Berrios-Peña, C. A. Fried, E. K. Yum, C. Tu and W. Leong, *J. Org. Chem.*, 1993, **58**, 4509–4510.
56. R. C. Larock and X. Han, *J. Org. Chem.*, 1999, **64**, 1875–1887.
57. Q. Huang, A. Fazio, G. Dai, M. A. Campo and R. C. Larock, *J. Am. Chem. Soc.*, 2004, **126**, 7460–7461.
58. A. Martins, U. Marquardt, N. Kasravi, D. Alberico and M. Lautens, *J. Org. Chem.*, 2006, **71**, 4937–4942.
59. V. Hornillos, A. W. van Zijl and B. L. Feringa, *Chem. Commun.*, 2012, **48**, 3712–3714.

60. O. Barberan, M. Alami and J. D. Brion, *Tetrahedron Lett.*, 2001, **42**, 2657–2659.
61. D. A. Petrone, H. A. Mlik, A. Clemenceau and M. Lautens, *Org. Lett.*, 2012, **14**, 4806–4809.
62. Y. Feng, Y. Wang, B. Landgraf, S. Liu and G. Chen, *Org. Lett.*, 2010, **12**, 3414–3417.
63. H. Kelgtermans, L. Dobrzanska, L. V. Meervelt and W. Dehaen, *Org. Lett.*, 2012, **14**, 5200–5203.
64. M. Prashad, Y. Liu, X. Y. Mak, D. Har, O. Repic and T. J. Blacklock, *Tetrahedron Lett.*, 2002, **43**, 8559–8562.
65. R. Jeyachandran, H. K. Potukuchi and L. Ackermann, *Beilstein J. Org. Chem.*, 2012, **8**, 1771–1777.
66. M. N. Reddy and K. C. K. Swamy, *Eur. J. Org. Chem.*, 2012, 2013–2022.
67. W. Liu, H. Cao, J. Xin, L. Jin and A. Lei, *Chem. – Eur. J.*, 2011, **17**, 3588–3592.
68. R. J. Li, S. F. Pi, Y. Liang, Z. Q. Wang, R. J. Song, G. X. Chen and J. H. Li, *Chem. Commun.*, 2010, **46**, 8183–8185.
69. L. F. Tietze, T. Hungerland, A. Düfert, I. Objartel and D. Stalke, *Chem. – Eur. J.*, 2012, **18**, 3286–3291.
70. E. Motti, F. Faccini, I. Ferrari, M. Catellani and R. Ferraccioli, *Org. Lett.*, 2006, **8**, 3967–3970.
71. E. Motti, N. D. Ca, R. Ferraccioli and M. Catellani, *Synthesis*, 2008, 995–997.
72. R. C. Larock and S. Babu, *Tetrahedron Lett.*, 1987, **28**, 5291–5294.
73. R. C. Larock and M. Y. Kuo, *Tetrahedron Lett.*, 1991, **32**, 56–572.
74. M. T. Stone, *Org. Lett.*, 2011, **13**, 2326–2329.
75. J. S. Mahanty, M. De, P. Das and N. G. Kundu, *Tetrahedron*, 1997, **53**, 13397–13418.
76. C. S. Cho, *J. Organomet. Chem.*, 2005, **690**, 4094–4097.
77. O. G. Schramm, T. Oeser, M. Kaiser, R. Brun and T. J. J. Müller, *Synlett*, 2008, 359–362.
78. R. Cinar, J. Nordmann, E. Dirksen and T. J. J. Müller, *Org. Biomol. Chem.*, 2013, **11**, 2597–2604.
79. A. Nikolaev, N. Nithiy and A. Orellana, *Synlett*, 2014, 2301–2305.
80. Z. Chen, J. Zhu, H. Xie, S. Li, Y. Wu and Y. Gong, *Chem. Commun.*, 2010, **46**, 2145–2147.
81. M. G. Shaibakova, G. A. Makhmudiyarov, L. M. Khalilov, E. A. Paramonov, A. G. Ibragimov and U. M. Dzhemilev, *Russ. J. Org. Chem.*, 2010, **46**, 428–432.
82. A. Deagostino, V. Farina, C. Prandi, C. Zavattaro and P. Venturello, *Eur. J. Org. Chem.*, 2006, 3451–3456.
83. R. P. Korivi and C. H. Cheng, *J. Org. Chem.*, 2006, **71**, 7079–7082.
84. J. Malm, P. Björk, S. Gronowitz and A. B. Hörnfeldt, *Tetrahedron Lett.*, 1994, **35**, 3195–3196.
85. M. G. Banwell, D. W. Lupton, X. Ma, J. Renner and M. O. Sydnes, *Org. Lett.*, 2004, **6**, 2741–2744.

86. S. Some, J. K. Ray, M. G. Banwell and M. T. Jones, *Tetrahedron Lett.*, 2007, **48**, 3609–3612.
87. M. J. Tomaszewski, A. Whalley and Y. J. Hu, *Tetrahedron Lett.*, 2008, **49**, 3172–3175.
88. K. Beydoun and H. Doucet, *Eur. J. Org. Chem.*, 2012, 6745–6751.
89. Z. Y. Gu, T. H. Zhu, J. J. Cao, X. P. Xu, S. Y. Wang and S. J. Ji, *ACS Catal.*, 2014, **4**, 49–52.
90. S. Gowrisankar, H. S. Lee, J. M. Kim and J. N. Kim, *Tetrahedron Lett.*, 2008, **49**, 1670–1673.
91. Z. Zheng and H. Alper, *Org. Lett.*, 2008, **10**, 829–832.
92. Y. Shen, C. Han, S. Cai, P. Lu and Y. Wang, *Tetrahedron Lett.*, 2012, **53**, 5671–5673.
93. K. Okuro and H. Alper, *Tetrahedron Lett.*, 2012, **53**, 2540–2542.
94. L. He, H. Li, H. Neumann, M. Beller and X. F. Wu, *Angew. Chem., Int. Ed.*, 2014, **53**, 1420–1424.
95. F. Zeng and H. Alper, *Org. Lett.*, 2010, **12**, 1188–1191.
96. F. Zeng and H. Apler, *Org. Lett.*, 2010, **12**, 3642–3644.
97. B. Roberts, D. Liptrot, T. Luker, M. J. Stocks, C. Barber, N. Webb, R. Dods and B. Martin, *Tetrahedron Lett.*, 2011, **52**, 3793–3796.
98. G. Qiu, G. Liu, S. Pu and J. Wu, *Chem. Commun.*, 2012, **48**, 2903–2905.
99. G. Qiu, Y. Lu and J. Wu, *Org. Biomol. Chem.*, 2013, **11**, 798–802.
100. C. Larsarp and H. Alper, *Org. Lett.*, 1999, **1**, 1619–1622.
101. L. Xue, L. Shi, Y. Han, C. Xia, H. V. Huynh and F. Li, *Dalton Trans.*, 2011, **40**, 7632–7638.
102. J. Salvadori, E. Balducci, S. Zaza, E. Petricci and M. Taddei, *J. Org. Chem.*, 2010, **75**, 1841–1847.
103. X. F. Wu, J. Schranck, H. Neumann and M. Beller, *Chem. - Eur. J.*, 2011, **17**, 12246–12249.
104. A. Arcadi, C. Asti, L. Brandolini, G. Caselli, F. Marinelli and V. Ruggieri, *Bioorg. Med. Chem. Lett.*, 1999, **9**, 1291–1294.
105. (a) X. F. Wu, M. Sharif, K. Shoaib, H. Neumann, A. Pews-Davtyan, P. Langer and M. Beller, *Chem. Eur. J.*, 2013, **19**, 6230–6233; (b) X. F. Wu, H. Neumann and M. Beller, *Chem. Eur. J.*, 2012, **18**, 12599–12602.
106. Q. Yuan and D. Ma, *J. Org. Chem.*, 2008, **73**, 5159–5162.
107. S. Tanimori, T. Nishimura and M. Kirihata, *Bioorg. Med. Chem. Lett.*, 2009, **19**, 4119–4121.
108. S. Tanimori, H. Kashiwagi, T. Nishimura and M. Kirihata, *Adv. Synth. Catal.*, 2010, **352**, 2531–2537.
109. W. S. Huang, R. Xu, R. Dodd and W. C. Shakespeare, *Tetrahedron Lett.*, 2013, **54**, 5214–5216.
110. X. Luo, E. Chenard, P. Martens, Y. X. Cheng and M. J. Tomaszewski, *Org. Lett.*, 2010, **12**, 3574–3577.
111. J. Yan, F. Zhou, D. Qin, T. Cai, K. Ding and Q. Cai, *Org. Lett.*, 2012, **14**, 1262–1265.

112. F. Zhou, J. Liu, K. Ding, J. Liu and Q. Cai, *J. Org. Chem.*, 2011, **76**, 5346–5353.
113. Q. Cai, J. Yan and K. Ding, *Org. Lett.*, 2012, **14**, 3332–3335.
114. J. H. Rigby and S. Laurent, *J. Org. Chem.*, 1998, **63**, 6742–6744.
115. D. E. Ames and A. Opalko, *Tetrahedron*, 1984, **40**, 1919–1925.
116. R. Bernini, S. Cacchi, G. Fabrizi and A. Sferrazza, *Synthesis*, 2008, 729–738.
117. Y. Yang, H. Huang, L. Wu and Y. Liang, *Org. Biomol. Chem.*, 2014, **12**, 5351–5355.
118. R. Grigg, H. Khalil, P. Levett, J. Virica and V. Sridharan, *Tetrahedron Lett.*, 1994, **35**, 3197–3200.
119. D. V. Kadnikov and R. C. Larock, *J. Org. Chem.*, 2004, **69**, 6772–6780.
120. C. S. Cho and J. U. Kim, *Tetrahedron Lett.*, 2007, **48**, 3775–3778.
121. S. R. Borhade and S. B. Waghmode, *Can. J. Chem.*, 2011, **89**, 1355–1363.
122. A. J. Ross, H. L. Lang and R. F. W. Jackson, *J. Org. Chem.*, 2010, **75**, 245–248.
123. T. Tsuritani, Y. Yamamoto, M. Kawasaki and T. Mase, *Org. Lett.*, 2009, **11**, 1043–1045.
124. H. Song, Y. Liu, Y. Liu and Q. Wang, *Org. Lett.*, 2014, **16**, 3240–3243.
125. L. E. Kaim, L. Grimaud and S. Wagschal, *Chem. Commun.*, 2011, **47**, 1887–1889.
126. J. Lu, X. Gong, H. Yang and H. Fu, *Chem. Commun.*, 2010, **46**, 4172–4174.
127. S. Torii, H. Okumoto and L. H. Xu, *Tetrahedron Lett.*, 1990, **31**, 7175–7178.
128. S. Torii, H. Okumoto and L. H. Xu, *Tetrahedron Lett.*, 1991, **32**, 237–240.
129. S. Torii, H. Okumoto, L. H. Xu, M. Sadakane, M. V. Shostakovsky, A. B. Ponomaryov and V. N. Kalimin, *Tetrahedron*, 1993, **49**, 6773–6784.
130. N. G. Kundu, J. S. Mahanty, P. Das and B. Das, *Tetrahedron Lett.*, 1993, **34**, 1625–1628.
131. J. A. Nieman and M. D. Ennis, *J. Org. Chem.*, 2001, **66**, 2175–2177.
132. F. Ye and H. Alper, *J. Org. Chem.*, 2007, **72**, 3218–3222.
133. C. Larksarp and H. Alper, *J. Org. Chem.*, 2000, **65**, 2773–2777.
134. A. Dos Santos, L. El Kaim, L. Grimaud and R. Ramozzi, *Synlett*, 2012, 438–442.
135. N. D. Cá, E. Motti and M. Catellani, *Adv. Synth. Catal.*, 2008, **350**, 2513–2516.
136. J. T. Reeves, D. R. Fandrick, Z. Tan, J. J. Song, H. Lee, N. K. Yee and C. H. Senanayake, *J. Org. Chem.*, 2010, **75**, 992–994.
137. S. Biswas and S. Batra, *Eur. J. Org. Chem.*, 2013, 4895–4902.
138. A. Huang, Y. Chen, Y. Zhou, W. Guo, X. Wu and C. Ma, *Org. Lett.*, 2013, **15**, 5480–5483.
139. L. Yu, X. Zhou, D. Wu and H. Xiang, *J. Organomet. Chem.*, 2012, **705**, 75–78.
140. H. Yuan, K. Li, Y. Chen, Y. Wang, J. Cui and B. Chen, *Synlett*, 2013, 2315–2319.

141. M. K. Ghorai, A. K. Sahoo and S. Kumar, *Org. Lett.*, 2011, **13**, 5972–5975.
142. Z. Liu and Y. Chen, *Tetrahedron Lett.*, 2009, **50**, 3790–3793.
143. H. Guo, J. Liu, X. Wang and G. Huang, *Synlett*, 2012, 903–906.
144. (a) M. Palucki, J. P. Wolfe and S. L. Buchwald, *J. Am. Chem. Soc.*, 1996, **118**, 10333–10334; (b) K. E. Torraca, S. I. Kuwabe and S. L. Buchwald, *J. Am. Chem. Soc.*, 2000, **122**, 12907–12908; (c) S. I. Kuwabe, K. E. Torraca and S. L. Buchwald, *J. Am. Chem. Soc.*, 2001, **123**, 12202–12206.
145. M. Palucki and N. Yasuda, *Tetrahedron Lett.*, 2005, **46**, 987–990.
146. Y. Wang and R. Franzén, *Synlett*, 2012, 925–929.
147. (a) B. Suchand, J. Krishna, B. V. Ramulu, D. Dibyendu, A. G. K. Reddy, L. Mahendar and G. Satyanarayana, *Tetrahedron Lett.*, 2012, **53**, 3861–3864; (b) B. V. Ramulu, L. Mahendar, J. Krishna, A. G. K. Reddy, B. Suchand and G. Satyanarayana, *Tetrahedron*, 2013, **69**, 8305–8315.
148. C. G. Saluste, S. Crumpler, M. Furber and R. J. Whitby, *Tetrahedron Lett.*, 2004, **45**, 6995–6996.
149. Y. Fang and C. Li, *J. Org. Chem.*, 2006, **71**, 6427–6431.
150. C. C. Malakar, D. Schmidt, J. Conrad and U. Beifuss, *Org. Lett.*, 2011, **13**, 1972–1975.
151. X. Y. Zhang, L. L. Fang, N. Liu, H. Y. Wu and X. S. Fan, *Chin. Chem. Lett.*, 2012, **23**, 1129–1132.
152. M. A. Ormar, J. Conrad and U. Beifuss, *Tetrahedron*, 2014, **70**, 5682–5695.
153. D. Rotili, D. Tarantino, V. Carafa, C. Paolini, J. Schemies, M. Jung, G. Botta, S. Di Maro, E. Novellino, C. Steinkühler, R. De Maria, P. Gallinari, L. Altucci and A. Mai, *J. Med. Chem.*, 2012, **55**, 8193–8197.
154. (a) Y. Luo, L. Hong and J. Wu, *Chem. Commun.*, 2011, **47**, 5298–5300; (b) X. Pan, H. Nie, Y. Luo, Y. Gao and J. Wu, *Org. Biomol. Chem.*, 2012, **10**, 8244–8250.
155. R. Grigg, V. Sridharan and C. Terrier, *Tetrahedron Lett.*, 1996, **37**, 4221–4224.
156. U. Anwar, R. Grigg, M. Rasparini, V. Savic and V. Sridharan, *Chem. Commun.*, 2000, 645–646.
157. P. H. Lee, D. Seomoon and K. Lee, *Org. Lett.*, 2005, **7**, 343–345.
158. F. Liron and P. Knochel, *Tetrahedron Lett.*, 2007, **48**, 4943–4946.
159. S. Braune, M. Pohlman and U. Kazmaier, *J. Org. Chem.*, 2004, **69**, 468–474.
160. Z. Lu, C. Hu, J. Guo, J. Li, Y. Cui and Y. Jia, *Org. Lett.*, 2010, **12**, 480–483.
161. K. Parthasarathy, M. Jeganmohan and C. H. Cheng, *Org. Lett.*, 2006, **8**, 621–623.
162. D. D. Hennings, S. Iwasa and V. H. Rawal, *J. Org. Chem.*, 1997, **62**, 2–3.
163. G. B. Bajracharya and O. Daugulis, *Org. Lett.*, 2008, **10**, 4625–4628.
164. (a) L. C. Campeau, M. Parisien, M. Leblanc and K. Fagnou, *J. Am. Chem. Soc.*, 2004, **126**, 9186–9187; (b) M. Parisien, D. Valette and K. Fagnou, *J. Org. Chem.*, 2005, **70**, 7578–7584; (c) L. C. Campeau, P. Thansandote and K. Fagnou, *Org. Lett.*, 2005, **7**, 1857–1860; (d) J. P. Leclerc, M. André

and K. Fagnou, *J. Org. Chem.*, 2006, **71**, 1711–1714; (e) L. C. Campeau, M. Parisien, A. Jean and K. Fagnou, *J. Am. Chem. Soc.*, 2006, **128**, 581–590; (f) M. Lafrance, D. Lapointe and K. Fagnou, *Tetrahedron*, 2008, **64**, 6015–6020.
165. E. D. Coy, L. E. Cuca and M. Sefkow, *Synth. Commun.*, 2011, **41**, 41–51.
166. T. Nishikata, H. Tsutsumi, L. Gao, K. Kojima, K. Chikama and H. Nagashima, *Adv. Synth. Catal.*, 2014, **356**, 951–960.
167. L. Zhang, M. Geng, P. Teng, D. Zhao, X. Lu and J. X. Li, *Ultrason. Sonochem.*, 2012, **19**, 250–256.
168. C. S. Nervig, P. J. Waller and D. Kalyani, *Org. Lett.*, 2012, **14**, 4838–4841.
169. K. C. Majumdar, A. K. Pal, A. Taher and P. Debnath, *Synthesis*, 2007, 1707–1711.
170. K. C. Majumdar, S. Chakravorty and N. De, *Tetrahedron Lett.*, 2008, **49**, 3419–3422.
171. K. C. Majumdar, A. Taher and P. Debnath, *Synthesis*, 2009, 793–800.
172. X. Xu, X. Xu, H. Li, X. Xie and Y. Li, *Org. Lett.*, 2010, **12**, 100–103.
173. E. Motti, N. D. Ca', D. Xu, A. Piersimoni, E. Bedogni, Z. M. Zhou and M. Catellani, *Org. Lett.*, 2012, **14**, 5792–5795.
174. (a) L. Mahendar, J. Krishna, A. G. K. Reddy, B. V. Ramulu and G. Satyanarayana, *Org. Lett.*, 2012, **14**, 628–631; (b) L. Mahendar and G. Satyanarayana, *J. Org. Chem.*, 2014, **79**, 2059–2074.
175. Z. Shen, Z. Ni, S. Mo, J. Wang and Y. Zhu, *Chem. - Eur. J.*, 2012, **18**, 4859–4865.
176. N. Barbero, R. SanMartin and E. Domínguez, *Green Chem.*, 2009, **11**, 830–836.
177. P. A. Turner, E. M. Griffin, J. L. Whatmore and M. Shipman, *Org. Lett.*, 2011, **13**, 1056–1059.
178. P. A. Turner, Samiunllah, J. L. Whatmore and M. Shipman, *Tetrahedron Lett.*, 2013, **54**, 6538–6540.
179. N. Marquise, P. J. Harford, F. Chevallier, T. Roisnel, V. Dorcet, A. L. Gagez, S. Sablé, L. Picot, V. Thiéry, A. E. H. Wheatley, P. C. Gros and F. Mongin, *Tetrahedron*, 2013, **69**, 10123–10133.
180. J. Hu, E. A. Adogla, Y. Ju, D. Fan and Q. Wang, *Chem. Commun.*, 2012, **48**, 11256–11258.
181. S. Cai, F. Wang and C. Xi, *J. Org. Chem.*, 2012, **77**, 2331–2336.
182. S. Cai, F. Wang and C. Xi, *Synthesis*, 2012, 1892–1897.
183. V. Kavala, C. C. Wang, D. K. Barange, C. W. Kuo, P. M. Lei and C. F. Yao, *J. Org. Chem.*, 2012, **77**, 5022–5029.
184. L. Liu, J. Hu, X. C. Wang, M. J. Zhong, X. Y. Liou, S. D. Yang and Y. M. Liang, *Tetrahedron*, 2012, **68**, 5391–5395.
185. Z. Y. Ge, X. D. Fei, T. Tang, Y. M. Zhu and J. K. Shen, *J. Org. Chem.*, 2012, **77**, 5736–5743.
186. M. A. Weidner-Wells, J. Altom, J. Fernandez, S. A. Fraga-Spano, J. Hilliard, K. Obemeng and J. F. Barrett, *Bioorg. Med. Chem. Lett.*, 1998, **8**, 97–100.

187. J. V. Suárez-Meneses, E. Bonilla-Reyes, E. A. Blé-González, M. C. Ortega-Alfaro, A. Vordero-Vargas and J. G. López-Cortés, *Tetrahedron*, 2014, **70**, 1422–1430.
188. G. Qabaja and G. B. Jones, *J. Org. Chem.*, 2000, **65**, 7187–7194.
189. J. Pandey, A. K. Jha and K. Hajela, *Bioorg. Med. Chem.*, 2004, **12**, 2239–2249.
190. M. Krzeszewski, O. Vakuliuk and D. T. Gryko, *Eur. J. Org. Chem.*, 2013, 5631–5644.
191. Y. Chen, Y. Lan, S. Wang, H. Zhang, X. Xu, X. Liu, M. Yu, B. F. Liu and G. Zhang, *Eur. J. Med. Chem.*, 2014, **74**, 427–439.
192. R. Singha, S. Roy, S. Nandi, P. Ray and J. K. Ray, *Tetrahedron Lett.*, 2013, **54**, 657–660.
193. K. Yang, J. Zhang, Y. Li, B. Cheng, L. Zhao and H. Zhai, *Org. Lett.*, 2013, **15**, 808–811.
194. D. Chernyak and V. Gevorgyan, *Org. Lett.*, 2010, **12**, 5558–5560.
195. D. K. Rayabarapu, P. Shukla and C. H. Cheng, *Org. Lett.*, 2003, **5**, 4903–4906.
196. T. Yao, D. Yue and r. C. Larock, *J. Org. Chem.*, 2005, **70**, 9985–9989.
197. A. C. Tadd, M. R. Fielding and M. C. Willis, *Chem. Commun.*, 2009, 6744–6746.
198. X. D. Fei, Z. Y. Ge, T. Tang, Y. M. Zhu and S. J. Ji, *J. Org. Chem.*, 2012, **77**, 10321–10328.
199. X. D. Fei, T. Tang, Z. Y. Ge and Y. M. Zhu, *Synth. Commun.*, 2013, **43**, 3262–3271.
200. K. Okuro, M. Furuune, M. Enna, M. Miura and M. Nomura, *J. Org. Chem.*, 1993, **58**, 4716–4721.
201. S. Inack-Ngi, R. Rahmani, L. Commeiras, G. Chouraqui, J. Thibonnet, A. Duchene, M. Abarbri and J. L. Parrain, *Adv. Synth. Catal.*, 2009, **351**, 779–788.
202. M. R. Kumar, F. M. Irudayanathan, J. H. Moon and S. Lee, *Adv. Synth. Catal.*, 2013, **355**, 3221–3230.
203. N. G. Kundu and M. Pal, *J. Chem. Soc. Chem. Commun.*, 1993, 86–88.
204. H. Y. Liao and C. H. Cheng, *J. Org. Chem.*, 1995, **60**, 3711–3716.
205. V. Subramanian, V. R. Batchu, D. Barange and M. Pal, *J. Org. Chem.*, 2005, **70**, 4778–4783.
206. R. C. Larock, M. J. Doty and X. Han, *J. Org. Chem.*, 1999, **64**, 8770–8779.
207. J. M. Zenner and R. C. Larock, *J. Org. Chem.*, 1999, **64**, 7312–7322.
208. M. Chakravarty and K. C. K. Swamy, *J. Org. Chem.*, 2006, **71**, 9128–9138.
209. K. Cherry, J. L. Parrain, J. Thibonnet, A. Duchene and M. Abarbri, *J. Org. Chem.*, 2005, **70**, 6669–6675.
210. R. C. Larock, S. Varaprath, H. H. Lau and C. A. Fellows, *J. Am. Chem. Soc.*, 1984, **106**, 5274–5284.
211. R. Joyeau, Y. Dugenet and M. Wakselman, *J. Chem. Soc. Chem. Commun.*, 1983, 431–432.
212. C. Hang, Q. Li, Y. Zhu and H. Katayama, *Synth. Commun.*, 2011, **41**, 3318–3324.

213. W. A. Nack, G. He, S. Y. Zhang, C. Lu and G. Chen, *Org. Lett.*, 2013, **15**, 3440–3443.
214. Q. Niu, H. Mao, G. Yuan, J. Gao, H. Liu, Y. Tu, X. Wang and X. Lv, *Adv. Synth. Catal.*, 2013, **355**, 1185–1192.
215. M. Lautens, E. Tayama and C. Herse, *J. Am. Chem. Soc.*, 2005, **127**, 72–73.
216. D. Solé, L. Vallverdú, X. Solans, M. Font-Bardia and J. Bonjoch, *J. Am. Chem. Soc.*, 2003, **125**, 1587–1594.
217. B. X. Tang, Y. H. Zhang, R. J. Song, D. J. Tang, G. B. Deng, Z. Q. Wang, Y. X. Xie, Y. Z. Xia and J. H. Li, *J. Org. Chem.*, 2012, **77**, 2837–2849.
218. T. Harayama, T. Sato, A. Hori, H. Abe and Y. Takeuchi, *Synlett*, 2003, 1141–1144.
219. E. Yamuna, M. Zeller and K. J. R. Prasad, *Tetrahedron Lett.*, 2011, **52**, 6030–6034.
220. S. De, S. Mishra, B. N. Kakde, D. Dey and A. Bisai, *J. Org. Chem.*, 2013, **78**, 7823–7844.
221. R. Grigg, V. Sridharan and A. Thayaparan, *Tetrahedron Lett.*, 2003, **44**, 9017–9019.
222. D. L. Priebbenow, S. G. Stewart and F. M. Pfeffer, *Org. Biomol. Chem.*, 2011, **9**, 1508–1515.
223. A. G. K. Reddy and G. Satyanarayana, *Tetrahedron*, 2012, **68**, 8003–8010.
224. E. M. Beccalli, G. Broggini, M. Martinelli, N. Masciocchi and S. Sottocornola, *Org. Lett.*, 2006, **8**, 4521–4524.
225. A. Nandakumar, D. Muralidharan and P. T. Perumal, *Tetrahedron Lett.*, 2011, **52**, 1644–1648.
226. A. Nandakumar, S. E. Kiruthika, K. Naveen and P. T. Perumal, *Org. Biomol. Chem.*, 2014, **12**, 876–880.
227. Y. Tokimizu, Y. Ohta, H. Chiba, S. Oishi, N. Fujii and H. Ohno, *Tetrahedon*, 2011, **67**, 5168–5175.
228. V. Fiandanese, G. Marchese, A. Punzi, F. Iannone and G. G. Rafaschieri, *Tetrahedron*, 2010, **66**, 8846–8853.
229. V. Fiandanese, S. Maurantonio, A. Punzi and G. G. Rafaschieri, *Org. Biomol. Chem.*, 2012, **10**, 1186–1195.
230. K. Brahma, B. Achari and C. Chowdhury, *Synthesis*, 2013, 545–555.
231. Q. Liao, L. Zhang, S. Li and C. Xi, *Org. Lett.*, 2011, **13**, 228–231.
232. G. Dai and R. C. Larock, *J. Org. Chem.*, 2003, **68**, 920–928.
233. M. Dell'Acqua, G. Abbiati and E. Rossi, *Synlett*, 2010, 2672–2676.
234. D. Yang, S. Burugupalli, D. Daniel and Y. Chen, *J. Org. Chem.*, 2012, **77**, 4466–4472.
235. F. Yang, J. Zhang and Y. Wu, *Tetrahedron*, 2011, **67**, 2969–2973.
236. L. Lin, Q. Wu, S. Huang and G. Yang, *Chin. J. Chem.*, 2012, **30**, 1075–1082.
237. A. K. Mandadapu, M. D. Dathi, R. K. Arigela and B. Kundu, *Tetrahedron*, 2012, **68**, 8207–8215.
238. W. J. Ang, C. H. Tai, L. C. lo and Y. Lam, *RSC Adv.*, 2014, **4**, 4921–4929.
239. C. C. Liu, R. P. Korivi and C. H. Cheng, *Chem. - Eur. J.*, 2008, **14**, 9503–9506.

240. (a) R. P. Korivi, Y. C. Wu and C. H. Cheng, *Chem. Eur. J.*, 2009, **15**, 10727–10731; (b) W. C. Shih, C. C. Teng, K. Parthasarathy and C. H. Cheng, *Chem. Asian J.*, 2012, **7**, 306–313.
241. S. Dhara, R. Singha, Y. Nuree and J. K. Ray, *Tetrahedron Lett.*, 2014, **55**, 795–798.
242. B. Wang, B. Lu, Y. Jiang, Y. Zhang and D. Ma, *Org. Lett.*, 2008, **10**, 2761–2763.
243. A. K. Verma, T. Kesharwani, J. Singh, V. Tandon and R. C. Larock, *Angew. Chem., Int. Ed.*, 2009, **48**, 1138–1143.
244. A. K. Verma, R. R. Jha, R. Chaudhary, R. K. Tiwari, K. S. K. Reddy and A. Danodia, *J. Org. Chem.*, 2012, **77**, 8191–8205.
245. R. R. Jha, A. K. Danodia and A. K. Verma, *Tetrahedron Lett.*, 2014, **55**, 4724–4730.
246. X. Pan, Y. Luo and J. Wu, *J. Org. Chem.*, 2013, **78**, 5756–5760.
247. H. Zhou, H. Jin, S. Ye, X. He and J. Wu, *Tetrahedron Lett.*, 2009, **50**, 4616–4618.
248. N. Okamoto, K. Sakurai, M. Ishikura, K. Takeda and R. Yanada, *Tetrahedron Lett.*, 2009, **50**, 4167–4169.
249. S. Dhiman, K. Pericherla, N. K. Nandwana, D. Kumar and A. Kumar, *J. Org. Chem.*, 2014, **79**, 7399–7404.
250. G. Pandey, S. Bhowmik and S. Batra, *Org. Lett.*, 2013, **15**, 5044–5047.
251. S. Hesse and G. Kirsch, *Tetrahedron*, 2005, **61**, 6534–6539.
252. G. Maestri, M. H. Larraufie, É. Derat, C. Ollivier, L. Fensterbank, E. Lacote and M. Malacria, *Org. Lett.*, 2010, **12**, 5692–5695.
253. Q. Cai, Z. Li, J. Wie, L. Fu, C. Ha, D. Pei and K. Ding, *Org. Lett.*, 2010, **12**, 1500–1503.
254. Y. Luo, X. Pan and J. Wu, *Org. Lett.*, 2011, **13**, 1150–1153.
255. X. Pan, Y. Luo and J. Wu, *Chem. Commun.*, 2011, **47**, 8967–8969.
256. X. Pan, Y. Luo, G. Liu, S. Pu and J. Wu, *Adv. Synth. Catal.*, 2012, **354**, 171–178.
257. X. Pan, Y. Luo and J. Wu, *Org. Biomol. Chem.*, 2012, **10**, 1969–1975.
258. M. Blanchot, D. A. Candito, F. Larnaud and M. Lautens, *Org. Lett.*, 2011, **13**, 1486–1489.
259. B. Liu, H. Gao, Y. Yu, W. Wu and H. Jiang, *J. Org. Chem.*, 2013, **78**, 10319–10328.
260. S. Dhara, M. Ghosh and J. K. Ray, *Synlett*, 2013, 2263–2265.
261. X. Yang, H. Liu, H. Fu, R. Qiao, Y. Jiang and Y. Zhao, *Synlett*, 2010, 101–106.
262. V. L. Truong and M. Morrow, *Tetrahedron Lett.*, 2010, **51**, 758–760.
263. A. V. Vypolzov, D. V. Dar'in, S. G. Ryazanov and P. S. Lobanov, *Chem. Heterocycl. Compd.*, 2011, **46**, 1481–1485.
264. J. Ju, R. Hua and J. Su, *Tetrahedron*, 2012, **68**, 9364–9370.
265. Q. Liu, Y. Zhao, H. Fu and C. Cheng, *Synlett*, 2013, 2089–2094.
266. X. Fan, B. Li, S. Guo, Y. Wang and X. Zhang, *Chem. – Asian J.*, 2014, **9**, 739–743.

267. C. J. Ball, J. Gilmore and M. C. Willis, *Angew. Chem., Int. Ed.*, 2012, **51**, 5718–5722.
268. N. Aljaar, J. Conrad and U. Beifuss, *J. Org. Chem.*, 2013, **78**, 1045–1053.
269. L. Gao, Y. Song, X. Zhang, S. Guo and X. Fan, *Tetrahedron Lett.*, 2014, **55**, 4997–5002.
270. C. li, W. T. Zhang and X. S. Wang, *J. Org. Chem.*, 2014, **79**, 5847–5851.
271. M. C. Willis, R. H. Snell, A. J. Fletcher and R. L. Woodward, *Org. Lett.*, 2006, **8**, 5089–5091.
272. C. S. Guy and T. C. Jones, *Synlett*, 2009, 2253–2256.
273. Z. J. Wang, J. G. Yang, F. Yang and W. Bao, *Org. Lett.*, 2010, **12**, 3034–3037.
274. S. Gupta, D. Koley, K. Ravikumar and B. Kundu, *J. Org. Chem.*, 2013, **78**, 8624–8633.
275. F. Wang, P. Zhao and C. Xi, *Tetrahedron Lett.*, 2011, **52**, 231–235.
276. T. Ma, W. Chen, G. Zhang and Y. Yu, *J. Comb. Chem.*, 2010, **12**, 488–490.
277. H. Suzuki and H. Abe, *J. Org. Chem.*, 1994, **59**, 6116–6118.
278. P. J. Manley and M. T. Bilodeau, *Org. Lett.*, 2004, **6**, 2433–2435.
279. L. Fu, X. Huang, D. Wang, P. Zhao and K. Ding, *Synthesis*, 2011, 1547–1554.
280. X. Zhang, X. Guo, L. Fang, Y. Song and X. Fan, *Eur. J. Org. Chem.*, 2013, 8087–8093.
281. A. C. Tadd, A. Matsuno, M. R. Fielding and M. C. Willis, *Org. Lett.*, 2009, **11**, 583–586.
282. K. Takenaka, N. Itoh and H. Sasai, *Org. Lett.*, 2009, **11**, 1483–1486.
283. T. Piou, L. Neuville and J. Zhu, *Org. Lett.*, 2012, **14**, 3760–3763.
284. R. Bernini, S. Cacchi, G. Fabrizi and A. Sferrazza, *Synthesis*, 2009, 1209–1219.
285. T. Zhao and B. Xu, *Org. Lett.*, 2010, **12**, 212–215.
286. R. F. Pellón, A. Martín, M. L. Docampo and M. Mesa, *Synth. Commun.*, 2006, **36**, 1715–1719.
287. R. F. Pellón, M. L. Docampo and M. L. Fascio, *Synth. Commun.*, 2007, **37**, 1853–1864.
288. C. Huang, Y. Fu, H. Fu, Y. Jiang and Y. Zhao, *Chem. Commun.*, 2008, 6333–6335.
289. X. Liu, H. Fu, Y. Jiang and Y. Zhao, *Angew. Chem., Int. Ed.*, 2009, **48**, 348–351.
290. X. Huang, H. Yang, H. Fu, R. Qiao and Y. Zhao, *Synthesis*, 2009, 2679–2688.
291. X. Zhang, D. Ye, H. Sun, D. Guo, J. Wang, H. Huang, X. Zhang, H. Jiang and H. Liu, *Green Chem.*, 2009, **11**, 1881–1888.
292. L. Yu, M. Wang, P. Li and L. Wang, *Appl. Organomet. Chem.*, 2012, **26**, 576–582.
293. M. U. Hung, B. S. Liao, Y. H. Liu, S. M. Peng and S. T. Liu, *Appl. Organomet. Chem.*, 2014, **28**, 661–665.
294. J. Zhou, L. Fu, M. Lv, J. Liu, D. Pei and K. Ding, *Synthesis*, 2008, 3974–3980.

295. L. Xu, Y. Jiang and D. Ma, *Org. Lett.*, 2012, **14**, 1150–1153.
296. D. Chen, Q. Chen, M. Liu, S. Dai, L. Huang, J. Yang and W. Bao, *Tetrahedron*, 2013, **69**, 6461–6467.
297. W. Xu and H. Fu, *J. Org. Chem.*, 2011, **76**, 3846–3852.
298. W. Xu, Y. Jin, H. Liu, Y. Jiang and H. Fu, *Org. Lett.*, 2011, **13**, 1274–1277.
299. H. Tian, H. Qiao, C. Zhu and H. Fu, *RSC Adv.*, 2014, **4**, 2694–2704.
300. T. Songsichan, J. Promsuk, V. Rukachaisirikul and J. Kaeobamrung, *Org. Biomol. Chem.*, 2014, **12**, 4571–4575.
301. D. S. Chen, G. L. Dou, Y. L. Li, Y. Liu and X. S. Wang, *J. Org. Chem.*, 2013, **78**, 5700–5704.
302. L. Chen, H. Fu and R. Qiao, *Synlett*, 2011, 1930–1936.
303. A. Maity, S. Mondal, R. Paira, A. Hazra, S. Naskar, K. B. Sahu, P. Saha, S. Banerjee and N. B. Mondal, *Tetrahedron Lett.*, 2011, **52**, 3033–3037.
304. Y. Ju, F. Liu and C. Li, *Org. Lett.*, 2009, **11**, 3582–3585.
305. A. Begouin, S. Hesse, M. J. R. P. Queiroz and G. Kirsch, *Synthesis*, 2006, 2794–2798.
306. D. Marosvölgyi-Haskó, A. Petz, A. Takács and L. Kollár, *Tetrahedron*, 2011, **67**, 9122–9128.
307. X. F. Wu, H. Neumann, S. Neumann and M. Beller, *Chem. – Eur. J.*, 2012, **18**, 8596–8599.
308. A. S. Suresh, P. Baburajan and M. Ahmed, *Tetrahedron Lett.*, 2014, **55**, 3482–3485.
309. K. P. Rao, A. K. Basak, P. K. Deb, S. Sharma and L. K. Reddy, *Tetrahedron Lett.*, 2013, **54**, 3694–3696.
310. T. Vlaar, E. Ruijter, A. Znabet, E. Janssen, F. J. J. de Kanter, B. U. W. Maes and R. V. A. Orru, *Org. Lett.*, 2011, **13**, 6496–6499.
311. T. Vlaar, P. Mampuys, M. Helliwell, B. U. W. Maes, R. V. A. Orru and E. Ruijter, *J. Org. Chem.*, 2013, **78**, 6735–6745.
312. T. Harayama, T. Sato, A. Hori, H. Abe and Y. Takeuchi, *Synthesis*, 2004, 1446–1456.
313. S. Goswami, A. K. Adak, R. Mukherjee, S. Jana, S. Dey and J. F. Gallagher, *Tetrahedron*, 2005, **61**, 4289–4295.
314. K. C. Majumdar, B. Chattopadhyay and S. Nath, *Tetrahedron Lett.*, 2008, **49**, 1609–1612.
315. G. Zhang, X. Zhao, Y. Yan and C. Ding, *Eur. J. Org. Chem.*, 2012, 669–672.
316. M. Nakamura, A. Aoyama, M. T. A. Salim, M. Okamoto, M. Baba, H. Miyachi, Y. Hashimoto and H. Aoyama, *Bioorg. Med. Chem.*, 2010, **18**, 2402–2411.
317. B. Gómez-Lor and A. M. Echavarren, *Org. Lett.*, 2004, **6**, 2993–2996.
318. H. Aoyama, K. Sugita, M. Nakamura, A. Aoyama, M. T. A. Salim, M. Okamoto, M. Baba and Y. Hashimoto, *Bioorg. Med. Chem.*, 2011, **19**, 2675–2687.
319. W. Cong, L. Zhao, X. Wu, J. Xu and H. Yao, *Tetrahedron*, 2014, **70**, 312–317.

320. R. Ferraccioli, D. Carenzi, O. Rombola and M. Catellani, *Org. Lett.*, 2004, **6**, 4759–4762.
321. R. Ferraccioli, D. Carenzi, E. Motti and M. Catellani, *J. Am. Chem. Soc.*, 2006, **128**, 722–723.
322. T. Furuta, Y. Kitamura, A. Hashimoto, S. Fujii, K. Tanaka and T. Kan, *Org. Lett.*, 2007, **9**, 183–186.
323. (a) L. Donati, S. Michel, F. Tillequin and F. H. Porée, *Org. Lett.*, 2010, **12**, 156–158; (b) L. Donati, P. Leproux, E. Prost, S. Michel, F. Tillequin, V. Gandon and F. H. Porée, *Chem. Eur. J.*, 2011, **17**, 12809–12819.
324. J. J. S. Lamna and J. M. Tour, *J. Am. Chem. Soc.*, 1994, **116**, 11723–11736.
325. C. Guo, K. Huang, B. Wang, L. Xie and X. Xu, *RSC Adv.*, 2013, **3**, 17271–17280.
326. S. K. Chattopadhyay, S. Maity, B. K. Pal and S. Panja, *Tetrahedron Lett.*, 2002, **43**, 5079–5081.
327. L. Stevenson, S. L. Pimlott and A. Sutherland, *Tetrahedron Lett.*, 2007, **48**, 7137–7139.
328. R. Grigg, I. Köppen, M. Rasparini and V. Sridharan, *Chem. Commun.*, 2001, 964–965.
329. K. Cherry, A. Duchene, J. Thibonnet, J. L. Parrain and M. Abarbri, *Synthesis*, 2005, 2349–2356.
330. Z. Xiang, T. Luo, K. Lu, J. Cui, X. Shi, R. Fathi, J. Chen and Z. Yang, *Org. Lett.*, 2004, **6**, 3155–3158.
331. K. C. Majumdar, S. Chakravorty and K. Ray, *Synthesis*, 2008, 2991–2995.
332. G. A. Ardizzoia, E. M. Beccalli, E. Borsini, S. Brenna, G. Broggini and M. Rigamonti, *Eur. J. Org. Chem.*, 2008, 5590–5596.
333. T. T. Jayanth, L. Zhang, T. S. Johnson and H. C. Malinakova, *Org. Lett.*, 2009, **11**, 815–818.
334. D. A. Petrone, H. Yoon, H. Weinstabl and M. Lautens, *Angew. Chem., Int. Ed.*, 2014, **53**, 7908–7912.
335. P. Thansandote, D. G. Hulcoop, M. Langer and M. Lautens, *J. Org. Chem.*, 2009, **74**, 1673–1678.
336. C. C. Liu, K. Parthasarathy and C. H. Cheng, *Org. Lett.*, 2010, **12**, 3518–3521.
337. R. G. Chary, G. Dhananjaya, K. V. Prasad, S. Vaishaly, Y. S. S. Ganesh, B. Dulla, K. S. Kumar and M. Pal, *Chem. Commun.*, 2014, **50**, 6797–6800.
338. V. R. Batchu, D. K. Barange, D. Kumar, B. R. Sreekanth, K. Vyas, E. A. Reddy and M. Pal, *Chem. Commun.*, 2007, 1966–1968.
339. F. Wang, H. Liu, H. Fu, Y. Jiang and Y. Zhao, *Org. Lett.*, 2009, **11**, 2469–2472.
340. Y. Shi, X. Zhu, H. Mao, H. Hu, C. Zhu and Y. Cheng, *Chem. - Eur. J.*, 2013, **19**, 11553–11557.
341. V. Kavala, C. C. Wang, Y. H. Wang, C. W. Kuo, D. Janreddy, W. C. Huang, T. S. Luo, C. H. He, M. L. Chen and C. F. Yao, *Adv. Synth. Catal.*, 2014, **356**, 2609–2626.

342. R. Adepu, A. Rajitha, D. Ahuja, A. K. Sharma, B. Ramudu, R. Kapavarapu, K. V. L. Parsa and M. Pal, *Org. Biomol. Chem.*, 2014, **12**, 2514–2518.
343. F. C. Jia, C. Xu, Q. Cai and A. X. Wu, *Chem. Commun.*, 2014, **50**, 9914–9916.
344. I. Ryu, K. Matsu, S. Minakata and M. Komatsu, *J. Am. Chem. Soc.*, 1998, **120**, 5838–5839.
345. G. Chouhan and H. Alper, *Org. Lett.*, 2008, **10**, 4987–4990.
346. G. Chouhan and H. Alper, *J. Org. Chem.*, 2009, **74**, 6181–6189.
347. A. Dieudonné-Vatran, M. Azoulay and J. C. Florent, *Org. Biomol. Chem.*, 2012, **10**, 2683–2691.
348. G. Shen, D. Chen, Y. Zhang, M. Sun, K. Chen, C. Jin, K. Li and W. Bao, *Tetrahedron*, 2012, **68**, 166–172.
349. Z. Xia, K. Wang, J. Zheng, Z. Ma, Z. Jiang, X. Wang and X. Lv, *Org. Biomol. Chem.*, 2012, **10**, 1602–1611.
350. Z. Y. Ge, Q. M. Xu, X. D. Fei, T. Tang, Y. M. Zhu and S. J. Ji, *J. Org. Chem.*, 2013, **78**, 4524–4529.
351. M. Harmata and N. Pavri, *Angew. Chem., Int. Ed.*, 1999, **38**, 2419–2421.
352. C. Bolm, H. Okamura and M. Verrucci, *J. Organomet. Chem.*, 2003, **687**, 444–450.
353. A. G. Pandey, M. J. McGrath, O. G. Mancheño and C. Bolm, *Synthesis*, 2011, 3827–3838.
354. D. Orain, A. C. Blumstein, E. Tasdelen and S. Haessig, *Synlett*, 2008, 2433–2436.
355. F. Lach, *Synlett*, 2012, 2639–2642.
356. D. Chen, J. Wu, J. Yang, L. Huang, Y. Xiang and W. Bao, *Tetrahedron Lett.*, 2012, **53**, 7104–7107.
357. R. Sekar, M. Srinivasan, A. T. M. Marcelis and A. Sambandam, *Tetrahedron Lett.*, 2011, **52**, 3347–3352.
358. R. Wang, W. Qian and W. Bao, *Tetrahedron Lett.*, 2012, **53**, 442–445.
359. T. A. J. Vijay, K. N. Nandeesh, G. M. Raghavendra, K. S. Rangappa and K. Mantelingu, *Tetrahedron Lett.*, 2013, **54**, 6533–6537.
360. S. E. Denmark and W. Pan, *Org. Lett.*, 2003, **5**, 1119–1122.
361. F. Hong, J. Xia and Y. Xu, *J. Chem. Soc., Perkin Trans. 1*, 1994, 1665–1666.
362. T. Castanheiro, M. Donnard, M. Gulea and J. Suffert, *Org. Lett.*, 2014, **16**, 3060–3063.
363. D. Yang, H. Liu, H. Yang, H. Fu, L. Hu, Y. Jiang and Y. Zhao, *Adv. Synth. Catal.*, 2009, **351**, 1999–2004.
364. D. Yang, H. Fu, L. Hu, Y. Jiang and Y. Zhao, *J. Comb. Chem.*, 2009, **11**, 653–657.
365. A. Cherepakha, V. O. Kovtunenko, A. Tolmachev, O. Lukin and K. G. Nazarenko, *Synth. Commun.*, 2011, **41**, 1977–1989.
366. W. J. Xiao and H. Alper, *J. Org. Chem.*, 1999, **64**, 9646–9652.

367. M. K. Parai and G. Panda, *Tetrahedron Lett.*, 2009, **50**, 4703–4705.
368. E. Abele, J. Popelis and Y. Visnevska, *Chem. Heterocycl. Compd.*, 2012, **48**, 1119–1121.
369. D. J. C. Prasad and G. Sekar, *Org. Biomol. Chem.*, 2009, 7, 5091–5097.
370. D. Ma, Q. Geng, H. Zhang and Y. Jiang, *Angew. Chem., Int. Ed.*, 2010, **49**, 1291–1294.
371. T. Dahl, C. W. Trone, B. Bang-Andersen, P. Nielsen and M. Jorgensen, *Angew. Chem., Int. Ed.*, 2008, **47**, 1726–1728.

CHAPTER 4

Macroheterocycle Synthesis

Macroheterocycles have many special properties and have been reported with applications in various areas. Procedures for preparing macroheterocycles have been developed as well; among them, the procedures based on using double-functionalized arenes as substrates will be summarized and discussed in this chapter.

4.1 1,2-Dihaloarenes

Oxcarbazepine (Trileptal) has become the most widely prescribed drug for the treatment of epilepsy, both in adults and children. Additionally, oxcarbazepine is also effective in the treatment of mood disorders and mania. In 2005, a five-step procedure was developed for its synthesis.[1] 2′-Aminoacetophenone and 1,2-dibromobenzene were used as starting materials. The key steps accomplished sequentially are palladium-catalyzed intermolecular α-arylation of ketone enolates and intramolecular *N*-arylation reactions (Scheme 4.1).

Taylor and co-workers prepared cyclic diarylborinic acids and applied them in the synthesis of benzannulated heterocycles in 2014. In the presence of a palladium catalyst, the reaction underwent two-fold Suzuki–Miyaura coupling with dihaloarenes or *gem*-dibromoolefins and the corresponding macroheterocycles were produced in good yields (Scheme 4.2).[2]

In 2014, a copper and iron co-catalyzed Ullmann coupling of diiodo- and dibromoarenes and diphenols to aryl ether macrocycles was developed.[3] In the presence of CuI (10 mol%) and $Fe(acac)_3$ (10 mol%) with K_2CO_3 as a base in DMSO at 110 °C, the corresponding macrocycles can be isolated in low to moderate yields. However, a long reaction time (7 days) is required. Li and co-workers developed a palladium-catalyzed Ullmann protocol and applied it to the synthesis of 6,7-dihydro-5*H*-dibenzo[*c*,*e*]azepines and biaryls.[4] In the presence of $Pd(OAc)_2$ and KOAc, intramolecular or intermolecular Ullmann

RSC Catalysis Series No. 24
Heterocycles from Double-Functionalized Arenes: Transition Metal Catalyzed Coupling Reactions
By Xiao-Feng Wu and Matthias Beller

Published by the Royal Society of Chemistry, www.rsc.org

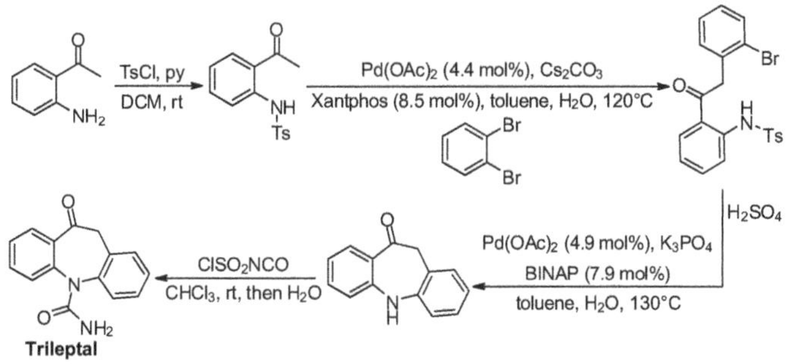

Scheme 4.1 Synthesis of Trileptal from 1,2-dibromobenzene.

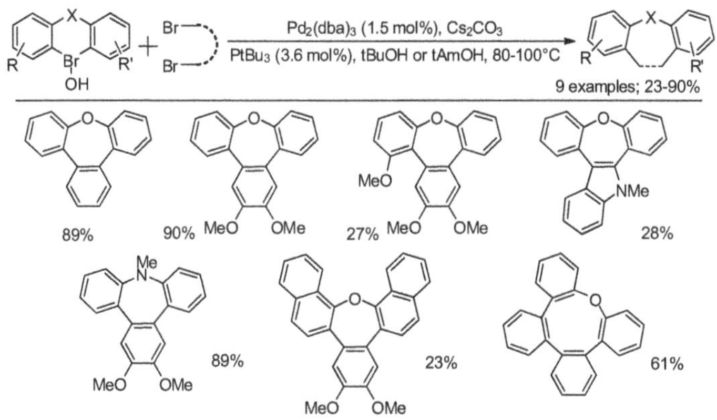

Scheme 4.2 Pd-catalyzed synthesis of macroheterocycles from cyclic diarylborinic acids.

> **Reaction Procedure** (Scheme 4.2): 10*H*-Dibenzo[*b*,*e*][1,4]oxaborinin-10-ol (19.6 mg, 0.1 mmol), $Pd_2(dba)_3$ (1.4 mg, 0.0015 mmol), *t*-$Bu_3PH(BF_4)$ (1.0 mg, 0.003 mmol) and Cs_2CO_3 (107.5 mg, 0.33 mmol) were added to an oven-dried Schlenk tube equipped with a magnetic stir bar. Please note, the Cs_2CO_3 used was from an older bottle and was likely to be hydrated. The flask was evacuated and refilled with argon three times. Toluene (2 mL) and the aryl halide (0.3 mmol) were added. The reaction was allowed to stir for 1 h at room temperature prior to heating to 100 °C for 16 h. The reaction was cooled to room temperature, diluted with EtOAc, filtered over Celite, concentrated *in vacuo* and purified by silica gel chromatography.

coupling of aryl halides proceeds efficiently under ligand-free and aerobic conditions to afford the corresponding 6,7-dihydro-5*H*-dibenzo[*c*,*e*]azepines and biaryls in moderate to excellent yields (Scheme 4.3).

The oxidative coupling of organozinc halides was also explored in the synthesis of macrocycles. Spring and co-workers combined a copper catalyst and oxygen with organozinc reagents; the coupled products were produced in high yields (Scheme 4.4).[5]

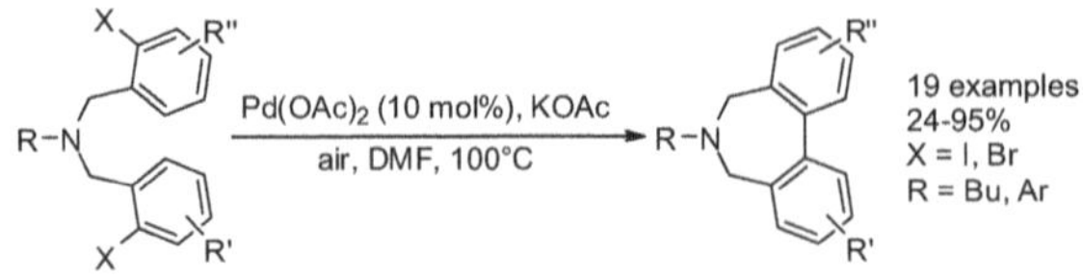

Scheme 4.3 Pd-catalyzed synthesis of dibenzo[*c*,*e*]azepines.

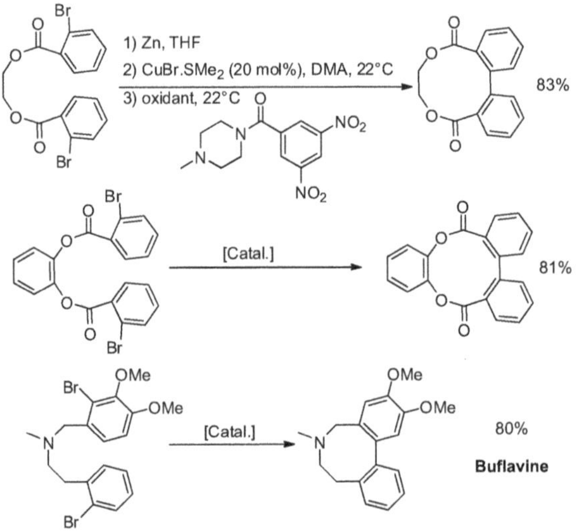

Scheme 4.4 Cu-catalyzed synthesis of buflavine.

Reaction Procedure (Scheme 4.4): Aryl bromide (1.0 mmol) in THF (2 mL) was added to Rieke zinc (4 mL, 5 g/100 mL suspension in THF). After the addition, the reaction mixture was heated at reflux and then concentrated *in vacuo*. The aryl zinc was dissolved in DMA (4 mL) and transferred *via* cannula onto a solid copper(I) bromide–dimethyl sulfide complex (20 mg, 0.1 mmol). Oxidant (147 mg, 0.5 mmol) in DMA (2 mL) was then added and the solution was kept stirring for 1 h at room temperature. The reaction mixture was filtered through a plug of silica eluting with hexane and EtOAc. The filtrate was concentrated *in vacuo* and the residue purified by flash column chromatography on silica gel.

4.2 2-Halophenol Derivatives

In 2010, Alper's group developed a new efficient synthetic method for the synthesis of 1,4-benzo- and pyrido-oxazepinones *via* a domino process through one-pot ring-opening–carboxamidation reaction sequences of *N*-tosylaziridines with 2-halophenols/pyridinol under phase-transfer conditions.[6] The method works efficiently with a range of *N*-tosylaziridines and 2-halophenols/pyridinol to provide facile access to a variety of 1,4-benzo- and pyrido-oxazepinones (Scheme 4.5a). Later on, Cai's group succeeded in replacing CO with isocyanide.[7] They established a new domino process for the synthesis of 1,4-benzoxazepine derivatives from *N*-tosylaziridines, 2-iodophenols and isocyanides. This process involves an aziridine ring-opening reaction with 2-iodophenol, followed by a palladium-catalyzed isocyanide-insertion reaction. A variety of desired products were formed in moderate to high yields (Scheme 4.5b).

More recently, Wu, Beller and co-workers reported a cascade procedure for the synthesis of 2,3-dihydrobenzodioxepinone from 2-bromophenols and epoxides.[8] The reactions undergo nucleophilic ring-opening of epoxides and subsequent palladium-catalyzed intramolecular alkoxylcarbonylation. Moderate to good yields of the versatile desired products were obtained in a regioselective manner (major product >90%) under mild conditions (Scheme 4.6).

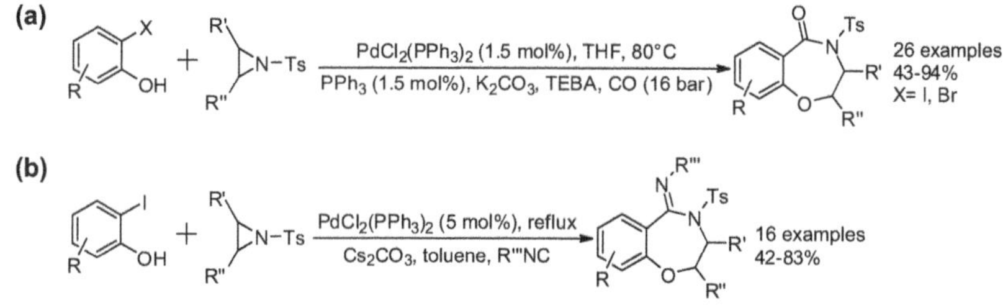

Scheme 4.5 Pd-catalyzed synthesis of oxazepinones.

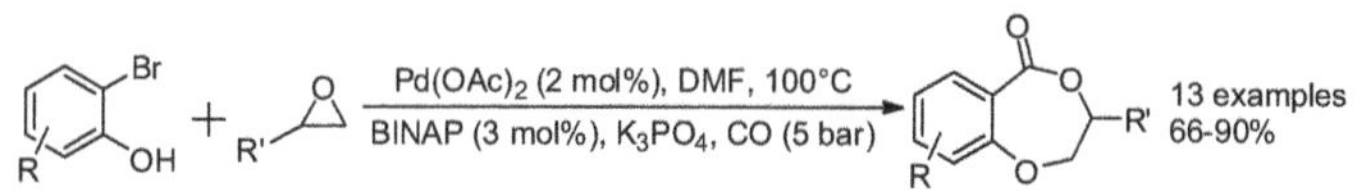

Scheme 4.6 Pd-catalyzed synthesis of 2,3-dihydrobenzodioxepinones.

Reaction Procedure (Scheme 4.6): A vial (6 mL) was charged with $Pd(OAc)_2$ (2 mol%) and Binap (3 mol%), K_3PO_4 (1 mmol) and a stir bar was added. Afterwards the vial was closed and purged with argon/vacuum three times. Then, DMF (2 mL), styrene oxide (0.75 mmol) and 2-bromo-phenol (0.5 mmol) were injected by syringe. The vial was placed in an alloy plate, which was transferred into an autoclave (300 mL) of the 4560 series from Parr Instruments under an argon atmosphere. After flushing the autoclave three times with CO, the pressure of CO was increased to 5 bar CO at ambient temperature. The reaction was performed for 16 h at 100 °C. After the reaction finished, the autoclave was cooled to room temperature and the pressure was carefully released. The solution was poured into water (20 mL) and extracted with Et_2O (20 mL) three times. After evaporation of the organic solvent, the residue was adsorbed on silica gel and the crude product was purified by column chromatography using *n*-pentane–ethyl acetate (20 : 1) as the eluent.

As heterogeneous catalysts are recyclable, palladium-complexed dendrimers supported on silica were prepared by Alper and Lu and evaluated as catalysts for intramolecular carbonylation reactions.[9,10] They showed that dendritic catalysts display high activity, affording oxygen, nitrogen, or sulfur-containing seven- or eight-membered ring fused heterocycles (dibenzoxazepinones and dibenzoxazocinones) in excellent yields. Moreover, this catalyst can be easily recovered by simple filtration in air and reused up to eight cycles with only a slight loss of activity. The catalyst was also applied in the carbonylative synthesis of twelve to eighteen-membered ring macrocycles. This process can tolerate a wide variety of functional groups, including halides, ethers, ketones, and esters. The heterogeneous dendritic catalysts facilitate excellent substrate reactivity, affording oxygen-, nitrogen-, or sulfur-containing tricyclic heterocycles in 70–92% yields. Importantly, these systems are easily recovered by simple filtration and can be reused several times with only a slight loss of activity.

In 2006, Yu and co-workers reported a palladium-catalyzed cascade carbometalation–cross-coupling of alkynes.[11] Dibenzoxapines containing tetrasubstituted exocyclic alkenes were prepared with high stereo- and regiocontrol (Scheme 4.7a). Employing a phosphine-free Pd(0) catalyst and water as the co-solvent afforded the target compounds in excellent yields. The practical utility of this novel method was demonstrated on the kilogram scale to afford the desired dibenzoxapine with over 70% yield. Later on, the group succeeded in preparing the same compounds by applying methylboronic acid as a coupling partner (Scheme 4.7b).[12] By replacing organoboronic acids with HCO_2Na, a reductive Heck reaction can occur and provides the corresponding trisubstituted exocyclic alkenes as the product (Scheme 4.7c).[13] A normal Heck reaction for the synthesis of the same product was reported as well.[14] The hydroboration of 2-[(2-ethynylphenyl)methoxy]-1-iodobenzene with bis(pinacolato)diboron in the presence of a palladium catalyst followed by consecutive oxidative addition, *cis*-cyclocarbopalladation, and *cis*-*β*-elimination could give a highly stereoselective

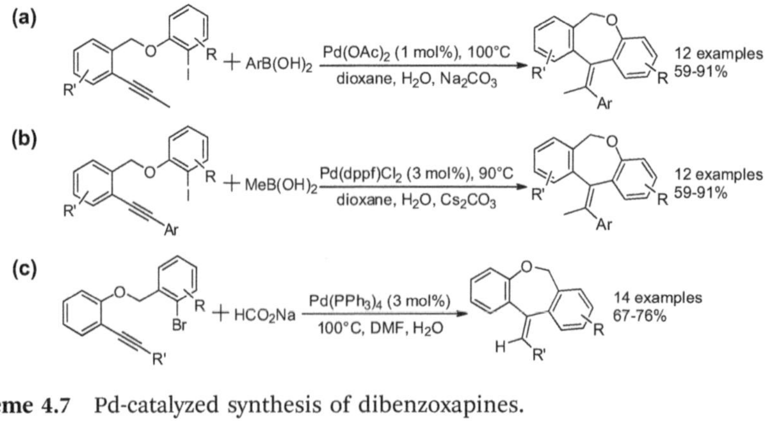

Scheme 4.7 Pd-catalyzed synthesis of dibenzoxapines.

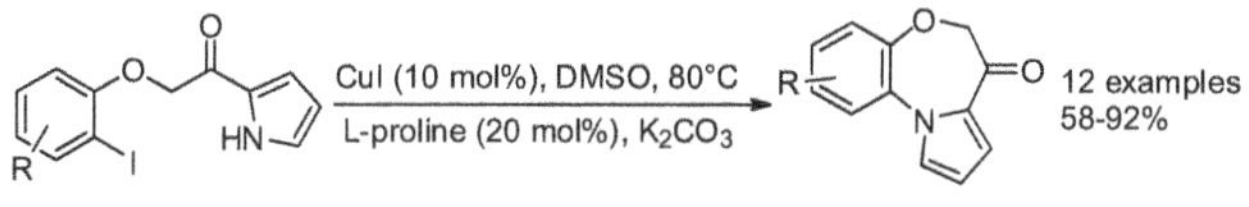

Scheme 4.8 Cu-catalyzed synthesis of pyrrolobenzoxazepinones.

Reaction Procedure (Scheme 4.8): A mixture of 2-(2-iodophenoxy)-1-(1*H*-pyrrol-2-yl)ethanone (1 mmol), CuI (0.1 mmol, 10 mol%), of L-proline (0.2 mmol, 20 mol%), and K_2CO_3 (2.5 mmol) in 3 mL DMSO under nitrogen, was heated at 80 °C, and monitored by TLC at the end of the reaction. After that, the reaction was cooled to room temperature, quenched with water (10 mL), and diluted with 10 mL of ethyl acetate. The layers were separated, and the aqueous layer was extracted with ethyl acetate (2×10 mL). The combined organic layers were dried over $MgSO_4$ and the solvent was removed *in vacuo* to afford a residue. The residue was purified by column chromatography on silica gel to afford the product.

exocyclic alkenylboronate ester. The subsequent cross-coupling of the exocyclic alkenylboronate ester with 2-bromo-*N*,*N*-dimethylacetamide in the presence of a palladium catalyst followed by LAH reduction gives (*E*)-doxepin in a fair yield (42%).

Gevorgyan and co-workers reported a mild, practical, and efficient method for the synthesis of symmetrical and unsymmetrical *o*-biphenols, *o*-phenol-naphthols, and *o*-binaphthols.[15] The method involves a Pd-catalyzed intramolecular C–H arylation of unsymmetrical bis-aryloxy silanes to give the seven-membered oxasilacycles, which, *via* a consecutive routine TBAF deprotection, furnish valuable unsymmetrical *o*-biphenols and *o*-binaphthols. In 2013, the intramolecular cyclization of di-*ortho*-bromoaryloxy acetal and *ortho*-bromoaryloxy acetal to the corresponding dibenzo[1,3]dioxepines was developed.[16]

Recently, a copper-catalyzed process for the construction of pyrrolobenzoxazepinones from 2-(2-iodophenoxy)-1-(1*H*-pyrrol-2-yl)ethanones was described.[17] A variety of pyrrolobenzoxazepinone derivatives were synthesized from the corresponding substrates by intramolecular Cu-catalyzed Ullmann *N*-arylations in good to excellent yields (Scheme 4.8).

4.3 2-Haloaniline Derivatives

As early as 1985, a palladium-catalyzed carbonylative procedure for the synthesis of 1,4-benzodiazepines was reported.[18] Starting from 2-haloanilines and amino acids, the desired seven-membered heterocycles were formed in moderate yields. More recently, an efficient, one-pot method for the synthesis of dibenzodiazepinone derivatives involving copper-catalyzed tandem C–N bond formation was reported.[19] Using various halo amide and 2-iodoaniline derivatives as substrates, an array of dibenzodiazepinone

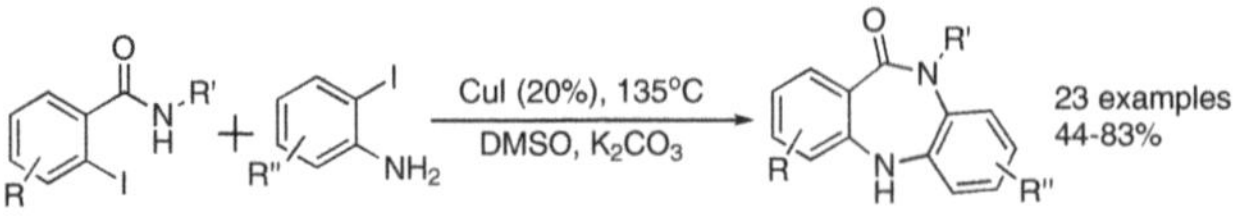

Scheme 4.9 Cu-catalyzed synthesis of dibenzodiazepinones from 2-haloanilines.

Reaction Procedure (Scheme 4.9): In an oven-dried, 10 mL round-bottomed flask equipped with a magnetic stirrer was added amide (0.5 mmol), 2-iodoaniline (0.5 mmol), CuI (20 mol%), K_2CO_3 (1.25 mmol) in anhydrous DMSO (1.5 mL). The reaction mixture was then stirred at 135 °C under a nitrogen atmosphere. After completion of the reaction, as determined by TLC, the reaction mixture was allowed to cool to room temperature. The crude reaction mixture was then purified by column chromatography without work-up using hexane–ethyl acetate as the eluent to yield the compound.

derivatives was produced in moderate to good yields (Scheme 4.9). Moreover, a dibenzodiazepinone derivative {2-(11-oxo-5*H*-dibenzo[*b*,*e*][1,4]diazepin-10(11*H*)-yl)benzonitrile} was utilized to synthesize the triazapentacyclic ring derivative.

In 2005, Fu and co-workers developed an efficient method for the preparation of medium- and large-sized nitrogen heterocycles *via* copper-catalyzed intramolecular *N*-arylation of phosphoramidates and carbamates.[20] Introduction of the phosphoryl group or *tert*-butoxycarbonyl at the *N*-termini can improve intramolecular cyclization under copper catalysis, and the phosphoryl (DIPP: diisopropylphospho) and *tert*-butoxycarbonyl can easily be removed under the mild conditions (Scheme 4.10).

Buchwald and co-workers developed a simple method for the preparation of medium ring heterocycles (7-, 8-, 9-, and 10-membered).[21] The process employs a Cu-catalyzed coupling of a *β*-lactam with an aryl bromide or iodide followed by intramolecular attack of a pendant amino group (Scheme 4.11). In some instances, the intermediate *β*-lactam was observable but can be converted to the aza-heterocycle by catalysis. Acetic acid was found to be superior to transition metal complexes as a catalyst for this ring-expansion process.

In 2006, Hii and co-workers reported the regioselective formation of medium-sized benzolactams *via* an intramolecular Heck reaction.[22] The reaction was found to be highly dependent on the catalytic conditions, particularly the structure of the substrate, as well as the nature of the ligand employed. The latter is found to affect the chemoselectivity, regioselectivity, and competitive double bond migration processes. Whereas six- to eight-membered heterocycles were obtained exclusively using the PPh_3 ligand (4 equiv.), biaryl coupling was preferred over 8-exo cyclization under "ligand-less" conditions. In the same year, an interesting palladium-catalyzed

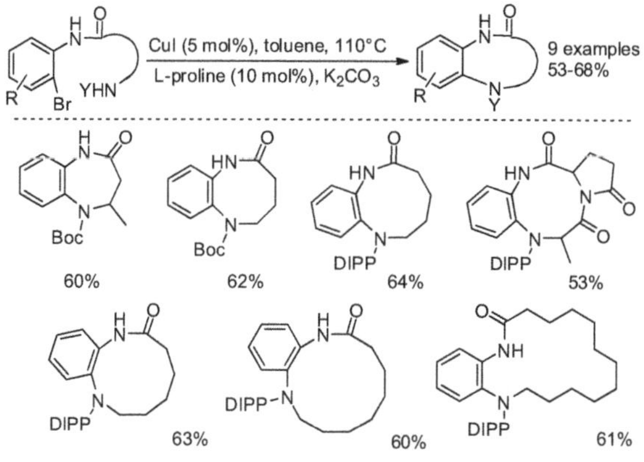

Scheme 4.10 Cu-catalyzed intramolecular synthesis of C–N coupling.

Reaction Procedure (Scheme 4.11): A Schlenk tube was charged with CuI (9.6 mg, 0.050 mmol, 5.0 mol%), 2-azetidinone (177 mg, 1.20 mmol), and K_2CO_3 (280 mg, 2.03 mmol), evacuated, and backfilled with argon. 2-Bromobenzylamine (1.00 mmol) and toluene (1.0 mL) were added under argon. The Schlenk tube was sealed with a Teflon valve, and the reaction mixture was stirred at 110 °C for 24 h in a preheated oil bath. After the resulting green-brown suspension was allowed to reach room temperature, dodecane (internal GC standard) and CH_2Cl_2 (3 mL) were added, and the supernatant solution was analyzed by GC to indicate >99% conversion of the aryl bromide starting material. The reaction mixture was filtered through a silica gel plug (0.5×0.5 cm) eluting with 10 : 1 CH_2Cl_2–MeOH (50 mL), the filtrate was concentrated, and the residue was purified by column chromatography on silica gel (CH_2Cl_2–MeOH 20 : 1) to provide the desired product.

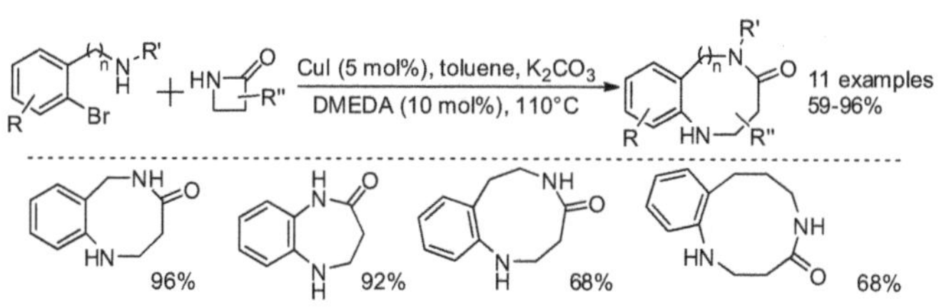

Scheme 4.11 Cu-catalyzed synthesis of macrocycles from *β*-lactams.

intramolecular coupling using the *o*-bromo- and *o*-iodo anilides of indoles and *N*-methyl indole was described.[23] The usability of this procedure was proved by the application of this in the short total synthesis of paullone and dimethyl paullone (Scheme 4.12).

Solé and co-workers explored the scope and limitations of Pd(0)-catalyzed intramolecular coupling reactions of amino-tethered aryl halides and ketones. Two different and competitive reaction pathways starting from ω-(2-haloanilino) alkanones, enolate arylation and addition to the carbonyl group have been observed, while ω-(2-halobenzylamino) alkanones exclusively underwent the enolate arylation process.[24] The dichotomy between ketone α-arylation and carbonyl-addition in the reactions of ω-(2-haloanilino) alkanones has been rationalized by the intermediacy of unprecedented four-membered azapalladacycles, from which X-ray data and chemical behavior were reported.

In 2010, Buchwald's group reported a Pd-catalyzed condensation of 2-bromostyrene and 2-chloroaniline derivatives yielding stable diphenylamine intermediates, which were selectively converted to five-, six-, or seven-membered heteroaromatics (indoles, carbazoles, acridines, and dibenzazepines).[25] The selectivity of these intramolecular transformations is uniquely ligand-controlled and offers efficient routes to four important classes of heterocycles from a common precursor (Scheme 4.13a). One year later, the group reported another protocol for the synthesis of dibenzodiazepines and their structural analogues.[26] In the presence of catalytic quantities of palladium, readily accessible precursors are cross-coupled with ammonia and then spontaneously undergo an intramolecular condensation to form the corresponding dibenzodiazepines in one step (Scheme 4.13b).

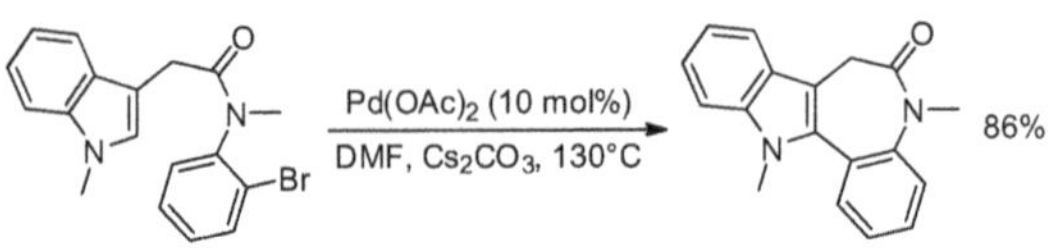

Scheme 4.12 Pd-catalyzed synthesis of dimethyl paullone.

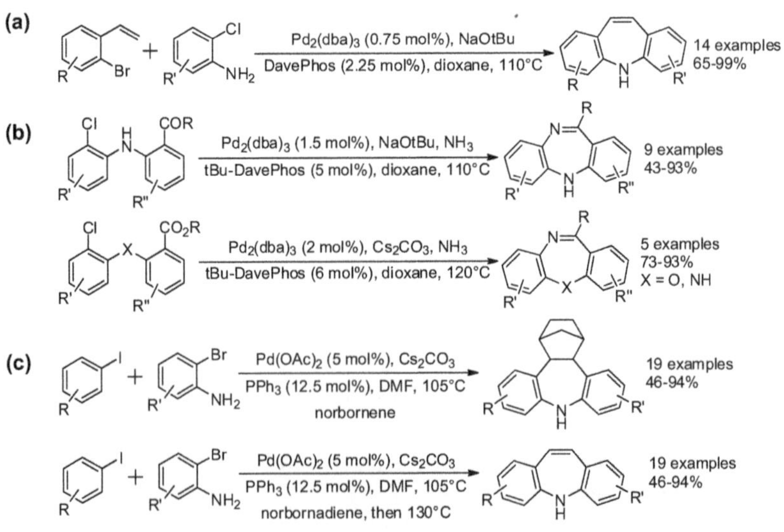

Scheme 4.13 Pd-catalyzed synthesis of dibenzoazepines and related compounds.

Reaction Procedure (Scheme 4.13a): DavePhos (8.8 mg, 0.0225 mmol), $Pd_2(dba)_3$ (6.9 mg, 0.0075 mmol), NaO*t*Bu (144 mg, 1.5 mmol) and the appropriate diarylamine intermediate (if solid) (1.0 mmol) were mixed in oven-dried Schlenk tube and degassed. The flask was refilled with argon (repeated 3–5 times) and degassed dioxane (1 mL) was added under an argon atmosphere (liquid intermediates were added with dioxane in the same manner). The tube was then placed in a preheated oil bath at 110 °C, and the mixture was stirred until completion of the reaction (followed by GC analysis). After cooling to room temperature, the solution was quenched with water (5 mL) and diluted with ethyl acetate (10 mL). The organic layer was separated and dried on Na_2SO_4, then filtered. After concentration under reduced pressure, the crude mixture was purified, either by silica gel chromatography, or by Biotage SP4 technique, eluting with hexanes–ethyl acetate mixtures to afford pure product.

(Scheme 4.13b): *t*BuDavePhos (17.1 mg, 0.05 mmol), $Pd_2(dba)_3$ (13.7 mg, 0.015 mmol), NaO*t*Bu (144 mg, 1.5 mmol), activated (flame dried) 4 Å molecular sieves (250 mg), and the appropriate diarylamine precursor (if a solid) (1.0 mmol) were mixed in oven-dried 20 mL borosilicate glass tube, which was sealed with a PTFE/silicon screw cap. The tube was evacuated and backfilled with argon (this sequence was repeated a total of 3 times), and a 0.5 M ammonia solution in 1,4-dioxane (10 mL, 5.0 mmol) was added *via* syringe under an argon atmosphere. Liquid precursors were previously dissolved in 10 mL of ammonia solution in an oven-dried borosilicate glass tube (under argon) and then transferred into the catalyst tube *via* syringe. The reaction mixture was then placed in a preheated oil bath at 85 °C, and mixture was stirred until completion of the reaction (2–5 h). At the end of this time, the tube was removed from the bath, allowed to cool to room temperature, the contents diluted with EtOAc (10 mL), and the mixture filtered through a plug of silica. The solution was then concentrated under reduced pressure and the residue purified on the Biotage SP4, eluting with hexanes–ethyl acetate–diethyl ether mixtures to afford the desired dibenzodiazepine.

(Scheme 4.13c): A Schlenk-type flask, equipped with a magnetic stir bar, was charged under nitrogen with Cs_2CO_3 (326 mg, 1.0 mmol), PPh_3 (14 mg, 0.055 mmol) and $Pd(OAc)_2$ (5 mg, 0.022 mmol) in DMF (5 mL). After 10 minutes with stirring, a DMF solution (5 mL) of the aryl iodide (0.48 mmol), the 2-bromoaniline (0.44 mmol) and norbornadiene (97 mg, 1.05 mmol) was added. The resulting mixture was stirred in an oil bath at 105 °C for 24 h and then at 130 °C for an additional 12–14 h. After cooling to room temperature, the mixture was diluted with EtOAc (30 mL) and extracted three times with a saturated solution of NaCl (25 mL). The organic layer was dried over anhydrous Na_2SO_4, the solvent was removed under reduced pressure and the products were isolated by flash column chromatography on silica gel using mixtures of hexane–EtOAc as eluent.

Meanwhile, Catellani, Derat and their co-workers developed the synthesis of dihydrodibenzoazepine and dibenzoazepine derivatives from 2-bromoanilines and aryl iodides.[27] The process was based on the sequential reaction of three components, an aryl iodide, a bromoaniline, and either norbornene or norbornadiene in the presence of palladium as the catalyst (Scheme 4.13c). Whereas the use of norbornene leads to dihydrodibenzoazepine derivatives, norbornadiene allows an additional step consisting of a retro-Diels–Alder reaction, thus leading to the parent dibenzoazepines.

4.4 1-Carbon-2-haloarenes

In 2011, a palladium-catalyzed C–O coupling was developed for the synthesis of benzannulated oxazepino- and pyranocarbazoles.[28] In the absence of a ligand, the desired product can be produced in excellent yields (Scheme 4.14).

Spring's group developed a copper-catalyzed procedure for the synthesis of tricyclic biaryl ether-linked aza-heterocyclic rings.[29,30] In the presence of catalytic quantities of copper(I), readily accessible acyclic precursors undergo an intramolecular carbon–oxygen bond-forming reaction facilitated by a "templating" chelating nitrogen atom. The methodology displays a broad substrate scope, is practical, and generates rare and biologically interesting tricyclic heteroaromatic products in moderate to excellent yields (Scheme 4.15).

In 2013, Yang and co-workers developed a simple and efficient method for the formation of functionalized dibenzo[*b*,*f*]oxepins and their analogues bearing both electron-donating and electron-withdrawing groups *via* a one-pot cascade reaction.[31] With the readily available starting materials 2-(2-hydroxyphenyl)-acetonitriles and 2-haloarylaldehydes, with the use of Cs_2CO_3 as the base and DMF as a solvent under copper-assisted/copper-free conditions, good yields of the desired products can be isolated (Scheme 4.16a). In the same year, Heo and co-workers developed an efficient one-pot procedure for the synthesis of dibenzoxepine lactam *via* a Cu-catalyzed etherification–aldol condensation cascade reaction.[32] The coupling reactions of isoindolin-1-ones with 2-bromobenzaldehydes in the presence of catalytic CuBr provided a wide range of dibenzoxepine lactams (Scheme 4.16b). The total synthesis of aristoyagonine was successfully achieved in a highly efficient manner using this protocol, which also afforded a variety of aristoyagonine derivatives.

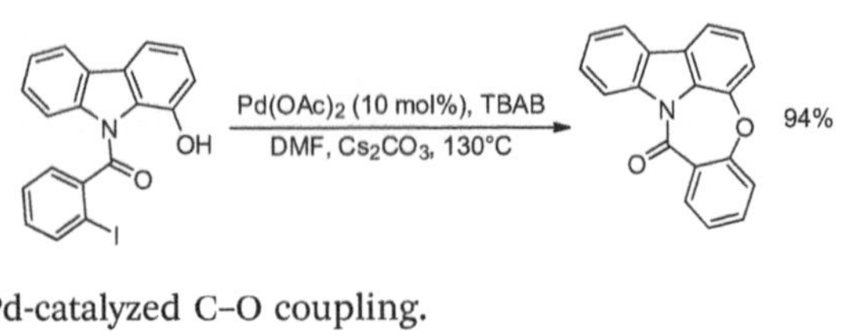

Scheme 4.14 Pd-catalyzed C–O coupling.

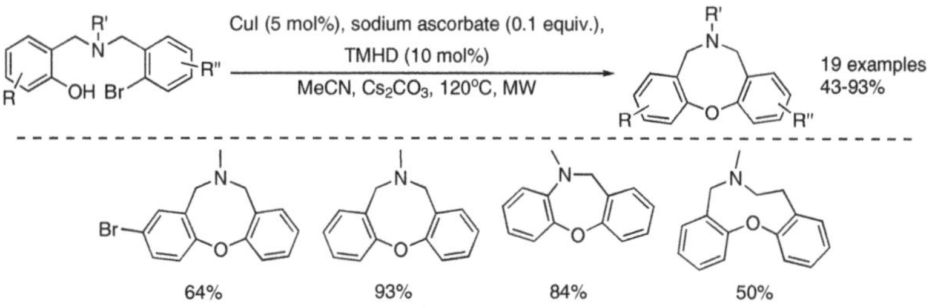

Scheme 4.15 Cu-catalyzed C–O coupling.

Reaction Procedure (Scheme 4.15): CuI (0.05 equiv.), sodium ascorbate (0.1 equiv.), Na_2SO_4 (2 equiv.), the acyclic substrate (1 equiv.), 2,2,6,6-tetramethyl-3,5-heptanedione (0.1 equiv.), Cs_2CO_3 (2 equiv.) and CH_3CN (0.190 mmol) were added sequentially to an oven-dried microwave tube. The mixture was then heated to 120 °C in the microwave (using the 'Standard method') for 5 hours. The mixture was filtered through a plug of Celite (eluting with CH_2Cl_2). The organic layer was washed with water and 1 M NaOH solution, dried (Na_2SO_4) and the solvent removed under reduced pressure. The residue was purified by column chromatography to afford the cyclic product.

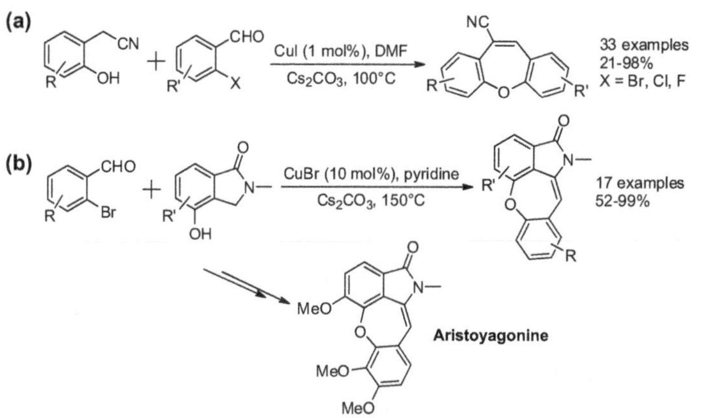

Scheme 4.16 Cu-catalyzed synthesis of dibenzoxepines.

Reaction Procedure (Scheme 4.16a): To a mixture of nitrile (0.53 mmol) and 2-bromobenzaldehyde (0.5 mmol) in a 25 ml two-necked reaction flask were added Cs_2CO_3 (1.5 mmol), CuI (0.005 mmol), and DMF (8 ml). The mixture was degassed with argon and then heated at 100 °C for 0.5–1 h. Most of the reactions were complete within 1 h. Once the reaction was completed (monitored by TLC), the mixture was poured onto crushed ice and then extracted with AcOEt. The combined organic layers were washed with brine and dried (Na_2SO_4). Concentration and purification by CC afforded products with desirable purities.

(Scheme 4.16b): A nitrogen-flushed microwave vial was equipped with a magnetic stir bar and charged with 4 Å molecular sieves (200 mg). The vial was flame dried for 10 min under high vacuum and purged with N_2. After cooling to room temperature, isoindolinone (0.3 mmol), 2-bromobenzaldehyde (0.6 mmol), CuBr (10 mol%), and Cs_2CO_3 (0.9 mmol) were added sequentially. The reaction mixture was suspended in pyridine (5 mL). Then, the reaction vial was sealed and placed into a heating block at 150 °C for 24 h. The mixture was cooled to room temperature, filtered through a Celite pad, and washed with acetone. The resulting solution was concentrated with a rotary evaporator, and the residue was purified by silica gel flash column chromatography (EtOAc–hexanes) to afford dibenzoxepine lactam.

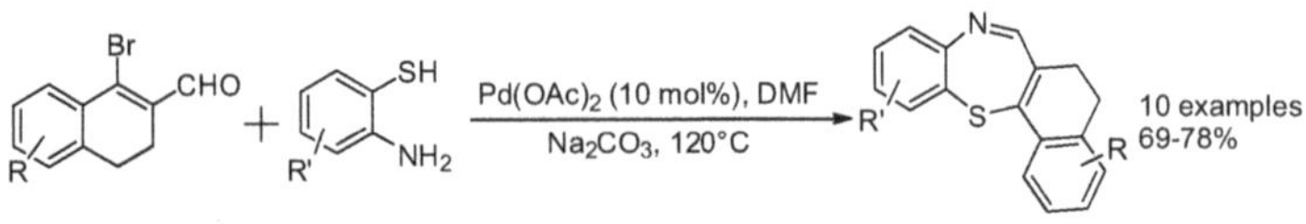

Scheme 4.17 Pd-catalyzed synthesis of benzothiazepines.

More recently, a simple and efficient reaction for the synthesis of steroid/nonsteroid fused benzo[*b*][1,4]thiazepines and 2-arylsubstituted benzo[*b*][1,4]thiazepines was developed.[33] By using $Pd(OAc)_2$ as the catalyst in DMF solvent with heating, a wide variety of steroidal/nonsteroidal β-bromovinyl aldehydes and 2-aminothiophenols undergo this reaction to give good yields of benzo[*b*][1,4]thiazepine derivatives (Scheme 4.17). A palladium catalyst was explored in benzoxazocine synthesis *via* intramolecular C–N coupling as well.[34]

In 1993, Corriu *et al.* studied the synthesis of nitrogen-containing heterocycles from (*Z*)-3-(tributylstanny1)allylamine, which was prepared by the reaction of *N*-(trimethylsily1)allylamine with 2 mol of *n*-butyllithium followed by treatment with chlorotributyltin and subsequent hydrolysis.[35] The unprotected (*Z*)-3-(tributylstannyl)allylamine underwent a palladium-catalyzed cross-coupling reaction with aromatic bromides affording a stereospecific preparation of substituted allylic amines with *Z* configuration of the carbon–carbon double bond. The reactions of *ortho*-functionalized aryl bromides offer a one-step preparation of 7-membered nitrogen heterocycles in high yields (Scheme 4.18).

Sharma's group developed the synthesis of dibenzoazepinones from 2-(2-bromophenyl)acetic acid esters and 2-aminophenyl boronate.[36] In the presence of $Pd(PPh_3)_4$ and Cs_2CO_3 in dimethoxyethane at 125–130 °C, the desired products were formed in moderate yields (Scheme 4.19).

Wallace and co-workers developed the synthesis of 5-substituted 6,7-dihydrodibenz[*c*,*e*]azepines, a class of secondary amine incorporating a centre-axis chirality relay, from 1-substituted *N*-(2-bromobenzyl)-1-phenylmethanamines *via* *N*-acylation and ring-closing intramolecular direct

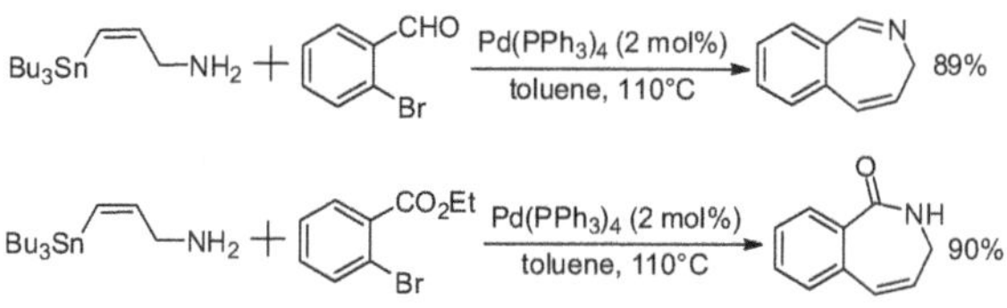

Scheme 4.18 Pd-catalyzed synthesis of heterocycles from (*Z*)-3-(tributylstanny1)allylamine.

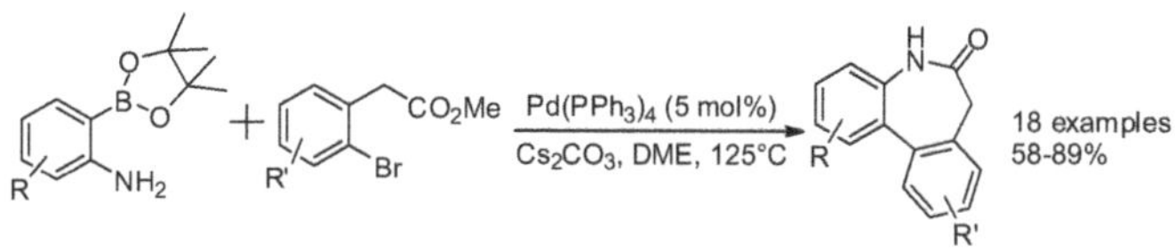

Scheme 4.19 Pd-catalyzed synthesis of dibenzoazepinones from 2-aminophenyl boronate.

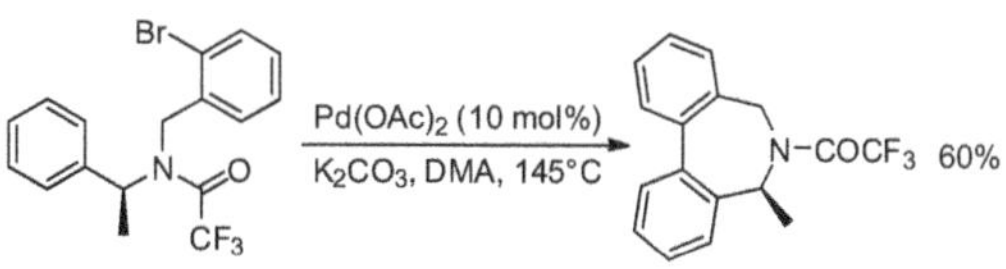

Scheme 4.20 Pd-catalyzed synthesis of 6,7-dihydrodibenz[*c*,*e*]azepines.

arylation in 2011 (Scheme 4.20).[37] The ring closure proceeds with high atropodiastereoselectivity.

A palladium-catalyzed ring-opening of cyclopropyl benzamides to benzo[*c*]azepine-1-ones was developed in 2013.[38] A variety of difficult to obtain benzo[*c*]azepine-1-ones were synthesized *via* this palladium-catalyzed, silver-promoted intramolecular cyclization of cyclopropyl benzamides in good yields (Scheme 4.21). Both aryl bromides and iodides were effective substrates for the transformation. Mechanistic studies indicated that the reaction proceeded through a cyclopropyl C(sp^3)–H cleavage step, followed by ring-opening, deprotonation, and reductive elimination.

An interesting palladium-catalyzed synthesis of 1,2,3,4-tetrahydro-5*H*-2-benzazepin-5-ones was developed in 2012 by Alper and Okuro.[39] The reaction proceeded *via* palladium-catalyzed intermolecular cyclocarbonylation of 2-iodobenzylamines with Michael acceptors. The desired products were isolated in moderate to good yields (Scheme 4.22). This methodology enables the direct preparation of highly functionalized 1,2,3,4-tetrahydro-5*H*-2-benzazepin-5-ones from readily available starting materials.

Narbonne *et al.* developed a one-pot catalytic method for the synthesis of dibenzo[*c*,*e*]azepines and their imine analogues in 2014.[40] The reaction was catalyzed by a joint palladium–norbornene organometallic catalyst. The complete diastereoselectivity observed originates from a chelated Pd(IV) complex *via* atroposelective aryl–aryl coupling. The desired dibenzo[*c*,*e*]azepines were isolated in good yields (Scheme 4.23).

In 1994, Negishi and Ma reported a palladium-catalyzed cyclization of allene-containing aryl and alkenyl halides.[41] In the presence of $PdCl_2(PPh_3)_2$, the corresponding seven- and eight-membered cycloalkane derivatives were isolated in good yields (Scheme 4.24a). In 2005, Grigg's group reported the synthesis of annelated benzazepines *via* intramolecular cyclization with allenes.[42] Moderate to good yields of the desired benzazepines were formed in the presence of palladium and indium catalysts (Scheme 4.24b). Recently, Ma and co-workers reported a $Pd(OAc)_2$-catalyzed reaction of 2,3-alkadienyl

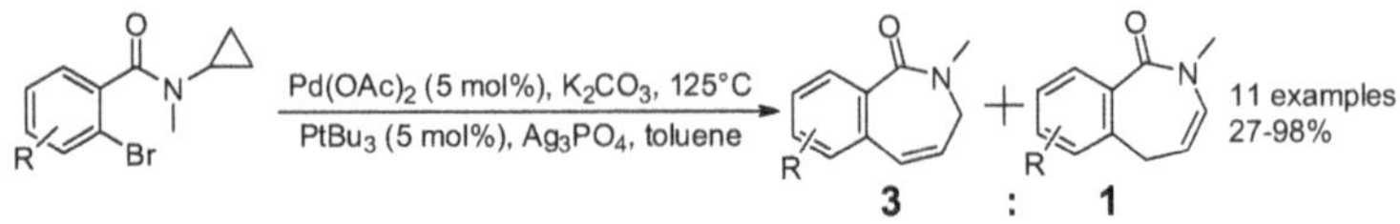

Scheme 4.21 Pd-catalyzed ring opening of cyclopropyl.

Reaction Procedure (Scheme 4.21): To a 5.0 mL microwave vial in a glovebox were added in the following order: $Pd(OAc)_2$ (5.6 mg, 0.025 mmol), $PtBu_3 \cdot HBF_4$ (7.2 mg, 0.025 mmol), K_2CO_3 (76.0 mg, 0.55 mmol), and Ag_3PO_4 (0.16 mmol, 69.1 mg). The vial was crimped shut. The benzamide (0.50 mmol) was weighed into a vial, dissolved in 0.5 mL of toluene and added to the reaction vial. The vial containing benzamide was rinsed with 0.5 mL of toluene for a total volume of 2.5 mL. This yellowish-orange solution was then heated to 125 °C in an oil bath for 16 h. The reaction turned black within 5 min. The reaction was cooled to ambient temperature, filtered over a cotton-Celite plug, and rinsed with 25 mL of ethyl acetate. It was then concentrated *in vacuo* to give the crude product. The crude product was then purified *via* column chromatography over silica gel (RediSep Rf Gold 24 g) using a solvent gradient of 10–50% ethyl acetate–hexanes to give isomers.

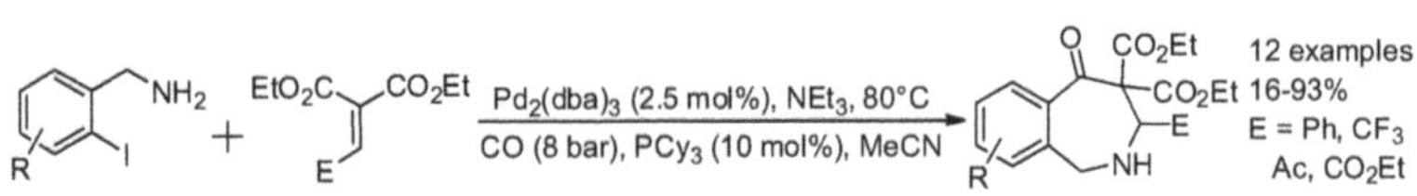

Scheme 4.22 Pd-catalyzed synthesis of 1,2,3,4-tetrahydro-5*H*-2-benzazepin-5-ones.

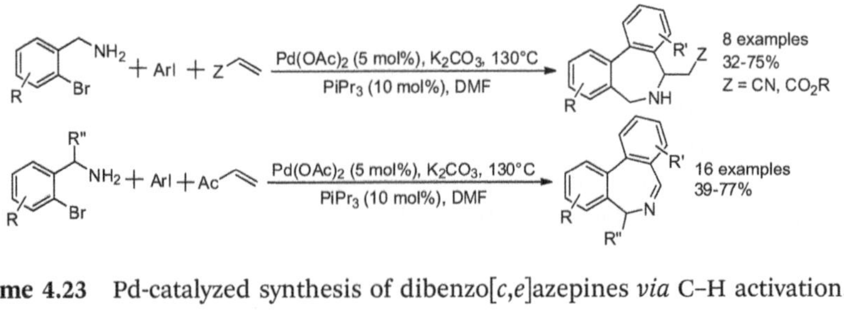

Scheme 4.23 Pd-catalyzed synthesis of dibenzo[*c*,*e*]azepines *via* C–H activation.

Reaction Procedure (Scheme 4.23): To a Schlenk-type flask was added under argon K_2CO_3 (83 mg, 0.66 mmol, 2.2 equiv.). At least 3 vacuum/argon cycles were made before adding a solution of DMF (3 mL) containing the aryl iodide (0.29 mmol, 1.1 equiv.), the (substituted) 2-bromobenzylamine (0.26 mmol, 1 equiv.), norbornene (12 mg, 0.18 mmol, 0.6 equiv.) and olefin (28 mg, 0.52 mmol, 2 equiv. for acrylonitrile or 38 mg, 0.29 mmol, 1.1 equiv. for butyl acrylate) and a solution of $Pd(OAc)_2$ (3.0 mg, 0.013 mmol, 0.05 equiv.) in 3 mL of DMF. A solution of triisopropylphosphine (10% on weight in hexane, 43 µL, 0.03 mmol, 0.1 equiv.) was then added. The same procedure could be adopted using hydrochloric salts of 2-bromobenzylamines by adding 1 more equiv. of base in the reaction vessel (thus 3.3 equiv. overall, 1 mmol, 138 mg). The resulting suspension was stirred with a magnetic bar at 130 °C until visible formation of palladium black (8–16 h). For 3 h, 4 additional equiv. of methyl acrylate (1.04 mmol, 90 mg) were added after full conversion and the mixture was kept at 130 °C for another 8 hours. The mixture was then allowed to cool to room temperature, diluted with EtOAc (30 mL), washed three times with a saturated K_2CO_3 solution (3×30 mL) and dried over $MgSO_4$. The solvent was removed under reduced pressure and the crude mixture was analyzed by ^1H NMR spectroscopy. The products were isolated by flash column chromatography on silica gel.

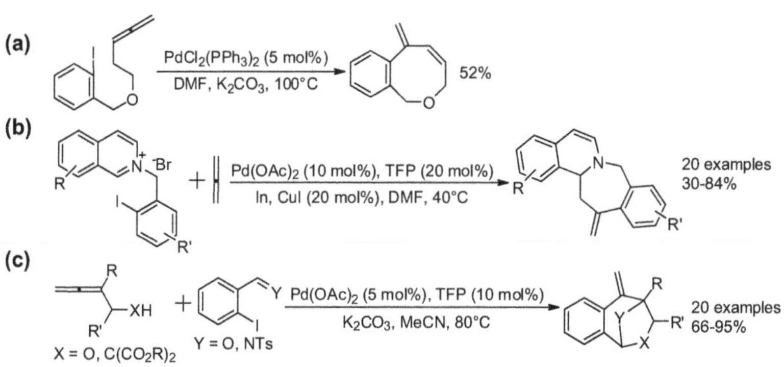

Scheme 4.24 Pd-catalyzed cyclization reactions of allenes.

malonates or 2,3-allenols with *o*-iodobenzaldehyde or its *N*-tosyl imine.[43] The reaction was performed in MeCN at 80 °C, and the desired oxa- or aza-bridged benzocycloheptane derivatives were produced in good yields in general (Scheme 4.24c). With the optically active 2,3-allenols, the absolute configurations of all three chiral centers can be conveniently established.

In 2004, Guy and co-workers reported the synthesis of five classes of heterocycles: dihydrodibenzo[*b*,*f*]azepine, -oxocine, and -thiocine and dibenzo[*b*,*f*]azepine and -oxepine using a strategy of alkylation followed by highly selective intramolecular Heck arylation reaction.[44] The hetero-tricyclic compounds were available in only two steps starting from

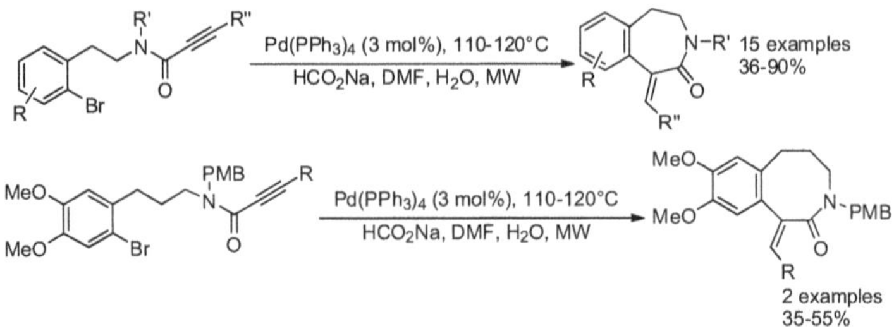

Scheme 4.25 Pd-catalyzed reductive intramolecular Heck reaction.

Reaction Procedure (Scheme 4.25): $Pd(PPh_3)_4$ (14 mg, 3 mol%) and HCOONa (41 mg, 1.5 equiv.) were loaded into a microwave instrument vial. The vial was evacuated and flushed with argon. Propynoic acid amide (0.4 mmol) dissolved in DMF (4.5 mL) was added, followed by distilled water (1.5 mL). The vial was sealed and irradiated with stirring at a minimum temperature of 110 °C at a maximum power level of 150 W for 15 min. Upon completion of the reaction time, the vial was cooled with a stream of air. After dilution with DCM, the organic phase was washed several times with brine, dried ($MgSO_4$) and concentrated under reduced pressure. The products were isolated by column chromatography on silica gel.

commercially available starting materials. The advantages of this strategy are (1) a very direct and convergent synthesis of tricyclic compounds and (2) the tolerance of various functional groups. Since then, this strategy has been studied by many other groups for the synthesis of oxocines,[45,46] naphthoxepines,[47] dibenzoazocinones,[48] benzoazocines,[49] benzazepines,[50] *etc.*[51] Eycken's group developed an efficient new procedure for the synthesis of 1-substituted 3-benzazepinones in 2007.[52] The medium-sized ring can be constructed with full regio- and stereoselectivity by applying a microwave-assisted reductive intramolecular Heck reaction (Scheme 4.25).

Cheng and co-workers reported the synthesis of isochromenones and oxepines *via* Pd-catalyzed cascade cyclization of alkynes and benzynes in 2012.[53] The reactions involve biscarbocyclization of alkynes and benzynes and C–H bond activation. Phenanthro[1,10-*bc*]-oxepine derivatives were prepared in good to excellent yields (Scheme 4.26a). They also developed the synthesis of seven-membered lactones *via* nickel- and zinc-catalyzed cyclization of 2-iodobenzyl alcohols with alkyl propiolates (Scheme 4.26b).[54] The catalytic reaction involves an unusual *E*/*Z* isomerization of a carbon–carbon double bond prior to ring closure and provides a convenient and unique method for the synthesis of seven-membered lactones.

In 2008, Fujii, Ohno and their co-workers developed an interesting method for the preparation of indole-fused 1,4-diazepines by copper-catalyzed domino three-component coupling–indole formation–*N*-arylation.[55]

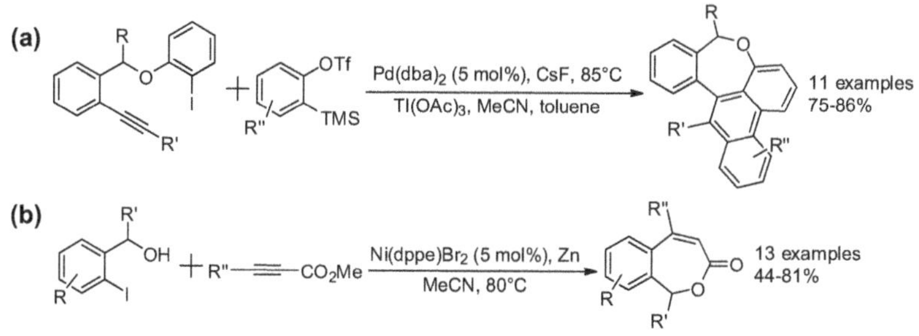

Scheme 4.26 Synthesis of seven-membered heterocycles from alkynes.

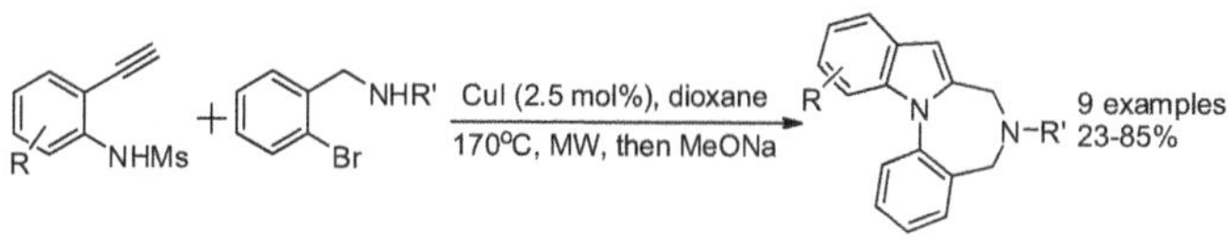

Scheme 4.27 Copper-catalyzed domino three-component synthesis of 1,4-diazepines.

Starting from simple 2-ethynylanilines and *o*-bromobenzylamines, complex indole-fused tetracyclic compounds were easily and directly synthesized in a single reaction vessel. This was the first example of a copper-catalyzed one-pot reaction including three catalytic cycles and the formation of four bonds. In detail, Mannich-type coupling of an ethynylaniline derivative with formaldehyde and *N*-substituted *o*-halobenzylamine, indole formation, and arylation of the indole nitrogen were included (Scheme 4.27).

Ma's group reported the synthesis of oxazepines *via* a copper-catalyzed intramolecular C–N coupling reaction.[56] Starting from 2-aminophenols and 2-bromobenzyl bromides, using CuI/L-proline as the catalyst system, the corresponding products were formed in good yields (Scheme 4.28a). This catalyst system was applied in the synthesis of 6,12*b*-diazadibenzo[*a*,*h*]azulen-7-ones and benzo[*f*]pyrrolo[1,2-*a*][1,4]diazepin-4-ones *via* intramolecular C–N coupling and provided the corresponding products in good to excellent yields (Scheme 4.28b).[57] A palladium catalyst was applied to this type of coupling reaction as well (Scheme 4.28c).[58] More recently, Laha and co-workers performed mechanistic studies on the palladium-catalyzed domino synthesis of 10,11-dihydro-5*H*-dibenzo[*b*,*e*][1,4]diazepines.[59] Starting from *o*-phenylenediamines and 2-bromobenzyl bromides or tosylates, the desired products were selectively produced (Scheme 4.28d). The reaction involves intermolecular mono-*N*-benzylation of the *o*-phenylenediamine through oxidative palladium insertion, primarily at the benzylic position of the 2-bromobenzyl bromide, followed by intramolecular *N*′-arylation. The intramolecular *N*′-arylation was demonstrated to be facilitated by coordination of a remote auxiliary nitrogen atom to the palladium atom. A variety of functional groups, including base-sensitive groups, were well tolerated

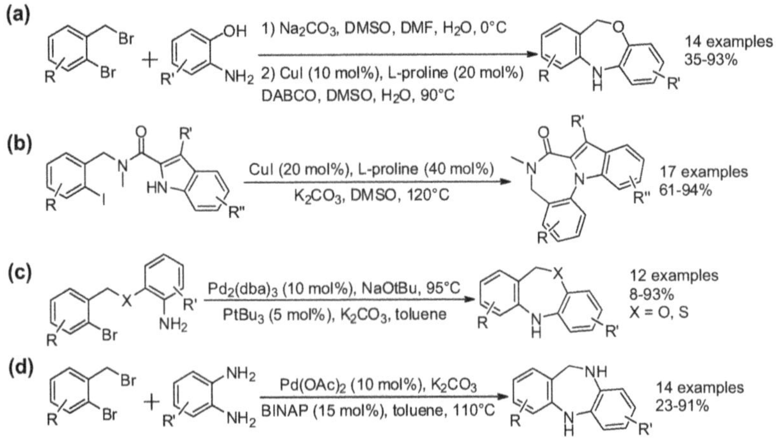

Scheme 4.28 Synthesis of heterocycles *via* C–N coupling.

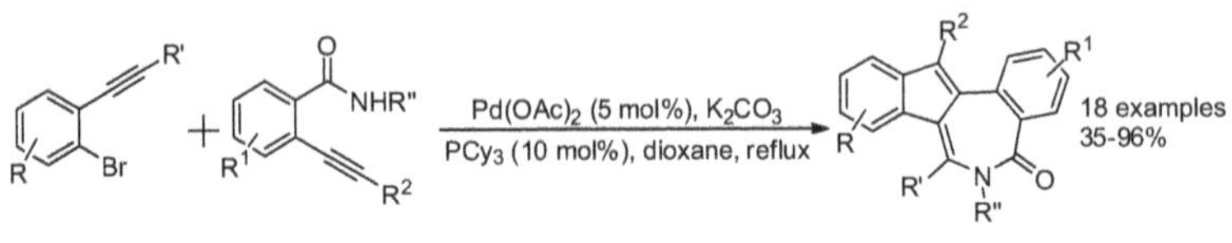

Scheme 4.29 Pd-catalyzed synthesis of indeno[1,2-*c*]-azepin-3(2*H*)-ones.

Reaction Procedure (Scheme 4.29): 2-Alkynylhalobenzene (0.24 mmol) was added to a mixture of $Pd(OAc)_2$ (0.01 mmol, 5 mol%), PCy_3 or PPh_3 (0.02 mmol, 10 mol%), K_2CO_3 (0.4 mmol), and 2-alkynylbenzamide (0.20 mmol) in 1,4-dioxane (2.0 mL). The reaction was stirred under reflux conditions. After completion of the reaction as indicated by TLC (usually 3–6 hours), the reaction was cooled and the mixture was purified immediately by flash chromatography on silica gel to give products.

under the optimized reaction conditions and various substituted dibenzo-[*b*,*e*][1,4]diazepines were formed in good to excellent yields.

Wu and Luo developed a novel and efficient route for the synthesis of indeno[1,2-*c*]-azepin-3(2*H*)-ones in 2011.[60] The reaction proceeded through a palladium-catalyzed tandem reaction of 2-alkynylhalobenzene with 2-alkynylbenzamide in the presence of PPh_3 or PCy_3. The indeno[1,2-*c*]azepin-3(2*H*)-ones, which incorporate both indene and unsaturated seven-membered ring lactam skeletons, were obtained in good to excellent yields (Scheme 4.29).

Zhu and co-workers developed a palladium- and copper-catalyzed procedure for the synthesis of medium- and large-sized ring-fused dihydroazaphenanthrenes and 1,4-benzodiazepine-2,5-diones.[61,62] A distinctly different catalytic property of palladium and copper catalysts was uncovered that leads to the development of a divergent synthesis of two different heterocyclic scaffolds from the same starting materials, simply by metal switching. Thus, starting from linear amide, palladium acetate triggers a domino intramolecular *N*-arylation–C–H activation–aryl–aryl bond-forming process to provide dihydroazaphenanthrenes, while copper iodide promotes only the intramolecular *N*-arylation reaction leading to 1,4-benzodiazepine-2,5-diones. In combination with the Ugi multicomponent reaction (Ugi-4CR) for the preparation of the linear amides, a two-step synthesis of either the 5,6-dihydro-8*H*-5,7*a*-diazacyclohepta[*jk*]phenanthrene-4,7-dione or 1,4-benzodiazepine-2,5-diones, by appropriate choice of metal catalyst, is subsequently developed from very simple starting materials.

Verkade's group developed a $Pd_2(dba)_3/P(i\text{-}BuNCH_2CH_2)_3N$-catalyzed C–N coupling for the one-pot synthesis of unsymmetrically substituted *trans*-4-*N*,*N*-diarylaminostilbenes and both symmetrically and unsymmetrically substituted *N*,*N*-diarylaminostyrene derivatives.[63] The procedure involves two or more palladium-catalyzed sequential coupling reactions (an amination and an intermolecular Heck reaction) in one-pot using the same catalyst system with two different aryl halides, including aryl chlorides and hetero aryl halides as the coupling partners. Interestingly, the same catalyst system can be used for both the amination and the intermolecular Heck coupling with a low catalyst loading. By using 2-bromophenylboronic acid and 2-bromophenethylamine as the substrates, 5,6-dihydro-7*H*-dibenz[*b*,*d*]azepine can be produced in 53% yield *via* a one-pot amination–Suzuki sequence in the presence of 4 mol% of $Pd(OAc)_2$ and 8 mol% of ligand.

4.5 Miscellaneous

In 2010, Alper and Zeng reported an interesting domino procedure for the synthesis of 1,4-benzothiazepin-5-ones from simple and readily accessible *N*-tosyl aziridines and *o*-iodothiophenols.[64] This process involves aziridines ring-opening with *o*-iodothiophenols, followed by palladium-catalyzed intramolecular carboxamidation. The scope and limitation of this transformation have been investigated in detail by using various aziridines and *o*-iodothiophenols; the corresponding benzothiazepinones were formed in good yields (Scheme 4.30a). The intramolecular cyclization of an aza Baylis–Hillman product *via* Heck coupling was established as well (Scheme 4.30b).[65]

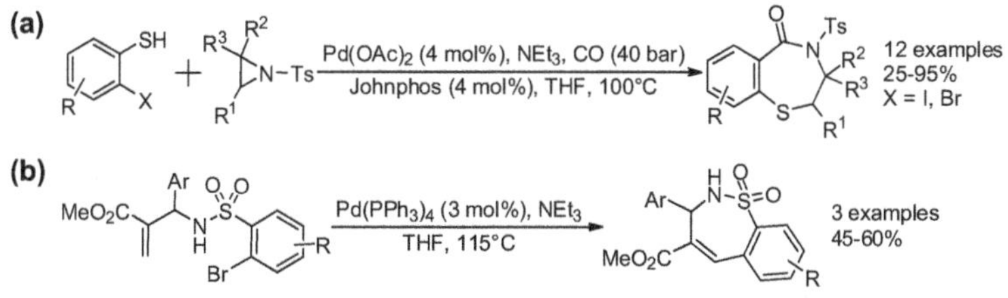

Scheme 4.30 Pd-catalyzed synthesis of sulfur-containing heterocycles.

Reaction Procedure (Scheme 4.30a): A mixture of *o*-iodothiophenols (0.50 mmol), *N*-tosyl aziridines (0.55 mmol), NEt_3 (1.50 mmol), $Pd(OAc)_2$ (0.020 mmol), Johnphos (0.020 mmol), and THF (6 mL) were sequentially added to a 45 mL glass-lined autoclave. After sealing, the autoclave was purged three times with carbon monoxide and pressurized with 40 bar of CO. The resulting mixture was stirred at 100 °C for 17 h. The autoclave was cooled to room temperature and then the excess carbon monoxide was released. The reaction mixture was filtered. The filtrate was concentrated in a rotary evaporator. The residue was purified by flash chromatography on silica gel using a mixture of hexanes and ethyl acetate (7 : 1 to 3 : 1) as the eluent to afford the products.

In summary, the main achievements in the synthesis of macrocyclic heterocycles from double functionalized arenes have been summarized and discussed. Compared with five- and six-membered heterocycles, macrocycles have been explored less. With the special reactivities and properties of macroheterocycles, especially the seven- and eight-membered cases, it will be interesting and necessary to work on this topic.

References

1. M. Carril, R. SanMartin, F. Churruca, I. Tellitu and E. Domínguez, *Org. Lett.*, 2005, **7**, 4787–4789.
2. E. Dimitrijevic, M. Cusimano and M. S. Taylor, *Org. Biomol. Chem.*, 2014, **12**, 1391–1394.
3. Q. Zhou, L. Su, T. Jiang, B. Zhang, R. Chen, H. Jiang, Y. Ye, M. Zhu, D. Han, J. Shen, G. Dai and Z. Li, *Tetrahedron*, 2014, **70**, 1125–1132.
4. M. Yu, R. Y. Tang and J. H. Li, *Tetrahedron*, 2009, **65**, 3409–3416.
5. X. Su, D. J. Fox, D. T. Blackwell, K. Tanaka and D. R. Spring, *Chem. Commun.*, 2006, 3883–3885.
6. G. Chouhan and H. Alper, *Org. Lett.*, 2010, **12**, 192–195.
7. F. Ji, M. F. Lv, W. B. Yi and C. Cai, *Adv. Synth. Catal.*, 2013, **355**, 3401–3406.
8. H. Li, A. Spannenberg, H. Neumann, M. Beller and X. F. Wu, *Chem. Commun.*, 2014, **50**, 2114–2116.
9. S. M. Lu and H. Alper, *J. Am. Chem. Soc.*, 2005, **127**, 14776–14784.
10. S. M. Lu and H. Alper, *Chem. – Eur. J.*, 2007, **13**, 5908–5916.
11. H. Yu, R. N. Richey, M. W. Carson and M. J. Coghlan, *Org. Lett.*, 2006, **8**, 1685–1688.
12. H. Yu, R. N. Richey, J. Mendiola, M. Adeva, C. Somoza, S. A. May, M. W. Carson and M. J. Coghlan, *Tetrahedron Lett.*, 2008, **49**, 1915–1918.
13. K. C. Majumdar, T. Ghosh and S. Ponra, *Tetrahedron Lett.*, 2013, **54**, 4661–4665.
14. C. Xue, S. H. Kung, J. Z. Wu and F. T. Luo, *Tetrahedron*, 2008, **64**, 248–254.
15. C. Huang and V. Gevorgyan, *Org. Lett.*, 2010, **12**, 2442–2445.

16. K. S. Masters, A. Bihlmeier, W. Klopper and S. Bräse, *Chem. - Eur. J.*, 2013, **19**, 17827–17835.
17. A. Huang, L. Feng, Z. Qiao, W. Yu, Q. Zheng and C. Ma, *Tetrahedron*, 2013, **69**, 642–646.
18. (a) M. Mori, M. Kimura, Y. Uozumi and Y. Ban, *Tetrahedron Lett.*, 1985, **26**, 5947–5950; (b) M. Mori, H. Kobayashi, M. Kimura and Y. Ban, *Heterocycles*, 1985, **23**, 2803–2806; (c) M. Mori, Y. Uozumi, M. Kimura and Y. Ban, *Tetrahedron*, 1986, **42**, 3793–3806.
19. S. D. Gawande, V. Kavala, M. R. Zanwar, C. W. Kuo, W. C. Huang, T. S. Kuo, H. N. Huang, C. H. He and C. F. Yao, *Adv. Synth. Catal.*, 2014, **356**, 2599–2608.
20. T. Yang, C. Lin, H. Fu, Y. Jiang and Y. Zhao, *Org. Lett.*, 2005, **7**, 4781–4784.
21. A. Klapars, S. Parris, K. W. Anderson and S. L. Buchwald, *J. Am. Chem. Soc.*, 2004, **126**, 3529–3533.
22. E. L. Cropper, A. J. P. White, A. Ford and K. K. Hii, *J. Org. Chem.*, 2006, **71**, 1732–1735.
23. J. G. Avila-Zárraga, A. Lujan-Montelongo, A. Covarrubias-Zúniga and M. Romero-Ortega, *Tetrahedron Lett.*, 2006, **47**, 7987–7989.
24. D. Solé, L. Vallverdú, X. Solans, M. Font-Bardia and J. Bonjoch, *J. Am. Chem. Soc.*, 2003, **125**, 1587–1594.
25. D. Tsvelikhovsky and S. L. Buchwald, *J. Am. Chem. Soc.*, 2010, **132**, 14048–14051.
26. D. Tsvelikhovsky and S. L. Buchwald, *J. Am. Chem. Soc.*, 2011, **133**, 14228–14231.
27. N. Della Cá, G. Maestri, M. Malacria, E. Derat and M. Catellani, *Angew. Chem., Int. Ed.*, 2011, **50**, 12257–12261.
28. K. Prabakaran, M. Zeller and K. J. R. Prasad, *Synlett*, 2011, 1835–1840.
29. P. Mestichelli, M. J. Scott, W. R. J. D. Galloway, J. Selwyn, J. S. Parker and D. R. Spring, *Org. Lett.*, 2013, **15**, 5448–5451.
30. J. L. Kenwright, W. R. J. D. Galloway, D. T. Blackwell, A. Isidro-Llobet, J. Hodgkinson, L. Wortmann, S. D. Bowden, M. Welch and D. R. Spring, *Chem. - Eur. J.*, 2011, **17**, 2981–2986.
31. Y. Wang, Y. Chen, Q. He, Y. Xie and C. Yang, *Helv. Chim. Acta*, 2013, **96**, 296–308.
32. H. S. Lim, Y. L. Choi and J. N. Heo, *Org. Lett.*, 2013, **15**, 4718–4721.
33. P. P. Kaishap, K. Shekarrao, P. Saikia, S. Gogoi and R. C. Boruah, *Tetrahedron Lett.*, 2014, **55**, 1927–1930.
34. (a) A. Neogi, T. P. Majhi, R. Mukhopadhyay and P. Chattopadhyay, *J. Org. Chem.*, 2006, **71**, 3291–3294; (b) A. Neogi, T. P. Majhi, B. Achari and P. Chattopadhyay, *Eur. J. Org. Chem.*, 2008, 330–336.
35. R. J. P. Corriu, B. Geng and J. J. E. Moreau, *J. Org. Chem.*, 1993, **58**, 1443–1448.
36. P. K. Deb, S. Sharma, A. Borude, R. P. Singh, D. Kumar and L. K. Reddy, *Tetrahedron Lett.*, 2013, **54**, 2916–2929.
37. C. A. Cheetham, R. S. Massey, S. L. Pira, R. G. Pritchard and T. W. Wallace, *Org. Biomol. Chem.*, 2011, **9**, 1831–1838.

38. C. L. Ladd, D. S. Roman and A. B. Charette, *Tetrahedron*, 2013, **69**, 4479–4487.
39. K. Okuro and H. Alper, *Synlett*, 2012, 2531–2533.
40. V. Narbonne, P. Retailleau, G. Maestri and M. Malacria, *Org. Lett.*, 2014, **16**, 628–631.
41. S. Ma and E. I. Negishi, *J. Org. Chem.*, 1994, **59**, 4730–4732.
42. L. A. T. Cleghorn, R. Grigg, C. Kilner, W. S. MacLachlan and V. Sridharan, *Chem. Commun.*, 2005, 3071–3073.
43. Q. Li, X. Jiang, C. Fu and S. Ma, *Org. Lett.*, 2011, **13**, 466–469.
44. L. A. Arnold, W. Luo and R. K. Guy, *Org. Lett.*, 2004, **6**, 3005–3007.
45. K. C. Majumdar, B. Chattopadhyay and B. Sinha, *Tetrahedron Lett.*, 2008, **49**, 1319–1322.
46. K. C. Majumdar and B. Chattopadhyay, *Synthesis*, 2009, 2385–2392.
47. K. C. Majumdar, I. Ansary, B. Sinha and B. Chattopadhyay, *Synthesis*, 2009, 3593–3602.
48. K. C. Majumdar, S. Samanta and B. Chattopadhyay, *Tetrahedron Lett.*, 2009, **50**, 4866–4869.
49. G. Satyanarayana and M. E. Maier, *Tetrahedron*, 2008, **64**, 356–363.
50. S. G. Stewart, C. H. Heath and E. L. Ghisalberti, *Eur. J. Org. Chem.*, 2009, 1934–1943.
51. G. Tasic, J. Randjelovic, N. Vusurovic, V. Maslak and S. Husinec, *Tetrahedron Lett.*, 2013, **54**, 2243–2246.
52. P. A. Donets and E. V. Van der Eycken, *Org. Lett.*, 2007, **9**, 3017–3020.
53. K. Parthasarathy, H. Han, C. Prakash and C. H. Cheng, *Chem. Commun.*, 2012, **48**, 6580–6582.
54. D. K. Rayabarapu and C. H. Cheng, *J. Am. Chem. Soc.*, 2002, **124**, 5630–5631.
55. Y. Ohta, H. Chiba, S. Oishi, N. Fujii and H. Ohno, *Org. Lett.*, 2008, **10**, 3535–3538.
56. L. Guo, B. Li, W. Huang, G. Pei and D. Ma, *Synlett*, 2008, 1833–1836.
57. H. J. Wang, Y. Wang, F. Camara, W. D. Paquette, A. J. Csakai and J. E. Mangette, *Tetrahedron Lett.*, 2011, **52**, 541–543.
58. B. J. Marglis, J. J. Swidorski and B. N. Rogers, *J. Org. Chem.*, 2003, **68**, 644–647.
59. J. K. Laha, K. S. S. Tummalapalli and A. Gupta, *Eur. J. Org. Chem.*, 2014, 4773–4779.
60. Y. Luo and J. Wu, *Chem. Commun.*, 2011, **47**, 11137–11139.
61. (a) G. Cuny, M. Bois-Choussy and J. Zhu, *Angew. Chem., Int. Ed.*, 2003, **42**, 4774–4777; (b) G. Cuny, M. Bois-Choussy and J. Zhu, *J. Am. Chem. Soc.*, 2004, **126**, 14475–14484.
62. A. Salcedo, L. Neuville, C. Rondot, P. Retailleau and J. Zhu, *Org. Lett.*, 2008, **10**, 857–860.
63. M. V. Nandakumar and J. G. Verkade, *Tetrahedron*, 2005, **61**, 9775–9782.
64. F. Zeng and H. Alper, *Org. Lett.*, 2010, **12**, 5567–5569.
65. A. Vasudevan, P. S. Tseng and S. W. Djuric, *Tetrahedron Lett.*, 2006, **47**, 8591–8593.

Subject Index

Page numbers in *italic* refer to schemes.